20th International Conference for Students and Young Scientists: Modern Techniques and Technologies

(MTT'2014)

IOP Conference Series: Materials Science and Engineering Volume 66

Tomsk, Russia
14 – 18 April 2014

ISBN: 978-1-63439-877-0
ISSN: 1757-8981

Printed from e-media with permission by:

Curran Associates, Inc.
57 Morehouse Lane
Red Hook, NY 12571

Some format issues inherent in the e-media version may also appear in this print version.

Printed by Curran Associates, Inc. (2015)

For permission requests, please contact the Institute of Physics
at the address below.

Institute of Physics
Dirac House, Temple Back
Bristol BS1 6BE UK

Phone: 44 1 17 929 7481
Fax: 44 1 17 920 0979

techtracking@iop.org

Additional copies of this publication are available from:

Curran Associates, Inc.
57 Morehouse Lane
Red Hook, NY 12571 USA
Phone: 845-758-0400
Fax: 845-758-2634
Email: curran@proceedings.com
Web: www.proceedings.com

20th International Conference for Students and Young Scientists: Modern Techniques and Technologies

(MTT'2014)

IOP Conference Series: Materials Science
and Engineering Volume 66

Tomsk, Russia
14 – 18 April 2014

ISBN: 978-1-63439-877-0
ISSN: 1757-8981

IOP Institute of Physics

TABLE OF CONTENTS

Preface

The active involvement of young researchers in scientific processes and the acquisition of scientific experience by gifted youth currently have a great value for the development of science. One of the research activities of National Research Tomsk Polytechnic University, aimed at the preparing and formation of the next generation of scientists, is the International Conference of Students and Young Scientists "Modern Techniques and Technologies", which was held in 2014 for the twentieth time.

Great experience in the organization of scientific events has been acquired through years of carrying the conference. There are all the necessary resources for this: a team of organizers – employees of Tomsk Polytechnic University, premises provided with modern office equipment and equipment for demonstration, and leading scientists – professors of TPU, as well as the status of the university as a leading research university in Russia. This way the conference is able to attract world leading scientists for the collaboration.

For the previous years the conference proved itself as a major scientific event at international level, which attracts more than 600 students and young scientists from Russia, CIS and other countries. The conference provides oral plenary and section reports. The conference is organized around lectures, where leading Russian and foreign scientists deliver plenary presentations to young audiences.

An important indicator of this scientific event is the magnitude of the coverage of scientific fields: energy, heat and power, instrument making, engineering, systems and devices for medical purposes, electromechanics, material science, computer science and control in technical systems, nanotechnologies and nanomaterials, physical methods in science and technology, control and quality management, design and technology of artistic materials processing. The main issues considered by young researchers at the conference were related to the analysis of contemporary problems using new techniques and application of new technologies.

IOP Conf. Series: Materials Science and Engineering **66** (2014) 011002 doi:10.1088/1757-899X/66/1/011002

ORGANIZING COMMITTEE OF THE CONFERENCE MTT'2014

- P. Chubik, Professor, Rector, chairman, TPU, Tomsk, Russia
- M. Sonkin, Professor, Vice-Rector for Research and Innovations, co-chair, TPU, Tomsk, Russia
- Dirk Vervake, Head of the international laboratory "Plastic and Organic Synthesis", TPU, Tomsk, Russia
- Kröning Hans-Michael Wilhelm Adolf, PhD, Professor, Saarland University, Germany
- Zhaohui Xi, Director of the Office of International cooperation and Exchanges, Harbin University of Science and Technology, China
- Sodnomdorj Dari, professor, Mongolian university of science and technology, Mongolia
- Kaushik Deb, department of computer science & engineering, Chittagong university of engineering & technology, Bangladesh
- Gromov Alexander, professor Aerospace Department Milan Polytechnic University, ITALY
- F. Gubarev, Head of the Division of Postgraduate Studies, TPU, Tomsk, Russia
- A. Sivkov, 1th Section Chairman, Professor, TPU, Tomsk, Russia
- S. Sylushkin, 2th Section Chairman, TPU, Tomsk, Russia
- N. Martyushev, 3th Section Chairman, TPU, Tomsk, Russia
- O. Muravliov, 4th Section Chairman, Professor, TPU, Tomsk, Russia
- G. Evtushenko, 5th Section Chairman, Professor, TPU, Tomsk, Russia
- B. Zenin, 6th Section Chairman, TPU, Tomsk, Russia
- V. Rudnitskiy, 7th Section Chairman, TPU, Tomsk, Russia
- A. Potylitsyn, 8th Section Chairman, Professor, TPU, Tomsk, Russia
- V. Kuleshov, 9th Section Chairman, Professor, TPU, Tomsk, Russia
- A. Zavorin, 10th Section Chairman, Professor, TPU, Tomsk, Russia
- M. Kukhta, 11th Section Chairman, Professor, TPU, Tomsk, Russia
- V. Vereshchagin, 12 th Section Chairman, Professor, TPU, Tomsk, Russia
- O. Sidorova, Academic Secretary of the Conference, TPU, Tomsk, Russia
- I. Sharkov, Department for development of publication activities, TPU, Tomsk, Russia

Tributes to colleagues

The Editors of the volume thank the organizers of the International Conference "Modern techniques and technologies":

- A. Sivkov, 1th Section Chairman, Professor, TPU, Tomsk, Russia
- S. Sylushkin, 2th Section Chairman, TPU, Tomsk, Russia
- N. Martyushev, 3th Section Chairman, TPU, Tomsk, Russia
- O. Muravliov, 4th Section Chairman, Professor, TPU, Tomsk, Russia
- G. Evtushenko, 5th Section Chairman, Professor, TPU, Tomsk, Russia
- B. Zenin, 6th Section Chairman, TPU, Tomsk, Russia
- V. Rudnitskiy, 7th Section Chairman, TPU, Tomsk, Russia
- A. Potylitsyn, 8th Section Chairman, Professor, TPU, Tomsk, Russia
- V. Kuleshov, 9th Section Chairman, Professor, TPU, Tomsk, Russia
- A. Zavorin, 10th Section Chairman, Professor, TPU, Tomsk, Russia
- M. Kukhta, 11th Section Chairman, Professor, TPU, Tomsk, Russia
- V. Vereshchagin, 12 th Section Chairman, Professor, TPU, Tomsk, Russia
- O. Sidorova, Academic Secretary of the Conference, TPU, Tomsk, Russia
- I.Sharkov, Department for development of publication activities, TPU, Tomsk, Russia

IOP Conf. Series: Materials Science and Engineering **66** (2014) 011004 doi:10.1088/1757-899X/66/1/011004

Conference Photographs

Схема эксперимента.
1. Инжектор микротрона; 2. Защита; 3. Поворотная
4. Фокусирующий магнит; 5. Коллиматор; 6. Соединит
фланцы; 7. Магнитно-индукционный датчик;
8. Экспериментальная камера; 9. Свинцовая защита; 10.
11. Очищающий магнит; 12. Майларовая пленка; 13.
14. Цилиндр Фарадея; 15. Видеокамера.

IOP Conf. Series: Materials Science and Engineering **66** (2014) 011005 doi:10.1088/1757-899X/66/1/011005

Peer review statement

All papers published in this volume of *IOP Conference Series: Materials Science and Engineering* have been peer reviewed through processes administered by the proceedings Editors. Reviews were conducted by expert referees to the professional and scientific standards expected of a proceedings journal published by IOP Publishing.

IOP Conf. Series: Materials Science and Engineering **66** (2014) 012001 doi:10.1088/1757-899X/66/1/012001

Synthesis of ultra dispersed graphite-like structures doped with nitrogen in supersonic carbon plasma flow

A Sivkov[1,a], A Pak[1,b], I Shanenkov[1,c], J Kolganova[1,d] and K Shatrova[1,e]

[1]Tomsk Polytechnic University, Tomsk, Russia

[a]sivkovaa@mail.ru, [b]ayapak@tpu.ru, cswordi@list.ru, [d]julia_kolganova@mail.ru, [e]kns2@tpu.ru

Abstract. In this paper the synthesis of ultra dispersed graphite-like structures doped with nitrogen by using a magneto plasma accelerator is shown. The synthesis of such structures is realized during supersonic carbon plasma jet flowing into the chamber filled with the nitrogen atmosphere. Plasma jet is generated by coaxial magneto plasma accelerator (CMPA) based on graphite electrode system. The CMPA is supplied from the pulsed capacitive energy storage with the maximum value of stored energy of 360 kJ. Electrical and energy parameters of plasma flow are recorded during experiment. According to X-ray diffraction (XRD) and high resolution transmission electron microscopy (HRTEM) methods obtained particles with well seen triangle sectors are graphite-like carbon structures doped with nitrogen atoms. This is confirmed by X-ray photoelectron spectroscopy (XPS) and Raman spectroscopy analyses in which carbon-nitrogen bonding configurations have been identified.

1. Introduction

The first introduction of the material with the stoichiometry C_3N_4 was made in 1922 [1]. After more than sixty years this material ran into renaissance and has become widespread, due at least partly to the paper published by Marvin L. Cohen [2]. The author made an assumption about a hypothetical phase of the carbon-nitrogen material which is isostructural to β-Si_3N_4, with the bulk modulus comparable to that of diamond. In 1996 Teter and Hemley predicted several carbon nitride polymorphs such as cubic, graphitic, pseudocubic etc. based on first-principles calculations [3]. In spite of numerous attempts concerning the synthesis of C_3N_4 [4], there is no unambiguous evidence about successful synthesis of ultra-hard crystalline carbon nitride solids [5]. At the same time graphitic carbon nitride with different stoichiometries has already been experimentally obtained. Up to now, several methods such as direct dynamic synthesis [6], synthesis using dynamic pressures [7], mechanochemical synthesis [8], solvothermal method [9], pyrolysis [10], are widely used. Also it has been claimed that the carbon nitride compound could be formed in thin amorphous films [11]. In the past few years there has been an increasing interest to graphitic carbon nitride materials due to their photocatalytic properties [12, 13]. Because of good prospects of carbon nitride as a promising complement to carbon materials, the synthesis process in the C-N system by using high-current coaxial magneto plasma accelerator was realized [14, 15]. In this paper the possibility of synthesis of ultra dispersed graphite-like structures doped with nitrogen via this versatile technique is presented.

2. Experimental

A sketch map of the system used for magnetoplasmadynamic synthesis is shown in figure 1. It consists of three main elements such as capacitive energy storage, coaxial magnetoplasma accelerator (1-7) and working chamber. Capacitive energy storage is divided into 24 sections with 1,2 mF capacitor banks (8). Maximum charging voltage of capacitor banks is 5 kV. Sectioning is made for varying the value of input energy. Such a way by using sectioning capacitive energy storage it's possible to supply the system with energy up to 360 kJ.

The z-pinch discharge current and the voltage between z-axis pinch accelerator electrodes were measured with using a Rogowski coil (R) and a resistive voltage divider (VD). The signals were registered in real time by a two-channel Tektronix TDS1012 oscilloscope (OSC).

The photoregistration system triggering is realized with pushing a button in the camera software PC suit. Camera generates a TTL signal, which by the way of the cascade amplification closes keys in the power circuit. Oscillographic testing is started with an appearance of the current signal front from the Rogowski coil.

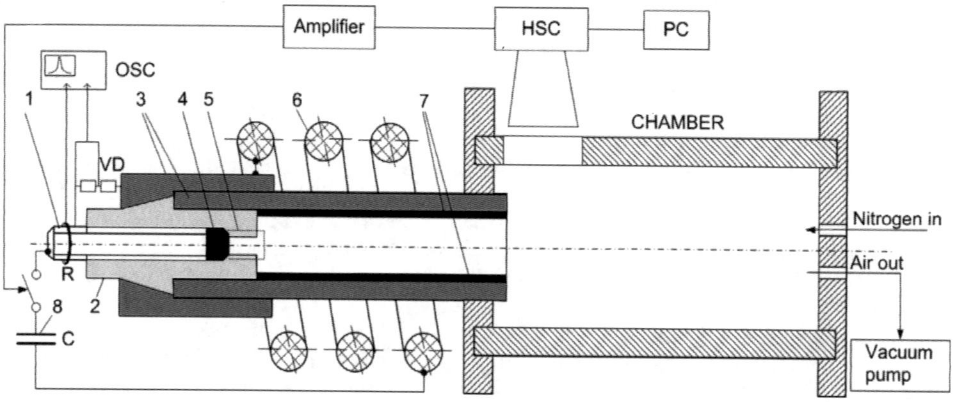

Figure 1. Sketch map of system: 1 – metal case of central electrode; 2 – insulator; 3 – metal case; 4 – graphite insert; 5 – plasma formation zone; 6 – inductor; 7 – graphite accelerating channel; 8 – capacitive energy storage; R -Rogowski coil; VD – voltage divider; HSC – high-speed camera.

Such a way powder sample was synthesized in the superhighspeed carbon plasma jet flowing into the chamber filled up with nitrogen atmosphere (N_2) at normal conditions. Before experiment working chamber was evacuated. The total capacity of energy storage was 6,0 mF (5 sections). The charging voltage was 3,5 kV. The precursor powder (a carbon black with purity of 99%) of 0.5 g was put into the plasma formation zone. Besides extra carbon ions are involved into the plasma flow due to erosion of accelerating channel. The typical work cycle of the system was up to 0.5 ms. The chamber was opened and the powder product was collected after an hour.

The final product (mass of 0.5 g) was analyzed by X-ray diffraction (XRD) with a Shimadzu XRD7000 diffractometer with a Cu $K\alpha$ radiation, high resolution transmitting electron microscopy (HRTEM) via JEOL JEM 2100F high resolution transmission electron microscope, X-ray photoelectron spectroscopy (XPS) using a Thermo Scientific ESCALAB 250 spectroscope and Raman spectroscopy using a Thermo Scientific Nikolet NXR 9650 Raman spectrometer.

3. Results and discussion

Figure 2 shows the electrical (a) and energy (b) discharge parameters during the working cycle. The discharge current and voltage were registered with using the Rogowski coil and resistive voltage divider, respectively. It's clearly seen that process time is about 270 μs and during this time the current maximum is over 150 kA. The presence of residual voltage (~ 400 V) can be connected with the recharge of capacitors. Using obtained current and voltage data the instantaneous power and energy were calculated.

The XRD pattern of the synthesized powder (1) and initial carbon black (2) used as a precursor is shown in figure 3. It can be noticed positions of diffraction peaks are typical for graphite-like structure. An analysis of XRD pattern was carried out by using database PDF2+.

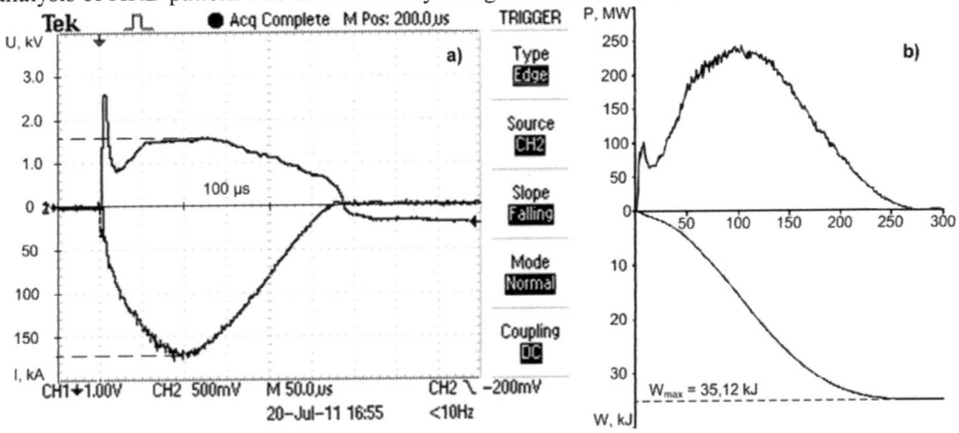

Figure 2. Electrical and energy discharge parameters: a) Current and voltage oscilloscope traces; b) Instantaneous power and energy.

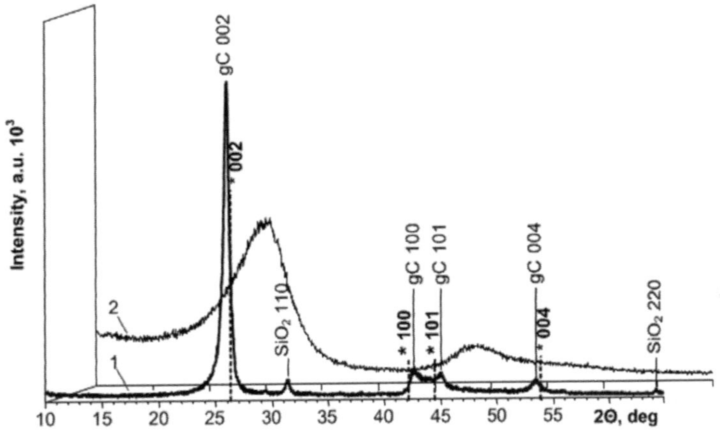

* - peak positions of etalon phase gC

Figure 3. XRD pattern of obtained powder: 1) Synthesized product; 2) Initial carbon black.

Following main crystalline phases such as graphite gC (space group P6-3mc (186)) and silicon oxide SiO_2 (space group P24/mnm (136)) were identified. The presence of small amount of silicon oxide in the powder can be explained with the features of the plasma formation zone construction, in particular, due to the using of fiberglass-filled plastic insulator. It is worth noting positions of diffraction peaks of gC are significantly different from the etalon model. It can be connected with differences in lattice parameters between etalon (a=2,470 Å; c=6,790 Å) and calculated gC phases. Lattice parameters of calculated gC are follow: a=2,448 Å; c=6,854 Å. These differences in lattice parameters, shifts of positions and broadening of the peaks for calculated phase may be attributed to the disorder induced by N incorporation into the carbon network.

To confirm the assumption about nitrogen incorporation into carbon network, an XPS method was used for obtaining the information about the chemical bonding nature of the powdered product. The survey scan XPS spectrum is shown in figure 4. It indicates the presence of mainly C and small amounts of O and N. The measured [N]/[C] ratio is 0.05. The presence of O may be due to surface absorption and oxidation [13].

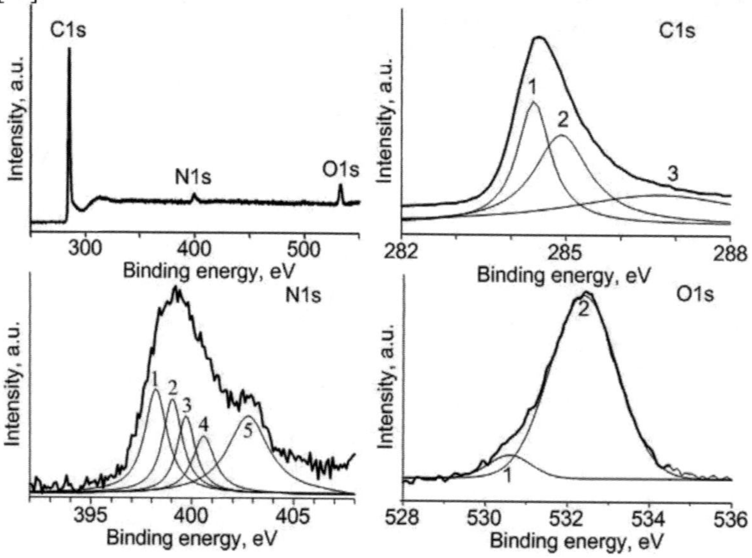

Figure 4. XPS spectrums of the obtained product.

The XPS spectra of C1s, N1s and O1s core level lines are also shown in figure 5. C1s consists of two main peaks with binding energy values at 284,4 and 285,2 eV, which are identified as sp^2 C-C and sp^3 C-C bonding, respectively [16-20]. The low and broadened band centered at 286.7 eV can be interpreted as sp^2 C in a C-N bonding configuration [18,19]. The N1s spectra can be decomposed in five Gaussian peaks. Peak identification may be as follows [16-20]: the *1* component (at 398,3 eV) has been assigned to a sp^3 C-N configuration; the component *2* (399,1 eV) correspond to C≡N bonding configuration; third component (399,8 eV) can be identified as C=N bonding configuration; *4* component (400,5 eV) assigned to a sp^2 C-N and the last one (402,5 eV) assigned to N=N type bonding. But it should be noted that there are some ambiguities in the literature concerning the structure and assignments of the binding energy of the N1s XPS spectra [21]. Deconvolution of O1s spectrum was carried out with using the NIST XPS database. According to it the first component (530,6 eV) is interpreted as C-O bindings and second one is Si-O bindings (532,4 eV). XPS results are in a good accordance with XRD data and confirm the presence of carbon, nitrogen and silicon oxide in the final product.

In figure 5 the Raman scattering spectrum of synthesized product in the 1000-2000 cm^{-1} region is shown. It's possible to see a series of sharp peaks in this region.

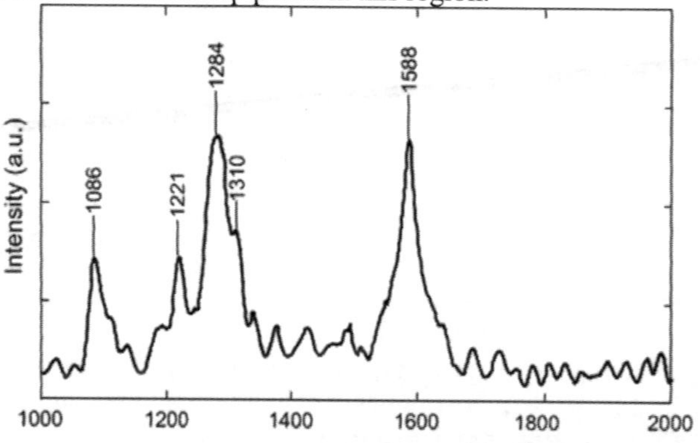

Figure 5. Raman spectrum.

Two main peaks can be identified as the so-called D and G peaks, attributed to disordered sp^2 microdomains and graphite-like sp^2 microdomains, respectively [22]. It should be noted that D-peak is situated in the region of lower frequencies (1284 cm^{-1}), which is different from data in the literature, where D-peak is situated around 1350-1360 cm^{-1} [22-24]. This may indicate the presence of high nitrogen concentration in the sample [25]. There exists good accordance in the $I_{(D)}/I_{(G)}$ ratio (~ 1,0), which is enhanced with increasing of nitrogen content [26]. According to the result that the intensity of D-peak is a little higher than that of G-peak, it is concluded that the symmetry in the sp^2 microdomains has been broken due to the incorporation of nitrogen into carbon network [25, 27]. The presence of three peaks in 1086 cm^{-1}, 1221 cm^{-1} and 1310 cm^{-1} is interesting. They are close to peaks of the calculated spectrum of graphitic C_3N_4: 1098 cm^{-1}, 1221 cm^{-1} and 1316 cm^{-1} [23], and may be assigned to nitrogen rich microdomains.

According to the HRTEM image (figure 6a) the typical particle has a natural faceted structure with inclusions of triangular sectors. Sectors are visible because the particle consists of crystalline and quasi-amorphous areas. The transition region between these areas in the radial direction is clearly visible in figure 6d. The set of these transitions forms triangular sectors. The measured interplanar distance in one of the sector is close to, but a little higher than that of graphite gC (d_{101}=3.395 Å) probably due to nitrogen incorporation and is equal to 3,50 Å. In the SAED pattern (figure 6b) it's possible to see that the first Debye ring consists of two closely spaced rings. It can be supposed that first ring is a superposition of diffraction maximums, which are attributed to families of gC–002 (d=3,395 nm) planes and probably planes of areas doped with nitrogen (d=3,550 nm). Also in dark-field TEM image (figure 6c) luminous triangle sectors corresponded to diffraction maximums of first Debye ring are clearly visible. So it can be assumed that these triangular sectors are nitrogen incorporated areas into the framework of crystalline graphite.

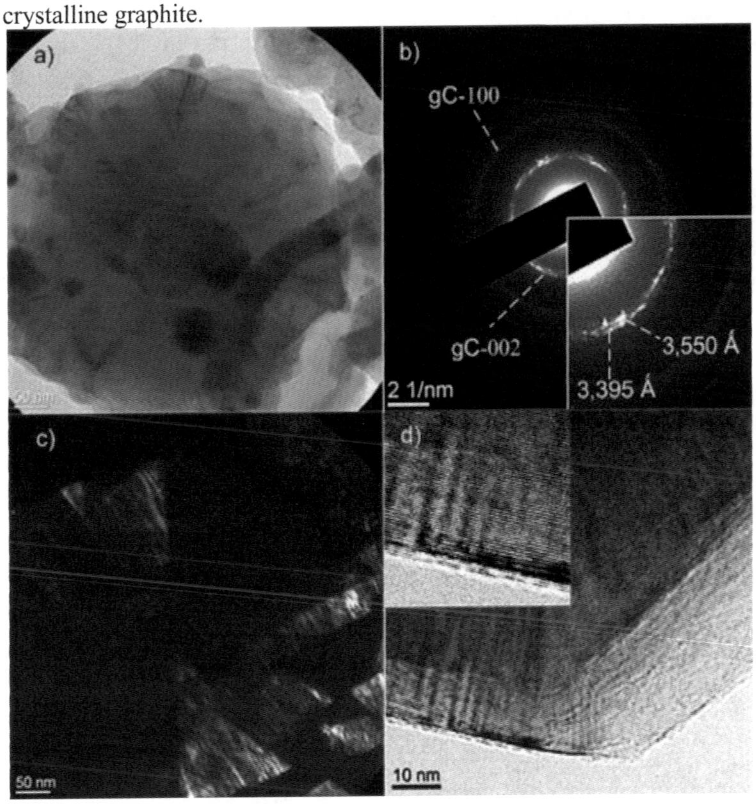

Figure 6. HRTEM-images of typical particle of synthesis product: a) Bright-field image; b) SAED; c) dark-field image; d) lattice image.

4. Conclusion

The ultra dispersed powder sample of graphite-like structures doped with nitrogen was synthesized in the supersonic carbon plasma jet flowing into the chamber filled up with the nitrogen atmosphere by using high current coaxial magneto plasma accelerator. The duration of pulsed power supply is 270 μs. During this time the value of energy equal to 35 kJ is supplied to z-pinch accelerator from the capacitive energy storage. According to HRTEM and XRD analysis the obtained particles with well seen triangle sectors are graphite-like carbon structures doped with nitrogen atoms. FT-IR, Raman and XPS analyses are in a good accordance with HRTEM and XRD data and confirmed the presence of carbon-nitrogen bonds in the powder.

References

[1] Franklin E C 1922 *J. Am. Chem. Soc.* **44** 486-509
[2] Cohen M L 1989 *Science* **245** 841-842
[3] Teter D and Hemley R J 1996 *Science* **271** 53-55
[4] Goglio G, Foy D and Demazeau G 2008 *Mater. Sci. Eng. R.* **58** 195-227
[5] Jiang X, Zhuang C, Zhao J and Jiang X 2012 *Diam. Related Mater.* **23** 44-49
[6] Ma H A, Jia X P Chen L X, Zhu P W, Guo W L, Guo X B, Wang Y D, Li S Q, Zou G T, Zhang G and Bex P 2002 *J. Phys.: Condens. Matter.* **14** 11269-73
[7] Shibata K and Sekine T 2006 *Solid State Commun.* **139** 443-472
[8] Zhao H Z, Chen X L, Jia C C, Zhou T, Qu X H, Jian J, Xu Y P and Zhou T 2005 *Mater. Sci. Eng.* B **122** 90-93
[9] Montigaud H, Tanguy B Demazeau G, Courjault S, Birot M and Dunogues J 1997 *C.R. Acad. Sci.* **325** 229-234
[10] Zhang J, Li Y, Zhu P, Huang D, Wu S, Cui Q and Zou G 2011 *Diam. Related Mater.* **20** 385-388
[11] Muhl S and Mendez J M 1999 *Diam. Related Mater.* **8** 1809-30
[12] Wang X, Maeda K, Thomas A, Takanabe K, Xin G, Carlsson J M, Domen K and Antonietti M 2009 *Nature Materials* **8** 76-80
[13] Xu X, Liu G, Randorn Ch and Irvine J T S 2011 *Inter. J. of hydrogen energy* **36** 13501-07
[14] Sivkov A A, Pak A Ya 2013 *Zhurnal Tekhnicheskoi Fiziki* **4** 85-91
[15] Sivkov A A, Pak A Ya, Nikitin D S, Rakhmatullin I A and Shanenkov I I 2013 *Nanotechnologies in Russia* **8** 489-494
[16] Li X, Zhang J, Shen L, Ma Y, Lei W, Cui Q and Zou G 2009 *Appl. Phys. A* **94** 387-392
[17] Boyd K J, Marton D, Todorov S S, Al-Bayati A H, Kulik J, Zuhr R A and Rabalais J W 1995 *J. Vac. Sci. Technol. A* **13** 2110-22
[18] Bhattacharyya S, Cardinaud C and Turban G 1998 *J. Appl. Phys.* **83** 4491-4500
[19] Wei S, Shao T and Ding P 2010 *Diam. Related Mater.* **19** 648-656
[20] Cappelli E, Trucchi D M, Kaciulis S, Orlando S, Zanza A and Mezzi A 2011 *Thin Solid Films* **519** 4059-63
[21] Dementjev A P, De Graaf A, Van de Sanden M C M, Maslakov K I, Naumkin A Vand Serov A A 2000 *Diam. Related Mater.* **9** 1904-07
[22] Guo J, Zhang J, Zhao Ch, Zhang J and Zhang Zh 2011 *Appl. Surf. Sci.* **258** 791-799
[23] Jiang J, Cheng W Zhang Ya, Zhu H and Shen D 2006 *J. Mater. Sci.* **41** 4117-21
[24] Ferrari A C, Rodil S E and Robertson J 2003 *Phys. Rev.* B **67** 155306
[25] Chowdhury A K M S, Cameron D C and Hashmi M S J 1998 *Thin Solid Films* **332** 62-68
[26] Rodil S E, Ferrari A C, Robertson J, Milne W I 2001 *J. Appl. Phys.* **89** 5425-30
[27] Dillion R O, Woollam J A and Kathanant V 1984 *Phys. Rev.* B **29** 3482-89

Acknowledgments

This work was supported in part by the State task to universities (no. 1998). We heartily thank Dr. Oleg Khasanov (Tomsk polytechnic University) for supporting HRTEM analysis.

Investigation of charge balance in ion accelerator TEMP-4M

I P Khailov[1] and V G Pak [1]
[1]*Tomsk, 634050, Russia*, National Research Tomsk Polytechnic University

E-mail: khailov@tpu.ru

Abstract. The paper presents the results of a study on the balance of charge in accelerator TEMP-4M operating in double-pulse mode with resistance load and ion diode. Crucially, it was found, that during the switching there is no losses of accumulated charge. It means, that all accumulated charge transferred to the load. However when the charge is transferred from the Marx generator to Blumlein line the half of accumulated charge is lost. Calibration of diagnostic equipment showed a good agreement between the calculated and experimental values of voltage and current. It means, that our diagnostic system is correct for registration parameters of the ion accelerator.
A distinctive feature of the ion accelerators with self-magnetically insulated diode is that there is no need to use additional energy source for the creation of an external magnetic field. That's why the efficiency of ion diodes with an external magnetic field is not more than 10-15%. The efficiency of energy conversion in self-magnetically insulated diodes will be determined by not only the efficiency of the diode, but the energy losses in the units of the accelerator. The aim of the researches is the analysis of the balance of charge in units of the ion beams pulsed generator and definition of the most significant channels of energy loss.

1. Experimental set-up and calibration of diagnostic equipment.

Investigations were conducted at the TEMP-4M accelerator [1, 2] set in a double pulse mode: the first pulse being negative (300–500 ns, 100–150 kV) and the second positive (150 ns, 200–250 kV). The accelerator consists of a capacitive storage—Marx generator, a nanosecond double forming line (Blumline line), and a vacuum ion diode with self-magnetically insulated of electrons and different. The flowchart of the double-forming line of accelerator TEMP-4M is shown on picture 1.

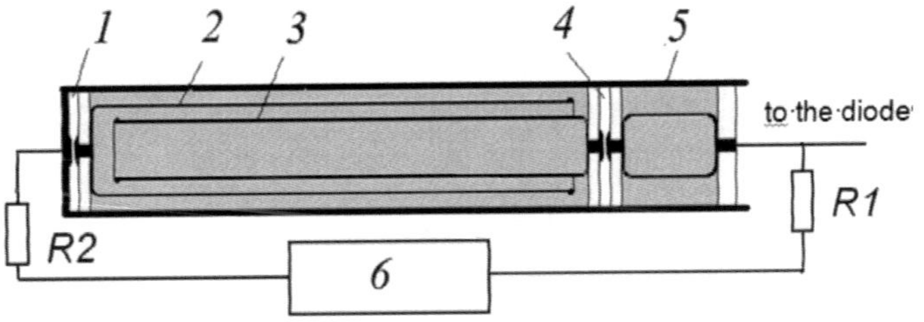

Figure 1. Flowchart of the double-forming line of accelerator TEMP-4M: 1 – main discharger (trigatron), 2 – the middle electrode, 3 – inner electrode, 4 – pre-discharger, 5 – airlock, 6 – cable delay line, R1 and R2 – resistance of the voltage divider and limiting resistor respectively.

The diode current and voltage were measured by a Rogowski coil [3], and a high frequency voltage divider, respectively. The shot-to-shot variation in the accelerating voltage for 100–200 pulses does not exceed 6–7% [4].

Analysis of the balance of the charge was made on the oscillograms of voltage and current, so the first stage of research was calibrated voltage divider and current sensors.

Figure 2 shows the results of the calibration of the voltage divider at the output of the double forming line when the accelerator operating with resistance load resistive load - 40 cm long tube with an aqueous solution of K2CO3.

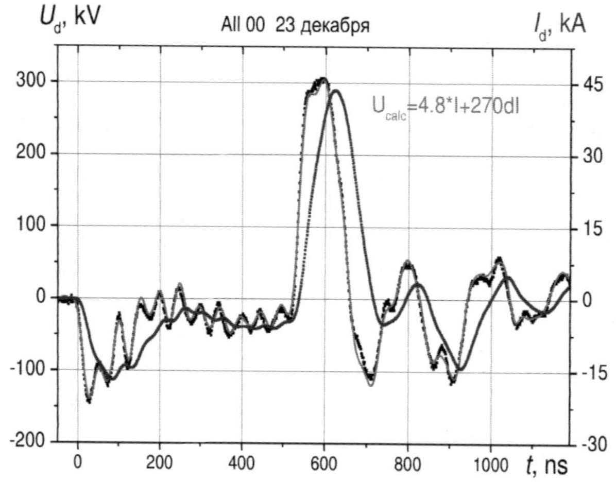

Figure 2. Experimental values of voltage (1, points) and current (2) on the output of Blumlein line and calculated values of the voltage (3, line)

Parasitic inductance of diode unite took into account in calculations (100 nH), which was determined in the short-circuit at the output of the double-forming line. The values of experimental and calculated voltage are in good agreement during the formation of the main voltage pulse and afterpulses. This confirms the correctness of measurement of current and voltage at the double-forming line output and the absence of parasitic leakage current in the diode unit.

2. Definition of capacity of Blumline line.

In the analysis of the energy balance in a double-forming lines we calculated energy in forming lines on the ratio $E= 0.5CU^2$. Therefore at the first stage of the research carried out calculation and measurement of the capacity of each forming line.

Actual capacity of Blumline line may differ from the calculated due to the error of calculation formulas, deviation of geometrical sizes along the length of the line, changes in the dielectric constant of water at high electric field intensity (70-80 kV / cm), etc. Therefore, the capacity of middle and inner lines were determined experimentally from the condition of balance of charge when Blumline line works in normal mode. While charging voltage of Blumline line was calculated by:

$$U_{mg} = \frac{1}{C}\int I_{mg}dt; \qquad (1)$$

where Umg, Img – charging voltage and current of DFL respectively, C – capacity of Blumline line.

Figure 3 shows the experimental and calculated values of the charging voltage of Blumline line[5].

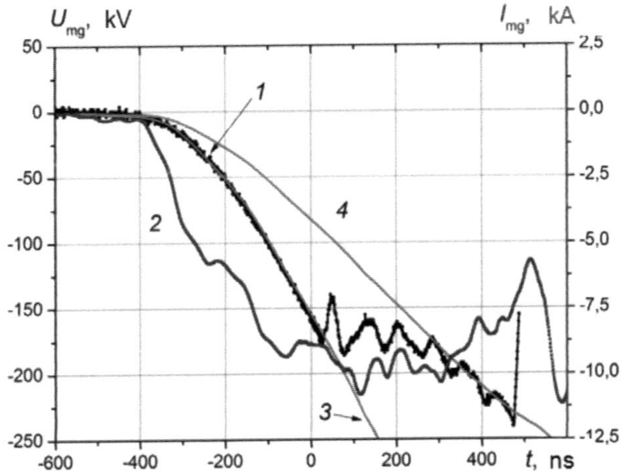

Figure 3. Waveform voltage (1 point) and current (2) at the output of pulsed generator and the calculated value of the voltage at C= 16.5 nF (3) and 30.5 nF (4)

Before pre-discharger is breakdown (t = 20 ns in Fig. 3) output voltage on pulsed generator coincides with the calculated ratio (1) with C = 16.5 nF. After the breakdown the parallel charging began in the inner and middle lines and output voltage of Marx generator (after completion of transition processes, t ≈ 280 NS) is equal with the calculated at C= 30.5 nF.

The capacitance value of forming lines can be determined experimentally from the analysis of transient processes for charging Blumlein line. To this end, the nitrogen pressure in both spark gaps has increased to levels above the pressure of self-breakdown. The equivalent circuit of the generator pulse voltage connected with Blumlein line is a serial oscillatory circuit, see figure 4

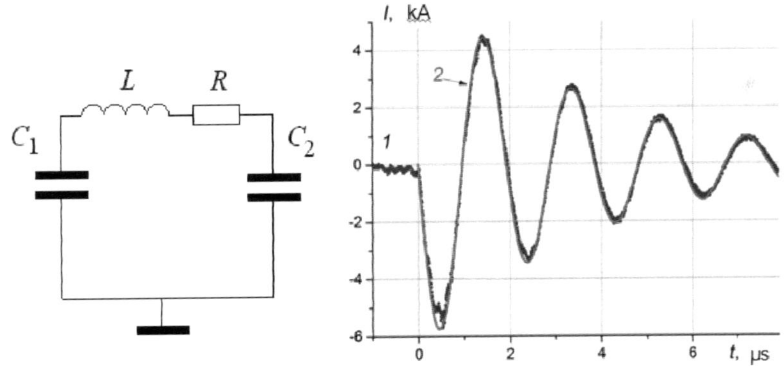

Figure 4. The equivalent circuit of Marx generator + Blumlein line: C1 - capacity of Marx generator, C2 - capacity of middle forming line, L is own inductance of Marx generator, R is the resistance losses. The waveform of the current measured by Rogowski coil of Marx generator in mode without breakdown of dischargers (1 point) and calculated current of serial oscillations (2 line).

Current of nonsingular oscillations in a serial resonant circuit is equal to:

$$I(t) = -I_0 \cdot e^{-\alpha \cdot t} \sin(\omega_d t) \qquad (2)$$

где $I_0 = \frac{U_0}{\sqrt{L/C}}$, $\alpha = \frac{R}{2L}$, $\omega_d = \sqrt{(\omega_0^2 - \alpha^2)}$, $\omega_0 = \frac{1}{\sqrt{LC}}$

For the scheme on figure 4:

$$C = \frac{C_1 \cdot C_2}{C_1 + C_2}$$

The best agreement between experimental and calculated values of the current of nonsingular oscillations obtained by LC = 9.3*10^{-14} s., R= 4 Ohm, I0 = 6.5 kA. These data help to determine the capacity of the middle forming line (Cm = 16.5 nF) using the value of self-inductance of Marx generator obtained in the analysis of transient processes in Marx generator working in short circuit mode (L = 7.5 uh). The obtained data allow us to calculate the reduction of tension in GIN for capacitors to discharge through the charging of resistance during charging double-forming line. The obtained data allow us to calculate the reduction of voltage in Marx generator due to capacitors discharging through the charging resistance during charging the Blumlein line. When the charging voltage of each stage 22.3 kV (for data Fig. 4) full output voltage of Marx generator (8 steps) must be 178 kV. Calculation of current of eigentones for a ratio (2) for these fig. 4 is executed at I0 = 6.5 kA that there corresponds U0 = 160 kV and a discharge of condensers for 10%.

To measure the capacities of the inner forming line the pressure was reduced in the preliminary gap. In this case current of a recharge of double-forming line is well described by a ratio (2) at C2 = 16.5 nF before preliminary gap is breakdown (t<0.5 microsec) and C2 = 30.5 nF after breakdown, see figure 5.

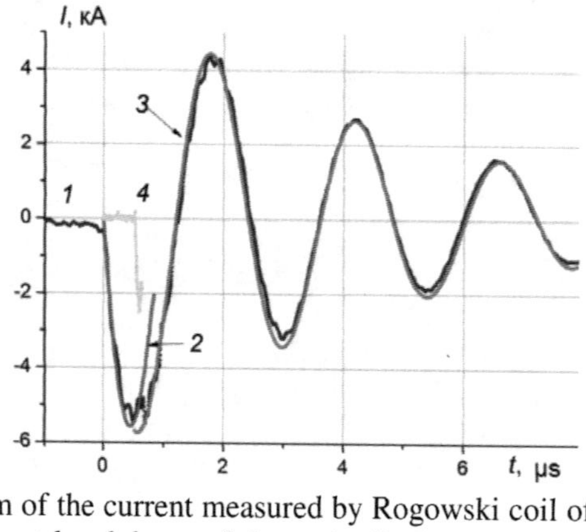

Figure 5. The waveform of the current measured by Rogowski coil of high voltage pulsed generator operating in mode without breakdown of the main discharger (1 point) and calculated current of serial oscillations till the pre-discharger is breakdown (2 line) and after the breakdown (3 line). Curve 4 - current of Rogowski coil on output of duble-forming lines after the preliminary discharger (first part)

Calculated and experimental values of the capacity of the internal forming line are the same. Middle forming line formed by body of accelerator and inner electrode. In the Blumlein line there is window for connection with Marx generator, therefore the experimental values of capacity of middle forming line there are less than calculated is executed.

3. The analysis of balance of charge.

At the first stage of researches calculation of an accumulated charge in Arkadyeva-Marx's generator on the following formula (2) was made:

Q=Суд*U (2)

Where Csp=50 nF, U=300 kV.

Calculations showed that Marx's generator accumulates 15 mC

Further the balance of a charge of Blumlein line was executed. Charge in Blumlein line counted by integration current from Marx generator to Blumlein line. Before the preliminary discharger is breakdown only inner forming line charging (C=14 nF), so the charge in Blumlein line is equal to multiplication of capacity of the inner line on a charge voltage. After breakdown of the preliminary discharger, both forming line are charge. At the same charge in the Blumlein line is the multiplication of sum of two capacities (24 nF) on a charge voltage. The balance of a charge of double-forming line is shown on figure 6.

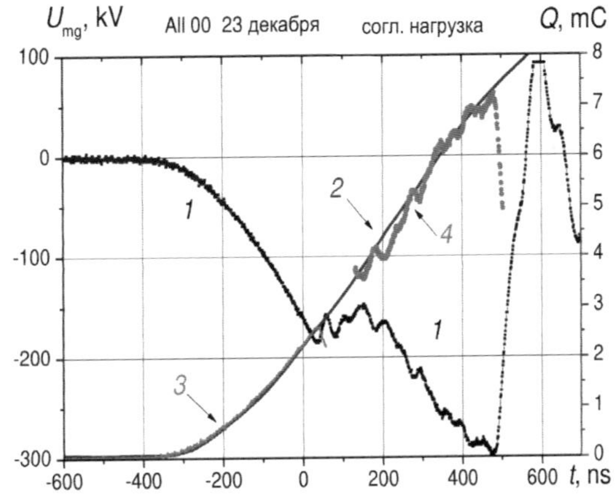

Figure 6. The oscillogram of a charge voltage of double-forming line (1), changing of a charge in Blulein line at calculation for integral of charging current (2), multiplication of a charge voltage on capacity 14 nF (3) and 24 nF (4).

The figure shows, that in the double-forming line the stored charge is equal to 7 mC. Losses of charge during switching the dischargers was not observed.

At the last stage researches of balance of a charge during the accelerator operating on the diode were carried out. Researches are executed on the strip focusing diode with self-magnetically insulated [6]. The results of researchers are presented on figure 7.

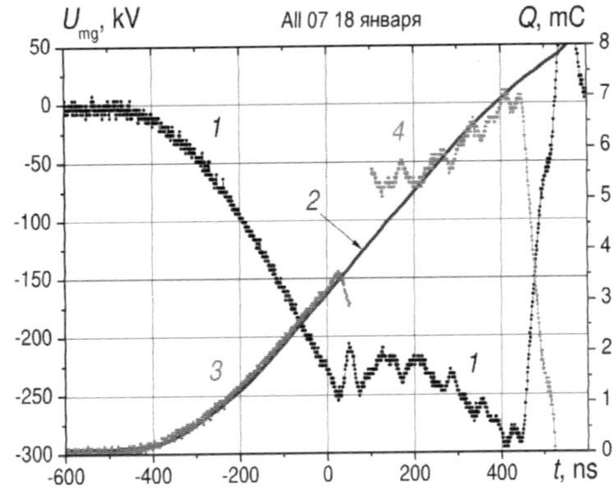

Figure. 7. The oscillogram of a charge voltage of double-forming line (1), changing of a charge in Blulein line at calculation for integral of charging current (2), multiplication of a charge voltage on capacity 14 nF (3) and 24 nF (4).

Conclusion.
The analysis of the balance of charge showed that double-forming line transmits the accumulated charge in a diode with a 100% efficiency, but a significant loss of charge occur during switching Marx generator. In this way, the main channel of loss of charge in the accelerator TEMP-4M is a Marx generator. Therefore, further work plan will be aimed at improving the transfer of charge from voltage impulse generator in the double line.

The work was supported by RFBR: 14-08-00449A «Investigation of ion beam divergence in self-magnetically insulated diodes» 2014-2015 yy.

References.
[1] Remnev G. E., Isakov I. F., Opekounov M. S., Kotlyarevsky G. I., Matvienko V. M., et al. Feb. 1997 "High-power Ion Sources for Industrial Application" *Surf. and Coatings Technol.* vol. **96** pp. 103–109.
[2] Remnev G. E., Isakov I. F., Pushkarev A. I., et al. 1999 "High Intensity Pulsed Ion Beam Sources and Their Industrial Applications" *Surf. and Coatings Technol.* vol. **114** pp. 206–212.
[3] Isakova Yu. I. December 2011 Diagnostic Equipment for the TEMP-4M Generator of High-current Pulsed Ion Beams *Journal of the Korean Physical Society* Vol. **59** No. **6** pp. 3531-3535.
[4] Pushkarev A. I., Isakova Yu. I., Khailov I. P. 2012 Shot-to-shot reproducibility of a self-magnetically insulated ion diode *Review of Scientific Instruments*, Volume **83** Issue **7** Article ID 073309 URL: http://link.aip.org/link/?RSI/83/073309
[5] Kumar D., Mitra S., Senthil K. et al. 2007 *Review of Scientific Instruments* v.**78** 115107
[6] Pushkarev A. I., Isakova Yu. I., Guselnikov V. I. 2011 Limitation of the electron emission in an ion diode with magnetic self-insulation *Phys. Plasmas* **18** 083109 URL: http://link.aip.org/link/?PHP/18/083109

Synthesis of a hybrid model of the VSC FACTS devices and HVDC technologies

Yu.S. Borovikov[1], A.S. Gusev[1], A.O. Sulaymanov[1], R.A. Ufa[1]

[1]*Tomsk, 634050, Russia,* National Research Tomsk Polytechnic University

E-mail: hecn@tpu.ru

Abstract. The motivation of the presented research is based on the need for development of new methods and tools for adequate simulation of FACTS devices and HVDC systems as part of real electric power systems (EPS). The Research object: An alternative hybrid approach for synthesizing VSC-FACTS and -HVDC hybrid model is proposed. The results: the VSC-FACTS and -HVDC hybrid model is designed in accordance with the presented concepts of hybrid simulation. The developed model allows us to carry out adequate simulation in real time of all the processes in HVDC, FACTS devices and EPS as a whole without any decomposition and limitation on their duration, and also use the developed tool for effective solution of a design, operational and research tasks of EPS containing such devices.

1. Introduction

Progress of power electronic allows to more efficient use of High-voltage direct current (HVDC) and Flexible Alternative Current Transmission Systems (FACTS) technologies in electric power systems (EPS). Voltage Source Converter (VSC) based on high-speed fully controlled power semiconductor switches is the main element of the most devices implementing these technologies [1, 2].

However, the introduction of technologies based on VSC opens up new possibilities for improving a controllability of EPS, as well as significantly complicates the design, operational and research tasks [3, 4]. Primarily this circumstance is connected with the specific operation of these devices: phase-phase operation of VSC; continuous high-speed operation of all possible normal, emergency and post-emergency operating conditions of EPS.

The mentioned specificity of HVDC and FACTS technologies excludes currently used in modern digital simulation tools of real EPS a processes decomposition and application of substantial simplifications of mathematical models of power equipment and EPS in general [3, 5, 6].

Thus, the mathematical model of EPS excluding these simplifications inevitably contains a rigid, non-linear system of differential equations of high order. According to the theory of discretization methods for differential equations the required model is ill-conditioned on the restrictive conditions of application of numerical integration and can't be adequately solved [7].

Consequently adequate modeling of three phase real EPS containing HVDC and FACTS devices using a purely numerical simulation becomes impossible [3, 5, 6].

Complex modeling approach, representing the hybrid simulation is an alternative for solving this problem.

2. The concept of the hybrid simulation tools and power systems

The basic provisions of the concepts:

- power equipment of EPS is described by mathematical models implemented via a hybrid coprocessors (HCP).
- methodically accurate with guaranteed instrumental error solution of differential equation systems carried out by means of the continuous implicit integration method.
- all types of commutation of power equipment and ensuring unlimited expandability of simulated EES carried out on a physical level model.
- the interconnection between the physical and mathematical modeling levels, control parameters of the simulated equipment and different switching elements are provided by means of appropriate voltage-current converters.
- model of automatic control and relay protection systems are implemented via a microprocessor units (MPU), containing a central processing unit (CPU) and functionally oriented peripheral processors (PP), as well as Server depending on the required composition of the input signals and the speed of algorithms execution.

Figure 1. Appearance of the experimental SHP

The structural diagram of the hybrid software and hardware simulation of EPS corresponding to above mentioned conception is shown in Figure 2:

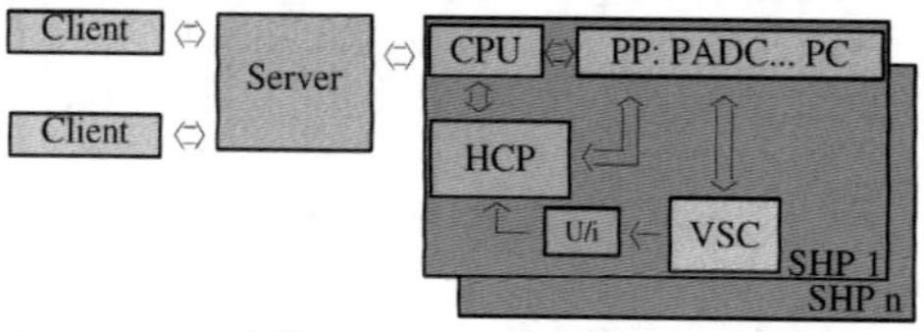

Figure 2. The structural diagram of the hybrid software and hardware simulation of EPS: SHP - specialized hybrid processor, CPU - central processing unit; PADC – processor of analog-to-digital converter; HCP – hybrid coprocessor; PC - switching processor; U / i - voltage-current converter; VSC - physical model of the power semiconductors VSC topology.

Developed according the described approach the concept and tools of the hybrid real time simulation of EPS allow to carry out:

- adequate simulation of all processes in VSC and three-phase power system as a whole without any decomposition and limitation on their duration;
- continuous methodically accurate solution in real time and on an unlimited range with guaranteed accuracy stiff nonlinear systems of differential equations of high order, adequately and accurately described a full spectrum of various processes in the power equipment and

power system as a whole, taking into account the automatic control and relay protection systems, preventive and emergency control of power systems;

- interconnection with various external software and hardware tools: operational information systems, SCADA system, etc.

The presented concept of model development of the traditional power equipment was tested and analyzed [8]. Only fragments of the development and research results of VSC model are presented within the format of this work.

3. VSC Simulation

The models of power switches according to the described concept are implemented on a physical level via the integrated microelectronic digital controlled analog switches (DCAS). To provide similarity of the model to a real device, the switching algorithms accounting input and output current and voltage values, as well as gate control signal should be implemented at the digital level. Thus, depending on the selected control algorithm, the DCAS can simulate any type of power electronic switches, including anti-parallel pair of different switches. The above mentioned algorithms implemented in MPU of SHP (Fig. 3).

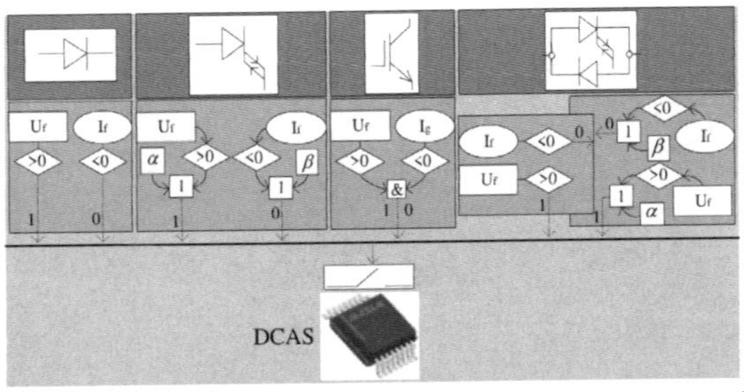

Figure 3. The control algorithms implementation scheme of power switches via DCAS.

According to the obtained DCAS characteristics (Fig.4) the switching time (t) is about $300ns$, while a switching time of IGBT is more then $2\mu s$ (Fig.5), as a result the DCAS can be considered an Ideal Switch.

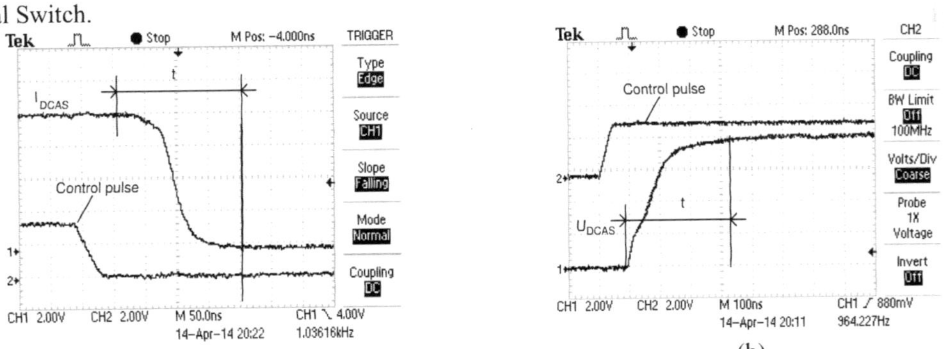

(a) (b)

Figure 4. The current (a) and the voltage (b) waveforms of the DACS commutation

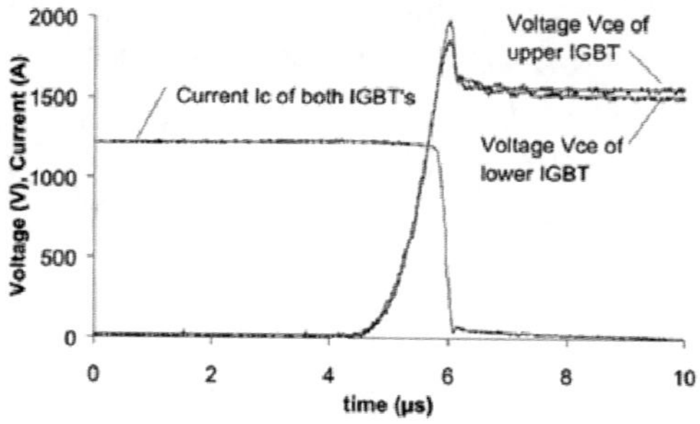

Figure 5. The current and voltage waveforms of real IGBT [9].

According to the shown in Fig. 6 schemes the IGBT (series 5SMY) and Ideal Switch commutation process is reproduced on the OrCAD software.

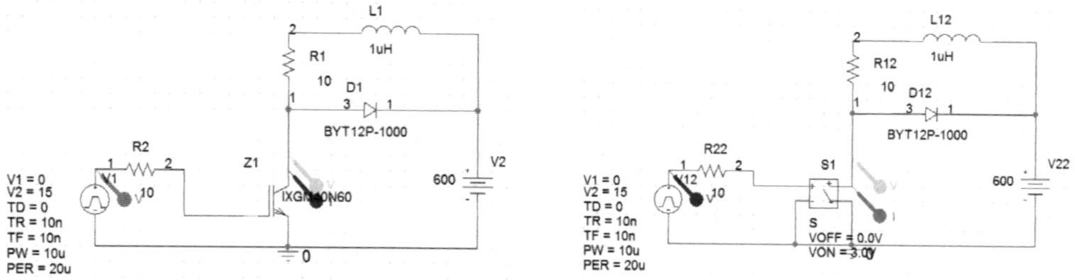

(a) (b)

Figure 6. Scheme of computer simulation of the IGBT (a) и Ideal Switch (b) commutating process

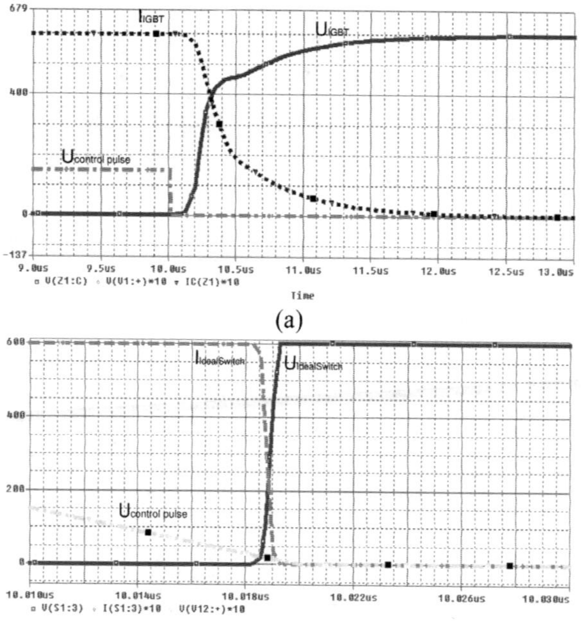

(b)

Figure 7. The current and the voltage waveforms of the IGBT (a) и Ideal Switch (b) turning on process

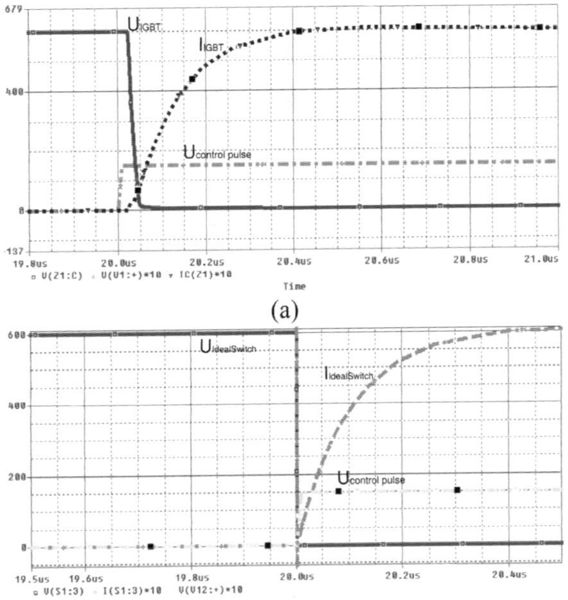

(a)

(b)

Figure 8. The current and the voltage waveforms of the IGBT (a) и Ideal Switch (b) turning off process

Application of this idealized VSC commutation is acceptable in the framework of solution the problem of modeling the real EPS. However, to research a process in the VSC, such as a control system and snubber circuits design and etc, the adequate commutation reproduction of real power semiconductors. For this purpose DCAS parameters should be adapted based on the analysis parameters and equivalent circuits of power semiconductors, taking into account a corresponding scaling. However, this task requires additional research.

4. Conclusion

- The presented concept and experimental implementation results confirm the required level of modeling adequacy of the power semiconductors and VSC based on them.
- An additional adaptation of the DCAS parameters will allow for adequate simulation of real power semiconductor switches and voltage converters based on them, as well as a full spectrum of quasi-steady-state and transient processes in power equipment and EPS as a whole with all possible normal, emergency and post-emergency processes in EPS as a whole containing devices based on VSC.
- The developed model allow to carry out the adequate simulation in real time of all the processes in HVDC, FACTS devices and EPS as a whole without any decomposition and limitation on their duration, and also use the develop tool for effective solution of the design, operational and research tasks of EPS containing such devices.

Acknowledgements

This work was supported by State Task «Science» 2.1318.2014 «Research and development of hybrid simulation complex of electric power system with active-adaptive network».

IOP Conf. Series: Materials Science and Engineering **66** (2014) 012003 doi:10.1088/1757-899X/66/1/012003

References

[1] Thepparat P, Retzmann D, Ogée E and Wiesinger M 2013 Smart Transmission System by HVDC and FACTS. // *Towards carbon free society through smarter grids*. Grenoble, France pp 1-6

[2] Chan-Ki K, Vijay K S, Gil-Soo J, Seong-Joo L, Seok-Jin L 2009 *HVDC Transmission: Power Conversion Application in Power System* (Singapore: Copyright) p 352

[3] Gnanarathna U, Gole A, Jayasinghe R 2011 Efficient Modeling of Modular Multilevel HVDC Converters (MMC) on Electromagnetic Transient Simulation Programs // *IEEE Transactions on Power Delivery* **vol 5 (1)** pp 316-324

[4] L'Abbate A, Fulli G 2010 Modeling and application of VSC-HVDC in the European transmission system. // *International Journal of Innovations in Energy Systems and Power* **vol 5 (1)** pp 8-16

[5] Forsyth P, Maguire T, Shearer D, Rydmell D 2009 Testing Firing Pulse Controls for a VSC Based HVDC Scheme with a Real Time Timestep < 3 μs. // *International Conference on Power Systems Transients* Kyoto, Japan pp 1-5

[6] Zhang Y; Gole A; Wu W, Zhang B, Sun H 2013 Development and Analysis of Applicability of a Hybrid Transient Simulation Platform Combining TSA and EMT Elements. // *IEEE Transaction on power system* **vol № 5 (1**) pp 357-366

[7] Hall G., Watt J.M. 1976 *Modern Numerical Methods for Ordinary Differential Equations* (Oxford University Press) p 350

[8] Prokhorov A, Borovikov Yu, Gusev A 2012 Real Time Hybrid Simulation of Electrical Power Systems: Concept, Tools, Field Experience and Smart Grid Challenges. // *International Journal of Smart Grid and Clean Energy* **vol 1 (1)** pp 67–78

[9] Siemens VSC Technology: Power Transmission and Distribution // EV-H2_01.98-1 http://www.technology.stfc.ac.uk/OWEN/workshop_4/pdfs/owen_siemens02.pdf

Estimation of winding insulation resistance to the corona discharges

A Leonov[1], V Red'ko[1] and E Soldatenko[1]

[1] Tomsk Polytechnic University, Tomsk, Russia

E-mail: leonov_ap@tpu.ru

Abstract. This article presents test results of enameled winding wires, characterizing an insulation electrical and mechanical strength. Standard and original test methods were used. Note that existing standard test methods do not estimate enamel insulation resistance to the electrical loads under winding operation of variable-speed drive. We show that estimation of wire corona resistance can be done by high frequency electrical impulse testing. Wire insulation plays the main role of reliability of insulation system.

1. Introduction

Widespread application of variable-frequency control (VFC) system of AC motors built on the base of pulse-width modulation (PWM) is the main development trend of automatic control system. Obvious advantages of VFC system applying go hand-in-hand with important disadvantages – detrimental effect on electrical insulation that could result in reducing sharply a winding lifetime – primarily, the turn insulation as the weakest winding unit. As a result, it acutely brings up an issue about testing methods and means of enamelled wire insulation, providing enough reliable estimate of its mechanical and electrical properties, as well as making assertion about applicability of one wire or another in variable-frequency drive (VFD) windings.

Currently, there are several methods normalized by different standards, specifications, etc. They make it possible to estimate the level of electrical and mechanical properties of enamel insulation. At the same time, the existing methods ignore aspects of winding operation at the VFC. It is known that PWM application causes sharp increase of electrical load level (overvoltage value is in excess of supply voltage value in 10 times; frequency – up to 10^{-4} s), leading to occurrence of partial discharges in winding [1, 2]. As a result, it raises a question about adequacy of existing test facilities for wires, being used in VSD windings [3 - 8].

2. Experimental results

This paper presents a done integrated research of electrical and mechanical properties for widely used winding wires by the existing standard methods. Mechanical properties were estimated by abrasion test, scratching an insulation by needle with ever-increasing load, and by calculation a number of wire insulation defects on unit length as-delivered condition and after winding on the metal rod. In addition, breakdown voltage of enamel insulation was determined as-delivered condition and after winding on the metal rod, equaling to double wire diameter. Obtained results are shown in Table 1 and indicate a satisfactory quality of all the test samples.

Table 1.

Test results of enameled wires

Wire mark	Average (test) value of double needle moves (diameter 0,4mm) to breakdown /min. permissible	Breakdown voltage, kV /min. permissible value by specifications, kV	Number of insulation defects on unit length, mm^{-1}	
			As-delivered condition	After mechanical stress
PET-200-1	275 / 30	6,8 / 5,3	0,00005	0,0001
PET-155	230 / 20	6,1 / 4,4	0,00012	0,00024
PETV-2	250 / 20	8,2 / 5,1	0,0001	0,0002
PETD-180	203 / 70	6,8 / 4,7	0	0,0001
PETD-2-K-180	393 / 70	8,3 / 4,5	0	0,0001
PEEA-155	33 / 20	4,2 / 2,7	0,0028	0,0036

Unfortunately, there are not enough domestic recommendations for estimating an enameled wire resistance to the corona discharges. This fact determined applying original methods for researching this issue. When performing the tests, the main idea is to place a test sample in the corresponding environment with surface (corona) discharges. Such conditions are obtained by using the electrode systems "wire-shot" (Fig. 1), "wire-plate" (Fig. 2). Taking into account the experience, the test voltage range can be from 4 to 5 kV. A lower voltage level provides surface discharges appearing; the top is not sufficient for immediate breakdown of the intact insulation. The criterion is the average time from a voltage input to a insulation breakdown.

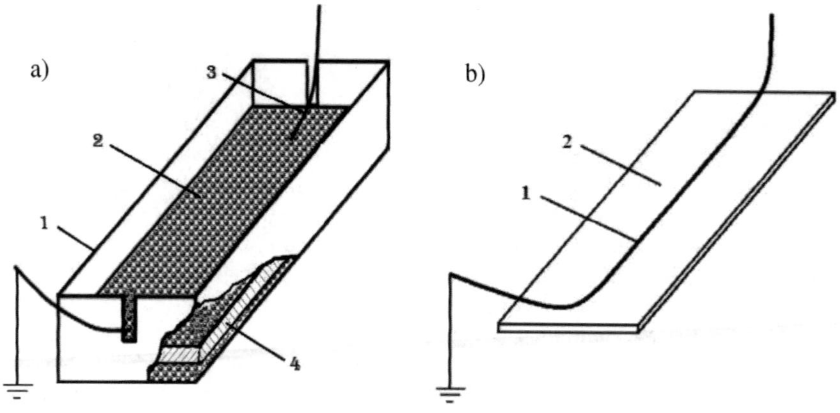

Fig. 1 Test sell: a - 1 – bath; 2 – steel shot; 3 – a grounded wire; 4 – metal electrode; b - 1 – a grounded wire; 2 – copper plate.

The paper presents a number of the tests in the above mentioned electrode systems. The objects have been wire marks: PETD2-K-180, PETD-180, PET-155 and PETV-2. The samples showing the greatest time to breakdown (PETD2-K-180, PETD-180) have additionally been tested after the winding on a metal rod and the tension of 5% and 10%. The results are summarized in Histogram 1 and 2. Thus, the wire PETD2-K-180 withstands better the electrical loads both in the as-delivery condition and after a mechanical impact.

In papers [9-12], a control-measuring system for the winding wire lifetime evaluation is proposed duplicating the conditions created by the frequency regulation system of an electric motor (temperature influence, frequency, voltage and rise time of pulses). In view of this, tests were conducted under the conditions of samples: an alternating voltage with the amplitude of 1200 V,

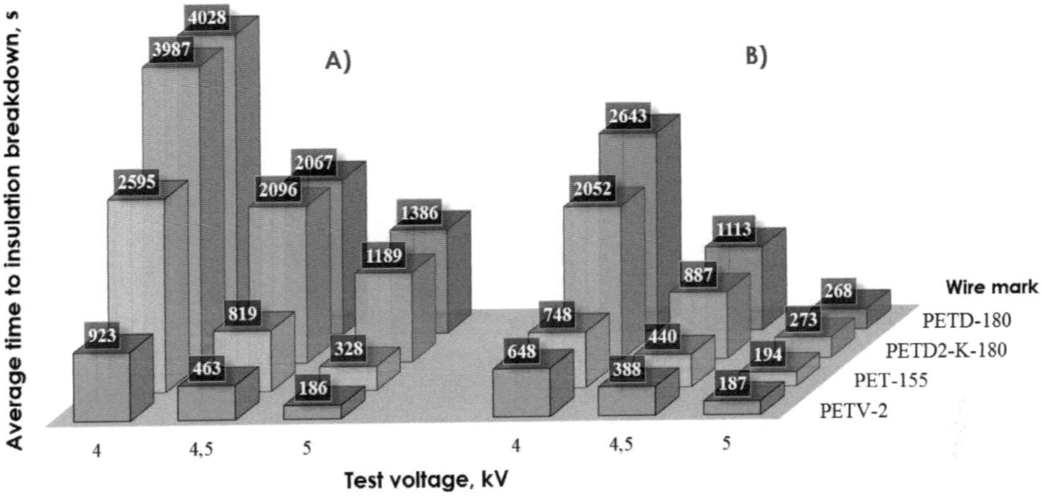

Histograms 1 – Average time to insulation breakdown (as-delivered condition):
a) electrode system "wire-shot"; b) electrode system "wire-plate".

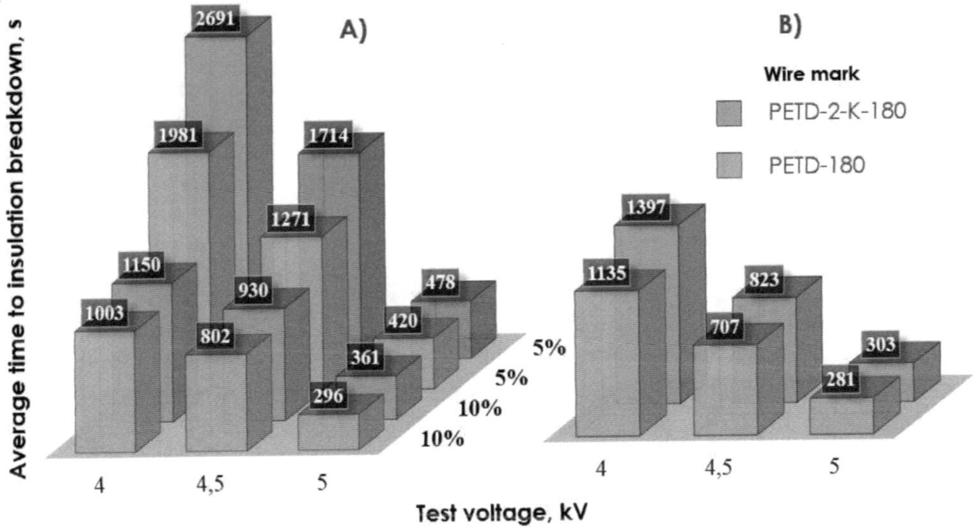

Histograms 2 – Average time to insulation breakdown (after mechanical stress):
a) electrode system: «wire-shot», after the winding;
b) electrode system: «wire-shot», after the tension.

frequency of 400 Hz with the quantization voltage frequency (modulation) of 5 kHz (the duration of the wave front), with the rate of pulse rise 4 ms and temperature corresponding to the heat resistance of enameled winding wires. The test samples are like a twisting wire with the operating zone of 125 mm.

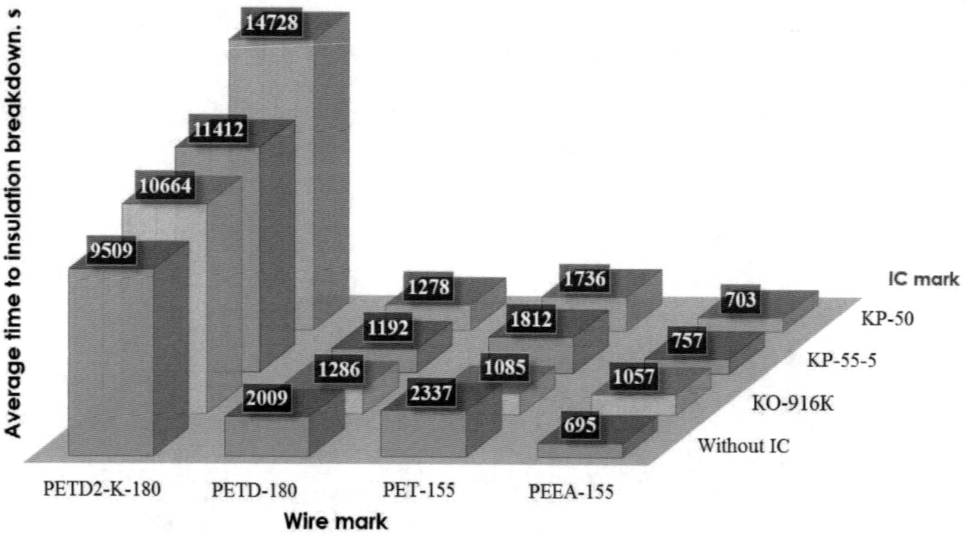

Histogram 3 – Average time to insulation breakdown by using HF test block

A corona resistance criterion is also the average time to an electrical insulation breakdown. Since the turn insulation system of AC windings is a complex of wire insulation and impregnation compound (IC), additional tests were carried out on high frequency test block by using a modern IC in the domestic market. These include impregnating varnish of KO-916K, impregnating compositions of KP-55-5 and KP-50. The results are shown in Histogram 3.

Histogram 3 shows that the PETD2-K-180 wire has the greatest corona resistance at the operating temperature. It enables to make recommendations for its use in the electric machines windings (in case of application of the system frequency control because of having the highest resistance to corona discharges). The wires having the thermal class F and wire PETD-180 are not much worth in size of the average time to the breakdown.

3. Conclusion

1. The obtained results clearly demonstrate the different degrees of enameled wires resistance to electrical loads.

Among the tested wire marks, these two ones PETD-180 and PETD2-K-180 have the greatest time to a breakdown. The value of the average time to a breakdown of their enamel insulation is almost comparable in the delivery state. This is due to the fact that the insulation of these wires is a two-layer composition. The inner layer of the wire PETD-180 is made of a polyester enamel, external - of polyamide-imide enamel. The inner layer of PETD2-K-180 is made of a trihydroxyethylcyanurate enamel containing a silica nanoparticles at least 0.1% vol., and external – of polyamide-imide enamel. Such compositions provide the necessary corona resistance properties of the insulation.

2. It is necessary to carry out a more accurate estimation of the insulation stability to the surface discharges after a mechanical stress (tension, winding on a metal rod). The results show that the greatest resistance to corona discharges after the tension and winding have wire PETD2-K-180.

3. The average time value to the enamel insulation breakdown of samples on a copper plate is lower than in a steel shot. This is due to the fact that the large surface area of the pad electrode promotes an ozone concentration increase and thereby the destruction rate increase of an enamel film. A low value of the average time to an insulation breakdown was obtained as a result of the field inhomogeneity at the samples attaching point. However, the obtained results on a pad electrode show the different levels of enamel wire resistance to electrical discharges.

4. The final estimation of the corona resistance is possible after testing high-frequency electrical pulses. In this case, the sample is exposed to the complex loads under operating conditions: temperature, electric voltage, corona discharges.

5. Application of impregnation compounds does not provide corona resistance, and in several cases deteriorate properties of wire insulation. Thus, wire insulation plays the main role of reliability of insulation system.

6. Application of corona resistance materials when designing the VFD is compulsory condition. And it is necessary to choose materials carefully, not reducing properties of other ones. It is connected with the fact that an increase of the repetitive surge of electrical impulses, generated by the voltage source inverter PWM, the operating loads level is increased, accelerating the aging process of an insulating system. The electrical aging leads to premature wearing and an insulation breakdown.

References

[1] Melfi M., Sung A.M.J., Bell S., Skibinski G.L. 1998 *Effect of Surge Voltage Risetime on the Insu-lation of Low Voltage Machines Fed by PWM Converters* (Industry Applications, IEEE Transac-tions on), pp 766– 775.

[2] Stone G.C., Culbert I.M., Lloyd B.A. 2007 *Stator insulation problems associated with low voltage and medium voltage PWM drives* (Cement Indus-try Technical Conference Record, IEEE, Apr/May), pp 187 – 192.

[3] Kaufhold M., Aninger H., Berth M., Speck J., Eberhardt M. 2000 *Electrical stress and failure mechanism of the winding insulation in PWM-inverter-fed low-voltage induction motors* (IEEE Transactions on Industrial Electronics), vol. **47** **(2)**, pp 396– 402.

[4] Bogh, D., Coffee, J., Stone, G., Custodio, J. 2006 *Partial discharge inception testing on low voltage motors* (Industry Applications, IEEE Transactions on, Jan.-Feb.), pp 148– 154.

[5] Fenger, M., Campbell, S.R., Pedersen, J. 2003 *Motor winding problems caused by inverter drives* (Industry Applications Magazine, IEEE, July-Aug.), pp 22– 31.

[6] Imai T., Kako N., Asano N., Masuda S., Katsuda M. 1984 *Enameled wires having resistance to overload and process for producing same* (US Patent) № **4 476 192**.

[7] Keane J.J., Pauze D.R. 1985 *Corona-resistant wire enamel compositions and conductors insulated therewith* (US Patent) № **4 503 124**.

[8] Hake E.J., Metzler A.D. 1999 *Electrical conductors coated with corona resistant, multilayer insulation system* (US Patent) № **5 861 578**.

[9] Goetter W.R., Vines L.D. 2007 *Use of nanomaterials in secondary electrical insulation coating* (US Patent) № **2007/0087116**.

[10] Cipelli L.C., Ichise H., Furlan F.C. 2005 *Pulsed voltage surge resistant magnet wire* (US Patent) № **2005/0118422**.

[11] Chih-Min Jang, Ru-Shi Liu, Chi-Ting Du, Yung-Chin Lin, Yao-Chung Tu, Wen-Cheng Wu, Chang-Chin Wu, Tsen-Hsu Lin, Yung Chieh Chang. 2002 *Pulsed voltage surges resistant enameled wires* (US Patent) № **2002/0197473** .

[12] Vojislav V. Divljakovic, Joseph A. Kline 200 *Pulse width modulation simulator for testing insulating materials*. (US Patent) № **6051980**.

Calculation of coaxial shunts superheat temperature

E V Bedareva [1], **P S Marinushkin** [2] **and L I Khudonogova**[1]

[1] Tomsk Polytechnic University, Tomsk, Russia
[2] Siberian Federal University, Krasnoyarsk, Russia

E-mail: bedareva@tpu.ru

Abstract
Coaxial shunt temperature field distribution has been obtained and analyzed using software package ANSYS. The main stages of 3D modeling of coaxial shunt temperature field in ANSYS are considered in the paper. It is determined that coaxial shunt temperature field distribution is nonuniform because of the material physical properties and free convection. The difference between theoretical temperature cylinder (60 °C) and experimentally observed temperature value (49 °C) of resistive pipe is caused by the fact that free convection and heat emission were not taken into account when calculating.

1. Introduction
Coaxial shunts are widely used as part of electric energy control devices and electrical power system test equipment, for precision AC and DC currents measurements over the wide dynamic and frequency ranges [1-5].

The main factor that determines the size of resistive element is load-carrying capacity. When choosing the overall dimensions, it is necessary to take into account resistive element superheat temperature [4], representing the difference between surface temperature of resistive element and ambient temperature.

The purpose of this paper is to carry out a thermal field modeling and calculate the value of superheat temperature of shunt prototype. This research will allow to determine critical operating modes and to find the maximum possible pulse current which will not lead to shunt destruction.

2. Coaxial shunt design
Figure 1 shows the structure of a coaxial shunt, consisting of an inner resistive cylinder made of manganin and the outer cylinder made of copper. Currents flow through the inner and outer cylinders in opposite directions. Thus magnetic field cancellation is achieved. The structure cooled by natural convection [1].

Simulations were performed for the coaxial shunt with the following parameters: resistive cylinder thickness – 1 mm; length – 90 mm; inner radius – 10 mm; the gap between cylinders – 1 mm; rated current – 1 kA; pulse duration – 1 s; transient period – 2,79 mks; superheat temperature – 40 °C. The inner cylinder is made of manganin, external – of copper, current and potential leads – brass.

3. Coaxial shunt thermal field modeling
Modeling of the shunt temperature field was carried out using finite element method (FEM) in software package ANSYS Mechanical.

Modeling process in ANSYS Mechanical consists of the following stages:
1) Importing a model created in a computer-aided design (CAD) system, or creating a solid model within ANSYS;
2) Gluing the parts;
3) Choosing the type of elements;
4) Specifying material properties;
5) Generating the mesh;

6) Applying boundary conditions and loads;
7) Defining dynamic analysis options;
8) Obtaining the solution;
9) Reviewing results.

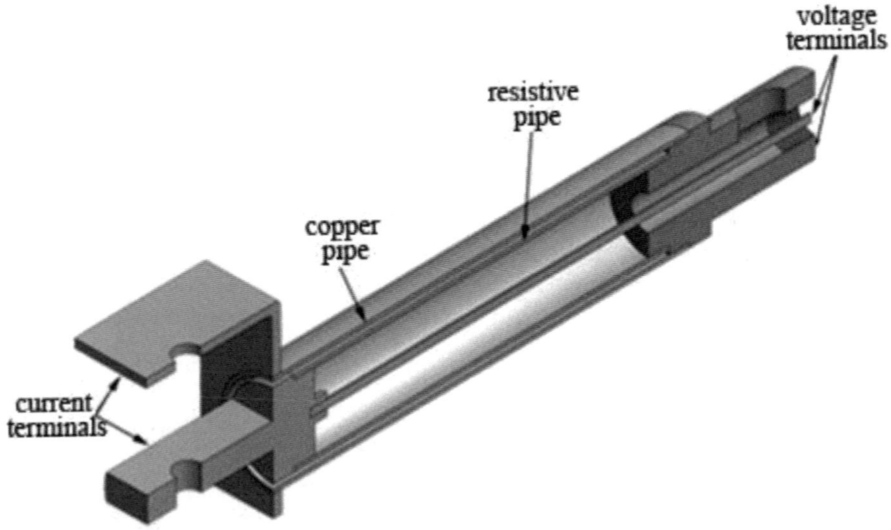

Figure 1. Coaxial shunt design

Geometrical parameters were calculated by means of the express calculator [6]. The solid model of a shunt was developed in Inventor Professional software package. Additional volumes representing areas filled with air were introduced into model of the shunt to account the heat exchange between parts.

To import the model from file (in SAT format) it is necessary to do the following operation: Utility>Menu > File > Import > SAT.

After importing the model has wireframe representation. To fix it and generate volumes the following operation should be done: Utility Menu > PlotCtrls > Style > SoliD Model Facets > Solid Model Facets > Normal Faceting.

For further modeling it is required to create common surfaces for neighbouring parts, which can be provide by gluing operation: Main Menu > Preprocessor > Modeling > Operate > Booleans > Glue > Volumes > Pick All. Contact phenomena between parts doesn't take into account.

Mesh generation is one of the most important stages for finding the solution. The choice of element type affects the set of degree of freedom, which, in his turn, determine the analysis type: structural, thermal, magnetic, electric, etc.

In this application, for 3D analysis, SOLID70 element for modeling air gapes and SOLID69 element for modeling metals are chosen. (Main Menu > Preprocessor > Element Type > Add/Edit/Delete >Thermal Mass > Solid > Scalar Tet 70, then Coupled Field > Scalar Tet 69).

For all materials, including in the prototype model, four physical properties are to be specified: thermal conductivity, thermal capacity, density, and resistivity. Inner pipe is made from manganin, and the outer one is made from copper. For current and voltage terminals brass is chosen. The specific material properties are defined by executing the command set: Main Menu > Preprocessor > Material Props > Material Models > required properties choosing.

The next step in the modeling process is the mesh generation. In this particular case tetrahedral shaped elements mesh were used. For meshing operations a meshing feature called smart element sizing ("SmartSize") was applied. It determined element size of 10 for both manganin and copper pipes, and size of 6 for the other parts of the prototype. The reason of the different element sizes for different parts is that pipes are the most essential part for finding the solution and obtaining the temperature field distribution. Mesh generation is performed by the following operations: Main Menu > Preprocessor > Meshing > MeshTool > set the "Element type number" > choose the material in accordance with the elements type > set the value of SmartSize parameter > set the parameter "Shape" = "Tet" (tetrahedral mesh) > set the parameter "Shape" = "Free" (Free meshing) > choose the part to be meshed in "Mesh Volumes" window.

The next stage is applying loads and boundary conditions. In the given case loads on elements are represented by:

1) Zero potential;
2) Current value of 1 kA;
3) Free convection.

Zero potential (A) and current (B) are applied on current terminals surfaces as shown in figure 2. Free convection (heat-transfer coefficient is 30 W/m^2·°C, ambient temperature is 20 °C) is applied on surfaces set off in turquoise, i.e. on all external surfaces.

Figure 2. Applying loads and boundary conditions

Process of setting the input current pulse consists of three load steps. The first one is characterized by the following parameters: initial time is 0.1 s and initial temperature is 20 °C. For the second step transient period is 2.79 mks, pulse duration is 1 s, and current value is 1 kA. In the third step current amplitude is 0 A, time to exposure completion is 5 s.

After all preprocessing stages, the performed task can be solved. To obtain the solution the following operation should be done: Main Menu > Solution > Solve > From LS Files. The order of load steps is specified in the options window.

In the postprocessing stage, obtained results visualization is implemented by the following commands: Main Menu > General Postproc > PlotResults > Contour Plot > Nodal Solu >"DOF solution" >"Temperature".

Figure 3 shows coaxial shunt temperature field distribution for the period of one second current pulse exposure.

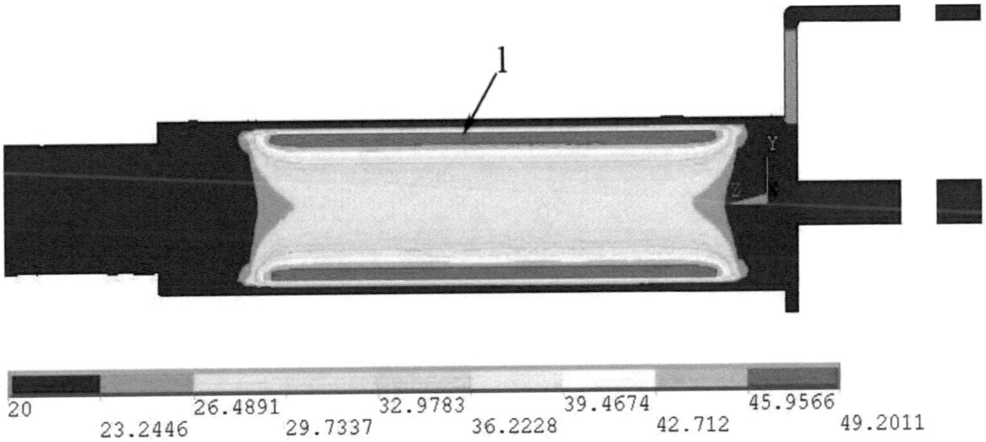

Figure 3. Coaxial shunt thermal field distribution at time $t = 1$ s

According to figure 3, when large-amplitude pulse current flows through the shunt, maximum superheat temperature occurs on the inner cylinder, decreasing towards the ends. Non-uniform thermal distribution of resistive element takes place because of the physical properties of materials, which current and voltage terminals are made from. The maximum temperature value is 49,2 °C.

Plot in figure 4, demonstrating temperature-time dependence for node 1 (figure 3), allows to estimate dynamic temperature changing of the resistive pipe in different time points. It is relevant to note that temperature-time dependence for other nodes of inner pipe is analogous to the one represented in figure 4.

When current pulse is stopped, shunt superheat temperature begins to decrease due to heat emission and heat exchange between other parts.

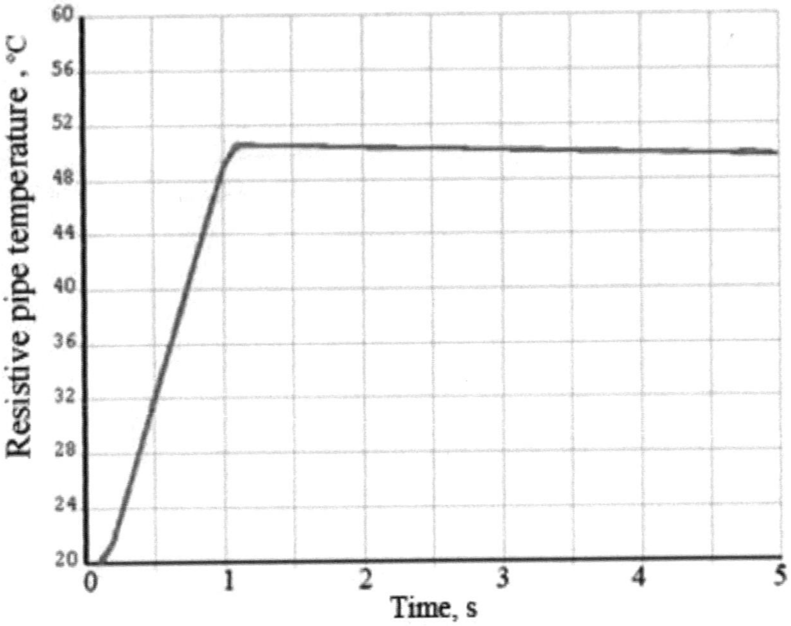

Figure 4. Resistive pipe temperature changing in time

4. Conclusion

In this paper coaxial shunt temperature field distribution has been obtained and analyzed using ANSYS. From the results obtained, the following conclusions can be made:

1) The theoretical temperature of a resistive cylinder (60 °C) given by the calculations doesn't coincide with experimentally observed temperature value (49,2 °C) because the theoretical calculations do not take into account natural convection and heat emission.

2) Ohmic shunt resistance does not change when temperature increases, hence, shunt behaves as a linear stable system.

3) Current pulse, effecting on the shunt, is not critical, so current amplitude can be therefore increased, and pulse duration can be extended as well.

4) Maximum superheat temperature occurs in the center of the inner pipe that caused by free convection and physical properties of the materials used in the model.

References

[1] Bedareva E, Tsymbalist E, Muravyov S and Baranov P 2013 Influence of methods of potential terminal connection on dynamic behavior of coaxial shunts *Bulletin of Tomsk Polytechnic University* **322** pp 159–164

[2] Muravyov S, Borikov V and Natalinova N 2009 A Computer system: measurement of welding surge current *Measurement and Control* **42** pp 44–47

[3] Kawamura T, Haginomori E, Goda Y and Nakamoto T 2007 Recent developments on high current measurement using current shunt *Transactions on electrical and electronic engineering* **2** pp 516–522

[4] Zarevich A, Muravyov S and Bedareva E 2013 Digital signal processing in pulse method for measuring frequency response function of high-current shunt *19th IMEKO TC4 Symposium* pp 496–500

[5] Muravyov S, Borikov V and Kaysanov S 2006 Computer system for measurement of welding process parameters *18th IMEKO World Congress* **2** pp 77–80

[6] Borikov V 2010 Computer-aided calculation of electrical and design parameters of coaxial shunts in LabVIEW professional development system *Prybory* **3** pp 42–46

IOP Conf. Series: Materials Science and Engineering **66** (2014) 012006 doi:10.1088/1757-899X/66/1/012006

Research of piezoelectric vibratory angular rate sensors with improved metrological characteristics

Pavel Marinushkin[1], **Elena Bedareva**[2]

[1]Siberian Federal University, Krasnoyarsk, 660074, Russia
[2]National Research Tomsk Polytechnic University,Tomsk, 634050, Russia

Email: marinushkin_ps@mail.ru

Abstract. Piezoelectric vibratory angular rate sensors are widely used in different applications. The current article presents and discusses a theoretical research of scale factor temperature drift of the piezoelectric gyroscope. Numerical simulation of scale factor temperature sensitivity using ANSYS finite element software package is also considered. The theoretical results are compared with the experimentally obtained values.

1. Introduction

In recent years areas related to the development of compact, reliable and cost-effective navigation and control systems for moving objects have been actively developing. These are: spatial orientation control of wells; orientation, navigation and control systems for unmanned aerial vehicles, autonomous underwater vehicles and mobile robots; stabilization systems for railroad transport. These systems use angular rate sensors (gyroscopes) to measure angular displacements of objects.

Analysis of publications and research work on the subject [1-3] has shown that despite of the fact that MEMS gyroscopes receive more and more wide application in the systems of navigation, stabilization and orientation of various objects, the problems concerning research and design of piezoelectric angular rate sensors are still actual. On a number of parameters (cost, power consumption) piezoelectric vibrating gyroscopes overcome micromechanical gyroscopes that make them more suitable for mobile, autonomous instrumentation [1].

Various designs and principles of operation of the piezoelectric vibrating gyroscopes were previously described in articles. However, above publications contain no detailed description of the dynamics and errors of the piezoelectric vibrating gyroscopes. Methods of calculation of sensitivity, zero offset and temperature drift are also not well presented.

Thus it is of great relevance to research and develop reliable and efficient modeling and design methods of the piezoelectric gyroscopes. As well as development of adequate models, providing a detailed study of the physical processes occurring in piezoelectric gyroscopes, and search for new design solutions.

One of the main problems of vibratory gyroscopes development is increasing of the gyroscope sensitivity to angular velocity. It is known, that the highest sensitivity of a vibratory gyroscope is achieved at coincidence of resonance frequencies of primary and secondary vibrations. That is why designs of piezoelectric vibratory gyroscopes based on axisymmetrical resonators look promising since rotational symmetry allows to achieve equality of frequencies of primary and secondary vibrations. Another important advantage of such gyroscopes is monolithic design of their resonators. Monolithic design of the resonator allows to minimize using of diverse materials and thus to increase stability of gyroscope characteristics in comparison with the gyroscopes based on conventional prismatic and bimorph resonators.

The design of such vibratory angular rate sensor containing thin-walled cylinder (resonator) made of piezoelectric ceramics is presented on figure 1.

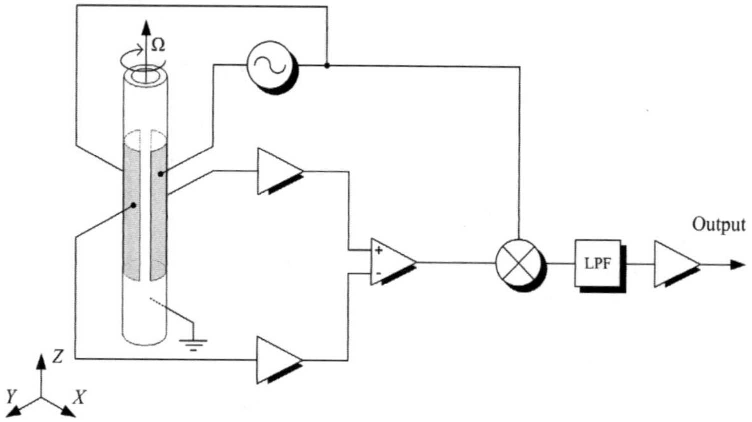

Figure 1. Tubular piezoelectric angular rate sensor.

Resonator has two pairs of electrodes on external surface, and a common electrode on inner surface. The principle of action of the piezoelectric vibratory gyroscope is based on transformation of energy of the primary vibrations excitated by means of the first pair of electrodes in plane *XZ* to energy of secondary vibrations in *YZ*-plane under action of Coriolis forces exerted by rotation of a gyroscope around *Z*-axis. The second pair of electrodes serves for detecting of vibrations in *YZ*-plane. An oscillator is connected between the driving electrodes for self-oscillating driving of resonator. When the excitation signal is applied to driving electrodes, resonator makes bending vibrations in *XZ*-plane. The signal due to the Coriolis force is synchronously detected by the synchronous detection circuit, low-pass filtered and then amplified by DC amplifier.

Simple 1-D analytical model of the vibratory piezoelectric gyroscope based on classical beam theory is considered in [4]. Disadvantage of this model is that shearing strain and rotational inertia are not taken into account when considering bending vibration of the resonator. This results in a higher predicted eigenfrequencies for a given set of boundary conditions. To increase the prediction accuracy Timoshenko beam model was used for the description of the gyroscope dynamics. Timoshenko beam theory considers effects of a shear strain and rotational inertia and thus can be used for describing the behavior of short beams (it's a better approximation of the real resonators). The obtained differential equations system was solved using MATLAB software.

Using the obtained modified model numerical estimations of the instability of the gyroscope scaling factor and shift of the resonator working frequencies in various temperature conditions were performed. It was found that the main factors affecting sensitivity of piezoelectric tubular vibratory gyroscope (figure 2) are temperature variations of dielectric permittivity ε_{33} and mechanical quality Q_m (37 % and 16 % respectively). The temperature variations of piezoelectric modulus d_{31} and compliance s_{11} have much less influence on sensitivity (10 % and 2 % respectively).

Technological imperfections, which are always present during the fabrication process of piezoelectric gyroscope, are of primary importance since they significantly reduce the measurement accuracy. Such technological imperfections include deviations from the ideal mass, stiffness and damping distributions. Resonators with imperfections show splitting of natural frequencies, location of the two mode shapes and different damping factors associated with each vibration mode. Another reason for the errors of piezoelectric gyroscopes is the presence of temperature-induced errors due to the temperature sensitivity of mechanical and electrical material properties of piezoelectric ceramics.

The change of material properties leads to imbalance and a change in the dynamic characteristics of the resonator.

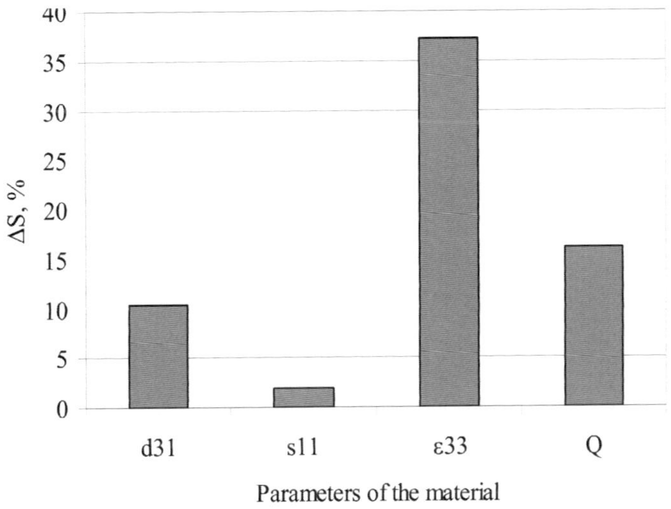

Figure 2. The influence of material properties variations on gyroscope sensitivity [5].

To evaluate the effect of technological temperature-induced errors on the output signal of the gyroscope, the finite elements method in three-dimensional formulation was used. Practical implementation of the method is executed using the package finite element simulation ANSYS v11. Numerical calculations for imperfect thin-walled cylindrical piezoelectric resonator (axial misalignment) were performed.

The transient analysis was performed to calculate the trajectories of the gyroscope resonator motion (figure 3, figure 4). These graphs show that the wall thickness variation leads to imbalance of the resonator. Thus, the displacement along the Y axis of the resonator under the influence of the Coriolis force is distorted by the projection of the drive on the same axis, which leads to an error in the measurement of angular velocity. As a result, at the output of the piezoelectric angular rate sensor the spurious output signal in the absence of angular speed Ω will be observed.

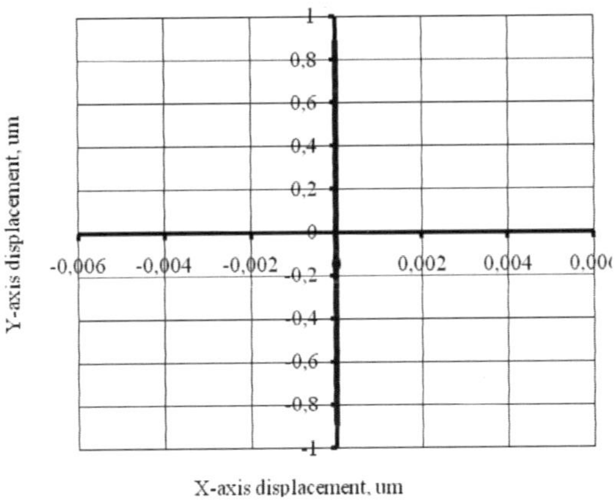

Figure 3. The trajectory of the ideal resonator motion (Ω = 0 rad/s).

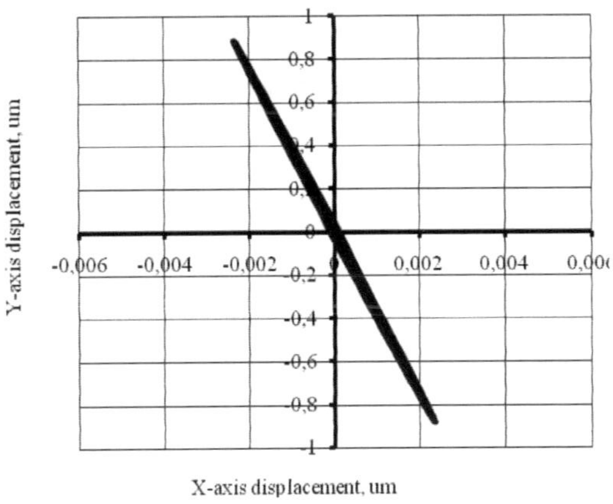

Figure 4. The trajectory of the imperfect resonator motion (Ω = 0 rad/s).

Temperature-induced position shift of the resonator vibration mode planes leads to the instability of the scale factor of an angular rate sensor [6]. The corresponding plot for two values of splitting of the natural frequencies of the resonator (0.3 Hz and 0.5 Hz) is shown in figure 5.

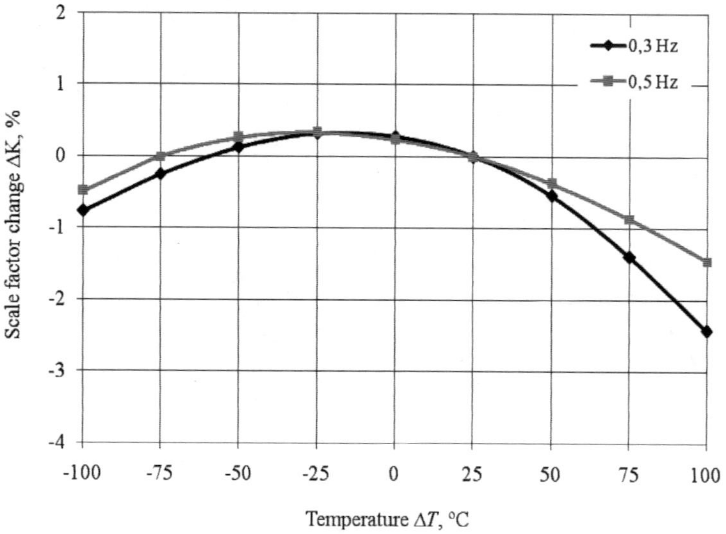

Figure 5. Temperature dependence of the scale factor of an angular rate sensor.

2. Results and discussion

Simulation results were verified experimentally. Characterization of the resonance frequency was conducted in laboratory conditions at room temperature. The resonators were excited by applying an AC voltage with amplitude not exceeding 100 mV in order to avoid resonators dissipative heating.

For the validation of the developed analytical model, results were compared with experiment and finite-element calculations (table 1). The closest match to the measured values of the resonant frequency provided numerical modeling. Also from the presented data it is visible that the proposed analytical model gives larger deviation of resonant frequency than numerical calculation but still provides a smaller error than the known analytical model [4].

Table 1. Results comparison.

Models to compare	Natural frequency, Hz	Deviation from the experimental results, %
Known analytical model [4]	13220	32.46
Proposed analytical model	12288	23.1
Proposed numerical model	10648	6.69
Experimental results	9980	–

Temperature-frequency characteristic of the resonator for the main flexural mode was measured over a -20 … +80 °C temperature range and compared to the analytical and numerical results (figure 6). From figure 6 it is visible that a quadratic parabola with a slope coefficient 9×10^{-3} $(^{\circ}\mathrm{C})^{-2}$ provides good agreement with the experimentally obtained curve. The difference between experimental and calculated data obtained within analytical and numerical models is 23.6% and 7% respectively.

Also it was found that the average value of the scale factor (magnitude of the drive signal is in the range 5 – 15 V) is 0.35 – 1.04 mV/deg/s (figure 7). Preliminary experiments showed that the value of zero drift is 0.8 deg/s. Obtained characteristics are in qualitative agreement with the results of analytical and finite-element modeling.

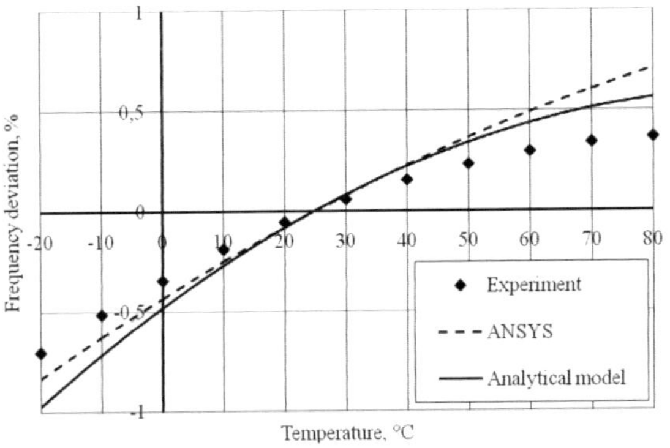

Figure 6. Temperature-frequency response of the piezoelectric resonator.

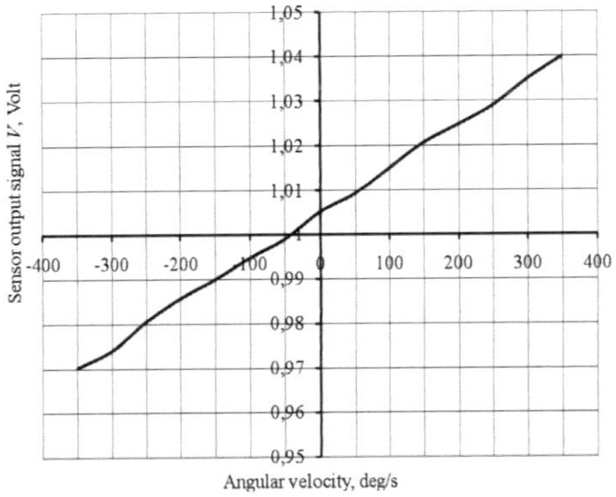

Figure 7. Rate response of the angular rate sensor.

Thus, experimental studies of the piezoelectric vibratory gyroscope prototype confirmed the accuracy of the theoretical studies and efficiency of the developed gyroscope as an angular rate sensor. The results obtained in this work can be used to develop error models of piezoelectric angular rate sensors of similar design, and provide an opportunity to improve metrological characteristics of position and deformation control instruments.

Acknowledgments
This work was financially supported by the Ministry of Education and Science of the Russian Federation in Siberian Federal University (Contract № 02.G25.31.0041).

References

[1] Tanaka M 2007 An industrial and applied review of new MEMS devices features *Microelectron. Eng.* **84** pp 1341–44
[2] Soria L, Pierro E, Garbone G, Contursi T 2009 Tuning fork microgyrometers: narrow gap vs. wide gap design *Journal of Sound and vibration* **322** pp 78–97
[3] Xi X, Wu Y, Wu X, Luan Q, Wu X 2013 A novel combined fused silica cylinder shell vibrating gyroscope *Sensors and Materials* **25** pp 323–339
[4] Yang J, Fang H 2003 A new ceramic tube piezoelectric gyroscope *Sensors and Actuators A* **107** pp 42–49
[5] Marinushkin P 2011 Modeling of temperature-induced errors of the piezoelectric vibratory gyroscope *IEEE Third School and Seminar on Fundamental Problems of Micro/Nanosystem Technologies MNST'2011* pp 19–21
[6] Wu Y, Xi X, Tao Y, Wu X, Wu X 2011 A study of the temperature characteristics of vibration mode axes for vibratory cylinder gyroscopes *Sensors* **11** pp 7665–77

IOP Conf. Series: Materials Science and Engineering **66** (2014) 012007 doi:10.1088/1757-899X/66/1/012007

Circuits for differential signal extraction in the lock-in amplifiers

P Baranov and T. Bukrina
National Research Tomsk Polytechnic University, Tomsk, 634050, Russia

E-mail: bpf@tpu.ru

Abstract. Confirmation of the metrological characteristics of measuring transducers provides maximum accuracy of the method of comparison with the standard instrument. Practical application of this technique is impossible without accurate instruments. As a measurement instrument to ensure the resolution of the order of several nanovolts when comparing two AC signals over a wide dynamic range, widely used lock-in amplifier with a differential input. The paper presents the results of the analysis of circuits based on operational amplifiers for comparing two signals and extraction differential signal in lock-in amplifier with a differential input. Common-mode rejection ratio for each of the circuits was estimated.

1. Introduction

Upon confirmation of the metrological characteristics of measuring transducers provides maximum accuracy of the method of comparison with the standard instrument. Practical application of this technique is impossible without accurate instruments of compare, resolution which largely determines the minimum measurement error. As a measurement instrument to ensure the resolution of the order of several nanovolts when comparing two AC signals over a wide dynamic range, widely used lock-in amplifier with a differential input (LIA).

The input stage of LIA providing simultaneous comparison of the two voltages and the extraction of the differential signal is the key terms of increased the dynamic range compared voltages and maximum resolution.

The paper presents the results of the analysis of circuits based on operational amplifiers for signal comparison and measurement of the differential signal in LIA.

2. Extraction circuits of differential signal

In modern LIA with a differential input used circuits based on operational amplifiers (op-amp) to extraction the differential signal [1-3].

Output voltage U_{out} of the op-amp will considered as a function of the differential U_d and common-mode U_{cm} voltages at the inputs of op-amp.

$$U_{out} = f\left(U_d, U_{cm}\right) = K_d U_d + K_{cm} U_{cm}, \qquad (1)$$

where K_d – the gain of the differential signal;

K_{cm} – common-mode signal gain.

Solving equation (1) relative to the differential signal U_d we are get:

$$U_d = \frac{U_{out}}{K_d} - \frac{U_{cm}}{K_{CMMR}} = \begin{cases} U_{out} / K_d \text{ if } U_{cm} = 0 \\ -U_{cm} / K_{CMRR} \text{ if } U_{out} = 0 \end{cases}, \tag{2}$$

where $K_{CMRR} = K_d / K_{cm}$ – common-mode rejection ratio (CMRR).

CMRR in comparison circuits based on the op-amp depends on two components: accuracy ratio of resistance in signal lines and the internal structure of the specific op-amp.

For different extraction circuits of the differential signal based on op-amps will evaluate CMRR. Since the value of CMRR determines the maximum resolution when comparing two similar signals. Analysis of the circuits will be carried out at DC with internal CMRR $\rightarrow \infty$.

2.1 Subtraction circuit using summing op-amp

Subtraction signals can be accomplished by adding to the inverted one of the input signals [4]. The circuits carries out this operation is shown in figure 1.

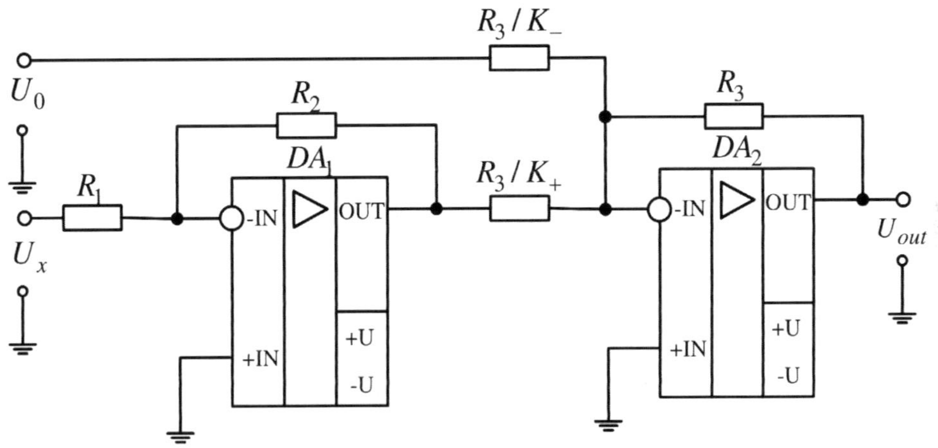

Figure 1. Subtraction circuit using summing op-amp

Op-amp DA_1 inverts the input voltage U_x, on condition of equality resistors R_1 and R_2 at the output voltage is:

$$U_{out} = K_+ U_x - K_- U_0. \tag{3}$$

Input signals represent as the sum of the common-mode and differential voltages in the inputs of op-amp DA_2 (4) and substitute in (3) to estimate the CMRR.

$$U_x = U_{cm} + \frac{U_d}{2} \text{ and } U_0 = U_{cm} - \frac{U_d}{2}. \tag{4}$$

$$\begin{gathered} U_{out} = (K_+ - K_-) U_{cm} + \frac{(K_+ + K_-)}{2} U_d; \\ K_{cm} = K_+ - K_-; \\ K_d = \frac{K_+ + K_-}{2}. \end{gathered} \tag{5}$$

From equation (5) CMRR is:

$$K_{CMRR} = \frac{1}{2} \cdot \frac{K_+ + K_-}{K_+ - K_-}. \tag{6}$$

If the resistance R_3/K_+ and R_3/K_- differ by value ΔK_d then:

$$K_+ = K_d + \frac{\Delta K_d}{2};\qquad K_- = K_d - \frac{\Delta K_d}{2}. \tag{7}$$

Then, substituting (7) in (6):

$$K_{cmrr} = \frac{K_d}{\Delta K_d}. \tag{8}$$

Consequently, the common-mode rejection ratio is inversely proportional to the difference between the resistances R_3/K_+ and R_3/K_-.

2.2 Subtraction circuit using one op-amp

Figure 2 shows the subtraction circuit using one op-amp.

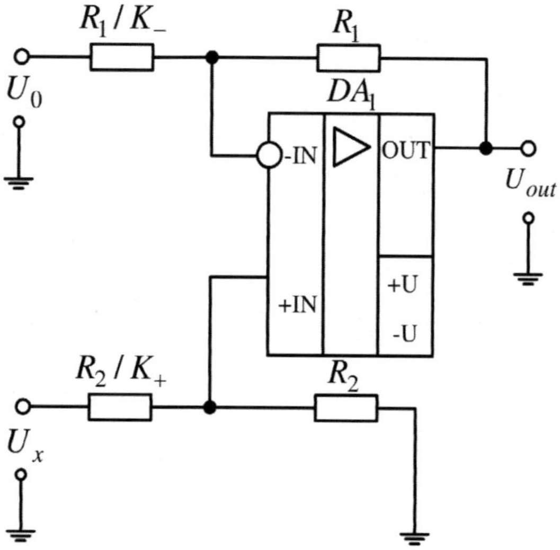

Figure 2. Subtraction circuit using one op-amp

Superposition is used to calculate the output voltage resulting from each input voltage:

$$U_{out} = \frac{(1+K_-)}{(1+K_+)} \cdot K_+ U_x - K_- U_0. \tag{9}$$

To estimate CMRR substituting (4) in (9):

$$U_{out} = \frac{(K_+ - K_-)}{(1+K_+)} U_{cm} + \frac{(K_+ + K_- + 2K_+K_-)}{2(1+K_+)} U_d;$$
$$K_{cm} = \frac{(K_+ - K_-)}{(1+K_+)};\tag{10}$$
$$K_d = \frac{(K_+ + K_- + 2K_+K_-)}{2(1+K_+)}.$$

From equation (10) CMRR is:

$$K_{CMRR} = \frac{K_d}{K_{cm}} = \frac{1}{2} \cdot \frac{(1+K_-)K_+ + (1+K_+)K_-}{(1+K_-)K_+ - (1+K_+)K_-}. \tag{11}$$

If the resistance R_2/K_+ and R_1/K_- differ by value ΔK_d then substituting (7) in (11):

$$K_{CMRR} \approx (1+K_d)\frac{K_d}{\Delta K_d}. \tag{12}$$

Consequently, at constant K_d common-mode rejection is inversely proportional difference resistances R_2/K_+ and R_1/K_- and directly proportionally to the gain of the differential signal. Thus, in the above circuits CMRR will depend on the accuracy of the ratio of resistances in the signal lines and on the sources output impedance of the compared signals.

2.3 Subtraction circuit using instrumentation amplifier

The instrumentation amplifier (IS) [5-6] is a precision amplifier unit with a differential input and closed-loop. IS provides a gain of the voltages difference between the two input signals, weakening any signals that are common to both inputs.

The most widely used IS circuit consisting of three op-amps (figure 3).

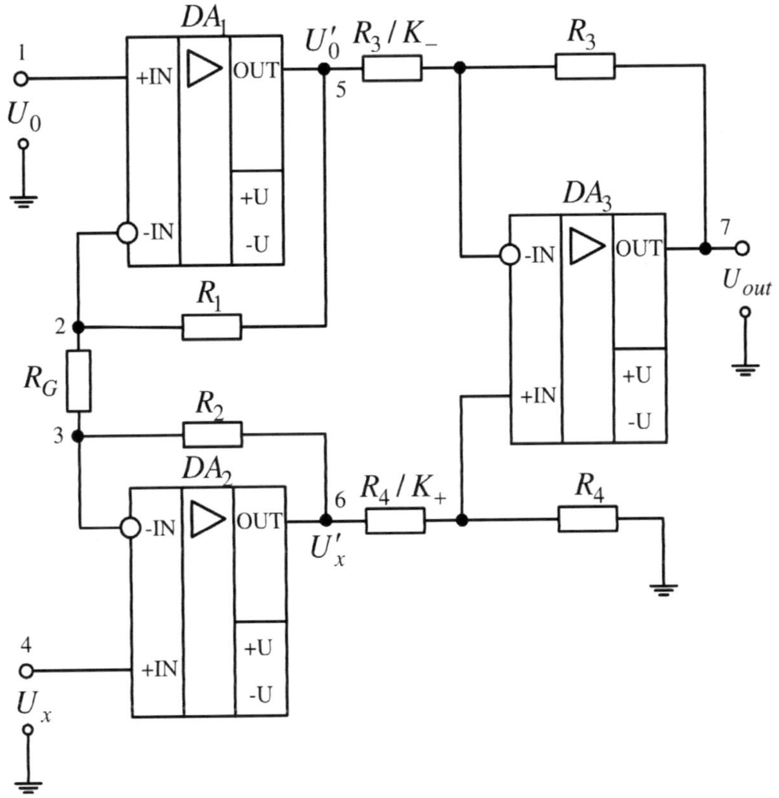

Figure 3. Instrumentation amplifier circuit consisting of three op-amps

IS constructed on a two-stage circuit, the first stage consists of op-amps DA_1, DA_2 and resistors R_1, R_2, R_G, has a the balanced input and high input impedance. The second stage consists of the op-amp

DA_3 and resistors R_3, R_4, R_5, R_6, forms the subtraction circuit on one op-amp, which is shown in figure 2. Using graph theory, we construct the signal graph for IS on the three op-amps (figure 4)

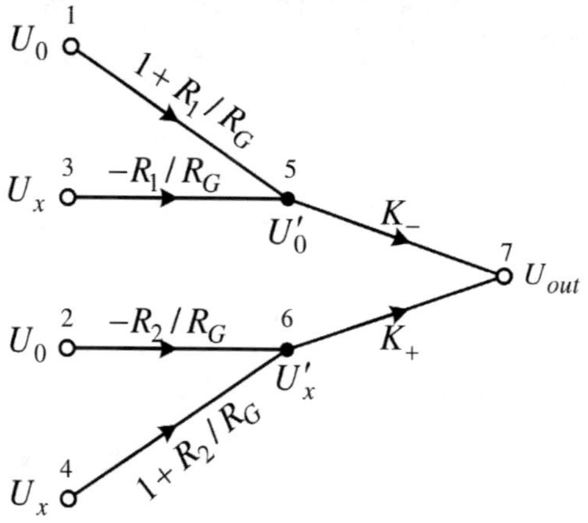

Figure 4. Signal graph for IS on the three op-amps

Amplifiers DA_1 and DA_2 we assume perfect, then on the basis of their inputs equipotential voltage U_0, simultaneously acts in the first and second nodes of the graph, and the voltage U_x, simultaneously acts in the third and fourth nodes. Using the principle of superposition, we find that the voltage U'_0 in fifth node of the graph is generated by transmission noninverting voltage U_0 of the first node and the inverting transmission voltage U_x from the third node circuit:

$$U'_0 = \left(1+\frac{R_1}{R_G}\right)U_0 - \frac{R_1}{R_G}U_x. \tag{13}$$

Similarly, we obtain an expression for the voltage U'_x:

$$U'_x = \left(1+\frac{R_2}{R_G}\right)U_x - \frac{R_2}{R_G}U_0. \tag{14}$$

The output voltage of the second stage and respectively output voltage of the IS:

$$U_{out} = \frac{(1+K_-)}{(1+K_+)}K_+U'_x - K_-U'_0. \tag{15}$$

Under the condition of equality K_+ and K_- output voltage of the IS:

$$U_{out} = (U_x - U_0)K_{d1} = (U_x - U_0)\left[1+\frac{(R_1+R_2)}{R_G}\right], \tag{16}$$

where K_{d1} - differential signal gain in first-stage.
To estimate CMRR substituting (4) in (15) considering expressions (13-14):

$$U_{out} = \frac{(K_+ - K_-)}{(1+K_+)}U_{cm} + \\ + \frac{(K_+ + K_- + 2K_+K_-)R_G + 2\left[K_+R_2 + K_-R_1 + K_+K_-(R_1+R_2)\right]}{2R_G(1+K_+)}U_d. \tag{17}$$

From equation (17) CMRR:

$$K_{CMRR} = \frac{\left(K_+ + K_- + 2K_+K_-\right)R_G + 2\left[K_+R_2 + K_-R_1 + K_+K_-\left(R_1 + R_2\right)\right]}{2R_G\left(K_+ - K_-\right)}. \quad (18)$$

If the resistance R_4/K_+ and R_3/K_- differ by ΔK_d and resistance R_1 and R_2 differ by value ΔR, substituting (7) and (19) into (18) we obtain:

$$R_1 = R + \frac{\Delta R}{2}; \quad R_2 = R - \frac{\Delta R}{2}. \quad (19)$$

$$K_{CMRR} = \frac{\left(K_d + K_d^2 - \Delta K_d^2\right)\left(R_G + 2R\right)}{2\Delta K_d R_G} + \frac{2\Delta R}{R_G}. \quad (20)$$

From expression (17) shows that the gain of common mode signal in the first stage is 1. Consequently CMRR will increase proportional gain the differential signal of the first stage.

CMRR does not depend on the sources output impedance of the compared signals in IS. Despite the fact, what CMRR depends on internal resistances of the first and second stages, they are practically identical for IS in integrally fabricated with laser trimmed resistances. This fact is a decisive advantage when choosing IS to compare two signals and extraction the differential signal in LIA as compared to the above discussed circuits.

Gain in IS set by changing the resistor R_G (figure 3). If an external resistor used, then its accuracy and temperature coefficient is directly influence on accuracy and drift of differential gain IS, also decrease CMRR (20). It is expedient to choose IS with programmable gain, since in such IS resistor R_G matched with the other elements on the nominal resistance and temperature coefficients, for maximum accuracy and linearity of the gain in operating frequency band.

3. Conclusion

Analysis circuits based op-amps for comparing two signals and extraction differential signal was performed. Common-mode rejection ratio for each of the circuits was estimated.

Proposed use instrumentation amplifier integrally fabricated with programmable gain for comparing two signals. Since such instrumentation amplifiers common-mode rejection is practically independent of the internal resistances and the resistor determines the gain of the differential signal matched with the other elements on the nominal resistance and temperature coefficients, for maximum accuracy and linearity of the gain in the range of operating frequencies.

References

[1] Callegaro L, D'Elia V 2002 *IEEE Transactions on Instrumentation and Measurement* **51** pp 632-635.
[2] Bengtsson L 2012 *Review of Scientific Instruments* **83** pp 1263-1271.
[3] Min M, Martens O, Parve T 2000 *Measurement* **27** pp 21-28.
[4] Horowitz P, Hill W 1989 *The Art of Electronics* (Second ed.), Cambridge University Press.
[5] AN1298.2 2009 *instrumentation amplifier application note* p 35.
[6] AN165 1988 *Integrated operational amplifier theory* p 10.

Experimental Investigations of Microwave Signal Attenuation in Radio Link within Geophysical Information Transmission

A S Goponenko, O V Stukach, V A Kochumeev, and A B Mirmanov
Tomsk Polytechnic University, Tomsk, Russia
S.Seifullin Kazakh Agro Technical University, Astana, Kazakhstan
E-mail: tomsk@ieee.org

Keywords: measurement-while-drilling, MWD, LWD, microwave pulse generator, power transmitter, Gunn diode generator.

Abstract. The paper describes the basic results of the project "Borehole Microwave" as researchers design microwave pulse signal transfer through a drilling pipe as a new communication channel. Methods of the telemetry information transfer are investigated. The "measurements while drilling" (MWD) mock-up of system based on the new concept of creation of telemetry equipment is developed. The experimental bench for electro-physical researches of various media in the drilling pipes is carried out. Investigations of measurement of attenuation of microwave pulse in drilling pipes are presented. Results show a possibility of operation in perspective field of investigations. The data transfer through a new communication channel in microwave band will open new possibilities of improving of the measurement-while-drilling equipment.

1. Research objectives

In the gas and petroleum drilling process, geophysical measurement system is usually used to provide real-time monitoring of the position and orientation of the bottom hole. Measurement systems provide borehole data collection, its transmission through communication channel, receiving information on the surface, processing and presentation of the results for the operator (Fig. 1). As a rule, any geophysical system consists of:
1 bottom equipment;
2 ground equipment;
3 communication channel;
4 accessories;
5 source of electrical energy (for systems with a wireless link).
Nowadays cable, hydraulic, acoustic, electromagnetic and combined channels are used in telemetry systems. Each of them has advantages and disadvantages, which are discussed in detail in [1, 2].
The communication channel primarily defines the structure of the whole system, informative capacity and reliability. The tasks of increasing channel capacity and reducing the resource using in geophysical work can not be solved within the existing systems and communication channels [3]. Therefore, the investigations of new ways of transmission and types of communication channels are urgent for improving geophysical equipment.

2. Radio link

Radio link now is used only in surface equipment of geophysical systems [4, 5]. But the use of wireless transmission of information directly into the borehole is not less preferable because of the possibility to work during drilling and less material using. Spreading the radio waves through the drill pipe is perspective but due to the lack of good receiving and transmitting equipment for telemetry it has not been studied before [6].

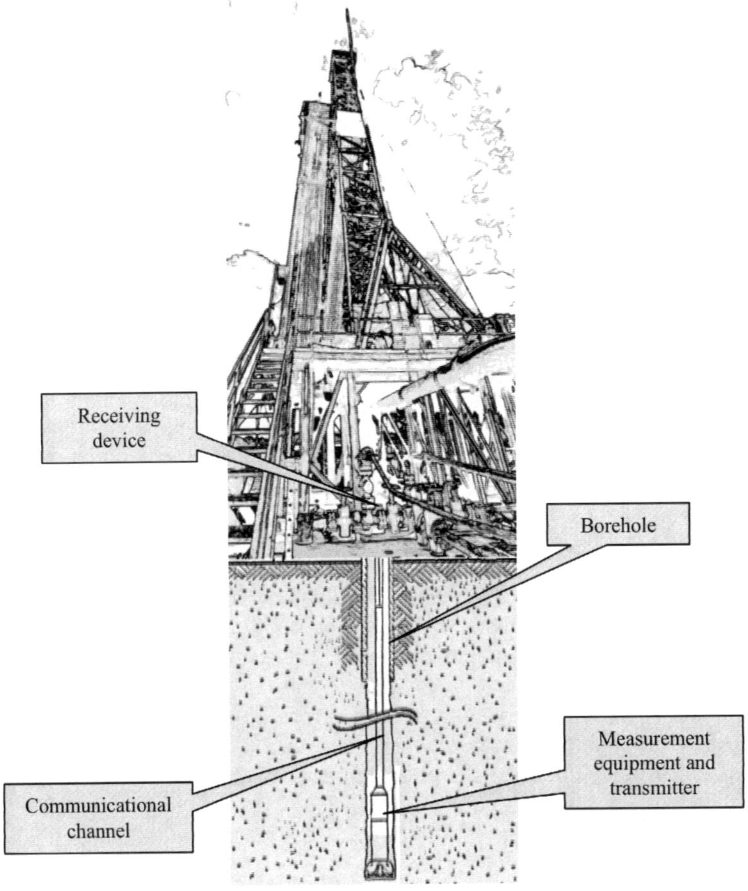

Fig. 1. The structure of geophysical measurement system

According to [3] ultra high frequency range of wavelengths may be the only suitable for the drill pipe as a communication channel. As a rule, channel is inhomogeneously filled with different media (rock, clay, sand) that have different permittivity and liquid mud with strong radio wave absorption. Therefore, receivers and transceivers should work in the band agreed with the channel parameters and provide the least signal attenuation in borehole. Signal absorption is significantly different for waves of different microwave lengths, especially in an inhomogeneous medium. Therefore, a study of attenuation and constructing models of the channel is necessary to develop requirements to receiver of telemetry system. In addition, it will allow assess the possible depth of the borehole.

3. Investigation of attenuation

Experimental setup was made to understand complex processes in the channel and evaluate microwave signal attenuation. It includes:
pulse generator Agilent 33250A;
DC power supply AIP B5.120/0.75;
oscilloscope Agilent DSO 1022A;

generator module based on Gunn diode;
panoramic SWR and attenuation meter R2-61.

Experimental setup is shown in Fig. 2.

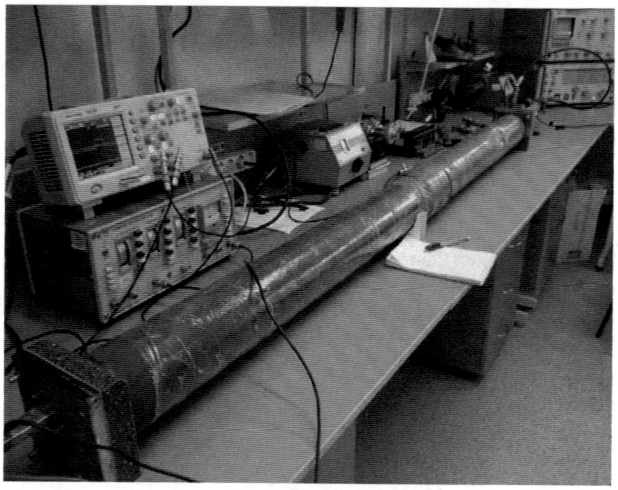

Fig. 2. Experimental setup

The steel drilling pipes with diameter of 100 mm and length of 1080 and 1920 mm were firmly fixed between the waveguide input and output of panoramic SWR and attenuation meter in the center. Measurements of wave attenuation depends frequency were carried out in the band of 8.5–9.5 GHz according to the procedure given in the instructions to the P2-61 equipment. The absolute value of coefficient S_{21} of the scattering matrix was measured. Pipes were filled in water and media with different dielectric permittivity to modeling the natural situation of downhole drilling and information transmission during drilling. Some results of measurements are shown in Fig. 3, 4.

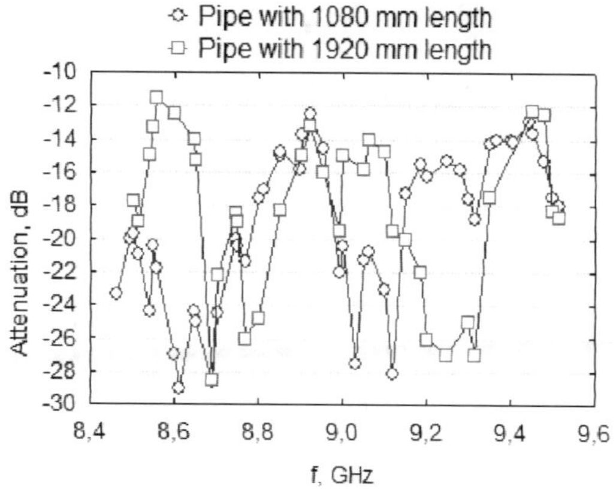

Fig. 3. Dependence of attenuation from frequency in the pipe

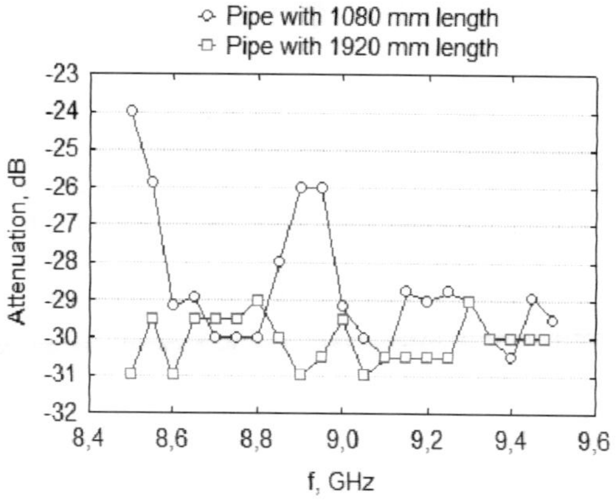

Fig. 4. Dependence of attenuation from frequency in the pipe with water

IOP Conf. Series: Materials Science and Engineering **66** (2014) 012008 doi:10.1088/1757-899X/66/1/012008

Preliminary investigations allows to conclude the following.
Radio link in drilling pipe differs from existing channels (cable, electromagnetics) with deep attenuation, more than 20 dB/m.
Attenuation depends on the length of the channel in the cable channel greatly than in a channel not filled in water.
Equability of attenuation according to frequency is higher in channel filled in water.
The behavior of radio link channel essentially differs from the behavior of electromagnetic channel, as addition concluded in [7].

4. Conclusion

Drilling pipe as a new channel of communication is a combination of the coaxial system and evanescent waveguide with an absorbing medium. Studying channel parameters, simulating and investigating microwave signal attenuation experimentally will lead to the new method of transmitting telemetry information, which in its turn will be helpful to increase the efficiency of geophysical works and their cost-cutting.

This research was supported by grant from Russian Foundation for Basic Researches (project № 13-02-98002).

References

[1] Molchanov A A and Abramov G S 2004. Wireless measurement systems for investigations of petroleum boreholes (theory and practice) Moscow, VNIIOENG p 516
[2] Prensky Stephen 2006 Recent advances in LWD/MWD and formation evaluation *World Oil* March issue pp 69-75
[3] Kochumeev V A, Mirmanov A B and Stukach O V 2012 Investigation of problem situations when designing the advanced geophysical information-measurement systems *Siberian Journal of Science* **3** p 99 (http://sjs.tpu.ru/journal/article/viewPDFInterstitial/361/321)
[4] Liu S, Sato M and Takahashi K 2004 Application of borehole radar for subsurface physical measurement *Journal of Geophysics and Engineering* 1 pp 221-227
[5] DeGauque P and Grudzinski R 1987 Propagation of electromagnetic waves along a drillstring of finite conductivity *SPE Drilling Eng* pp 127 -134
[6] Goponenko A S, Kochumeev V A, Pushkarev V P, Mirmanov A B and Stukach O V Experimental Investigations of Characteristics of X-Band Pulse Generator Based on Gunn Diode 2013 *International Siberian Conference on Control and Communications* Krasnoyarsk p 1
[7] Chuprov V P, Belkov A V, Mishin Yu S and Shibanov S N 2011 Results of investigation of signal attenuation in electromagnetic communication channel *Karotazhnik* **203** p 11

Modelling coils system for generating homogeneous magnetic field

V Ogay, P Baranov and A. Stepankova
National Research Tomsk Polytechnic University, Tomsk, 634050, Russia

E-mail: bpf@tpu.ru

Abstract. Magnetometers are used for measuring the characteristics of magnetic field and magnetic properties of the material. A changeable source of a homogeneous magnetic field is necessary for verification and calibration of magnetometers. Often the Helmholtz coil is used for generating magnetic homogeneous field, but homogeneous field area generated by Helmholtz coils is confined to a small volume in the center of the coils. The paper describes result of modeling a coils system to generate a homogeneous magnetic field with increased volume in comparison to Helmholtz coils.

1. Introduction

Magnetometers are used for measuring the characteristics of magnetic field and magnetic properties of the material.

Depending on lockable physical quantities, there are magnetometers determining the field strength gradient field, the direction of the field, magnetic flux, magnetic flux density, coercive force, magnetic permeability, magnetic susceptibility, magnetic moment.

The main metrological characteristics of magnetometers are:

- Sensitivity – is the minimum value of the magnetic field induction, fixed measuring instrument. This parameter is expressed in most of cases in nT (nanoTeslas);
- Resolution – is determined by the minimal difference of induction, which can register the instrument.

There are magnetometers that measure the absolute values of the field characteristics and their relative changes in time or in area. A changeable source of a homogeneous field is necessary for verification and calibration of magnetometers [1-4].

2. The coils system for generating homogeneous field

The system of solenoid coils or Helmholtz coils [5], shown in figure 1 is often used for generating homogeneous field.

The Helmholtz coils are two concentric rings of radius R arranged at a distance R from each other as shown in figure 1. The coils are connected in series with the current source to create a homogeneous field. According to the law of Biot-Savart-Laplace, the resulting field of the two coils is equal to the vector sum of the fields generating by each coil. The axial field of one coil of a predetermined point can be calculated according to the equation (1):

$$B(z) = \frac{\mu_0 NIR^2}{2(R^2 + (z-h)^2)^{2/3}}, \tag{1}$$

where μ_0 – magnetic permeability of vacuum, H/m;
N – the number of turns of the each coil;
I – current through the coils, A;
z – coordinate of the point, m;
h – the distance of the coil center from the beginning of the coil center coordinates, m.

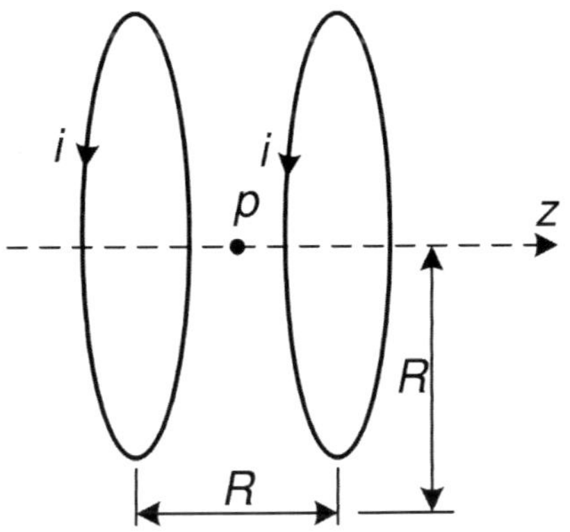

Figure 1. Helmholtz coils

Axial field of two coils can be calculated by formula (2):

$$B(z) = \frac{1}{2}\mu_0 NIR^2 \left\{ \left[R^2 + \left(z + \frac{R}{2} \right)^2 \right]^{-\frac{3}{2}} + \left[R^2 + \left(z - \frac{R}{2} \right)^2 \right]^{-\frac{3}{2}} \right\}. \tag{2}$$

Formula (2) can be expanded in a Taylor series:

$$B(z) = B(0) + \frac{1}{2}B^{(2)}(0)z^2 + \frac{1}{24}B^{(4)}(0)z^4 + \dots \tag{3}$$

It is easy to show that the choice of the distance between the coils is equal to their radius, similar to the condition $B^{(2)}(0) = 0$.

Homogeneous field area generating by Helmholtz coils is confined to a small volume in the center of the coils. For example, field uniformity ≤ 1% for the two coils with the radius 55 mm is achieved only for distances up to 17 mm from its center. The radiuses of the coils are often increased to increase the amount of homogeneous field. However, the same effect can be achieved by increasing the number of coils in the system [6].

Coil system offered in [6] consists of three coils of different radius with a certain ratio of number of turns in the coils. Manufacturing such a system needs a complicated frame. The system of three coils of the same radius is offered in [7]. The system of three coils requires the following conditions:

$$\begin{aligned} B^{(2)}(0) &= 0; \\ B^{(4)}(0) &= 0. \end{aligned} \tag{4}$$

Further increase of the homogeneity of the field can be increased by increasing the number of coils in the system [7]. In this case, the conditions (4) are added to a growing number of even derivatives equal to zero in the center of the coil system. The axial field for multi-coil system can be calculated using the equation (5):

$$B(z)=\frac{1}{2}\mu_0 NIR^2\left\{\begin{matrix}\left[R^2+z^2\right]^{-\frac{3}{2}}+k\left[R^2+(z+a)^2\right]^{-\frac{3}{2}}+\\+k\left[R^2+(z-a)^2\right]^{-\frac{3}{2}}+p\left[R^2+(z+b)^2\right]^{-\frac{3}{2}}+...\end{matrix}\right\}. \quad (5)$$

The first term in the expression (5) corresponds to the coil coordinate z = 0, the other terms correspond to the remaining pairs of coils arranged at the same distance from the origin and having the same number of turns.

For the coils system with the even number of coils, the first coil member of equation (5) is zero. In calculating such a system of coils, it is necessary to find n/2 of unknown distances to the center of the coils and (n-2)/2 of the unknown number of turns. The system of nonlinear differential equations (6) is compiled and calculated to find the unknown parameters:

$$\begin{cases}B^{(2)}(0)=0;\\B^{(4)}(0)=0;\\...\\B^{(2(n-1))}(0)=0.\end{cases} \quad (6)$$

The number of coils has been chosen to be eight, based on the requirements for the volume of the region of homogeneous axial field and acceptable difficulty of manufacturing the coils. The system of equations has been obtained for an integer equal to the number of turns and the distance between the centers of the coils that are multiples of 0.5 mm.

The calculation results given in Table 1.

Table 1. Parameters of the eight-coil system to create a homogeneous magnetic field.

Number coil	The number of turns	The distance from the coil center to the origin, (mm)
1	10	13,5
2	12	42,5
3	17	80,5
4	34	146
5	10	–13,5
6	12	–42,5
7	17	–80,5
8	34	–146

The calculated value of the magnetic induction in the center of the coil system at a current of 50 mA is 22,445 μT.

3. Modeling of the coil system

Software COMSOL Multiphysics 4.3b is used for modeling coils system. Figure 2 shows the geometry of the coils system created in COMSOL.

The calculation of the magnetic field produced in the module mf as only it allows you to use the section Multi-Turn Coil Domain coils for modeling coils in 3D. The coils have closed geometry with a circular cross-section perpendicular to the axis z, so, Circular type of coil was chosen. The direction of current flow in this case is modeled by specifying the border (reference edge), along which the current flows (Figure 3).

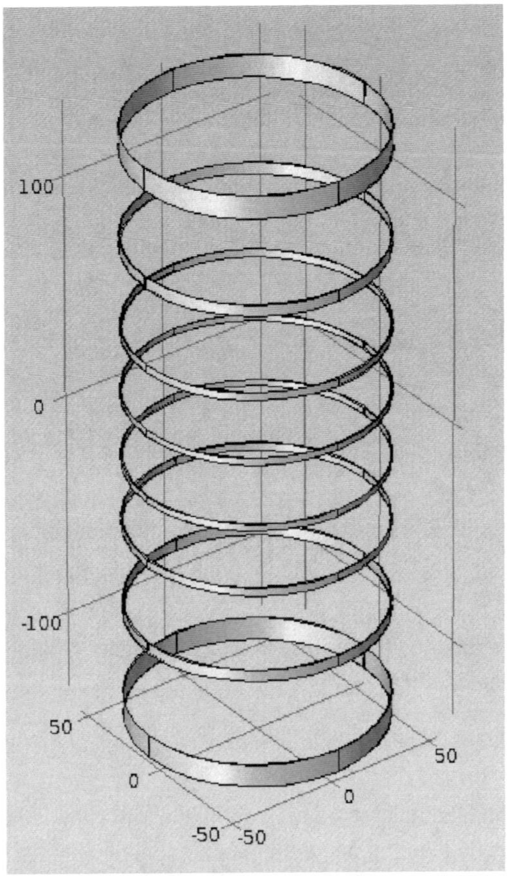

Figure 2. The system of 8 axial coils in COMSOL

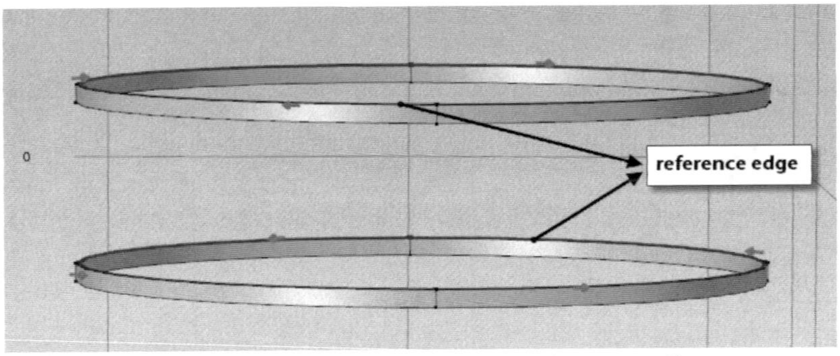

Figure 3. The assignment of directing the current flow

Modeling results of the coils system at the DC current of 0.05 A are shown in figure 4 and figure 5. Inhomogeneity of the magnetic field was calculated by formula:

$$\delta(z) = \frac{B(z) - B(0)}{B(0)} \cdot 100\,\% \tag{7}$$

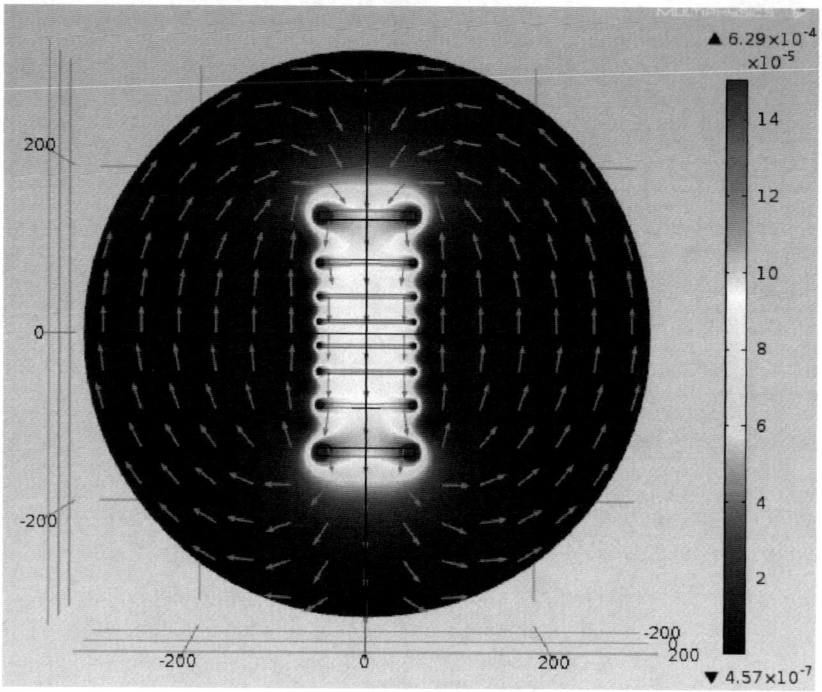

Figure 4. The distribution of the magnetic field

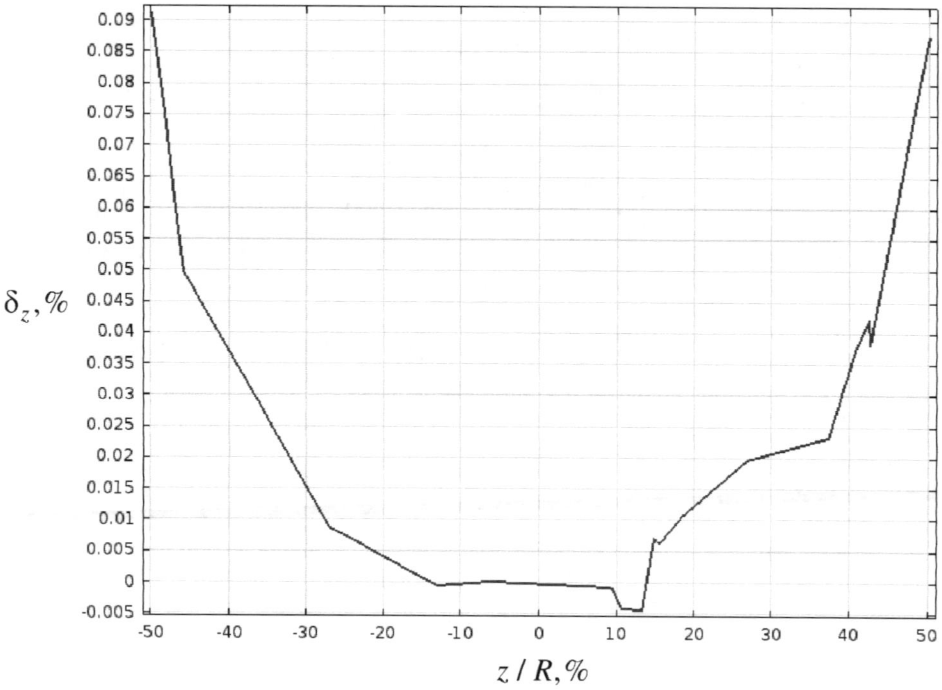

Figure 5. The inhomogeneity of the magnetic field

4. Conclusion

A changeable source of a homogeneous magnetic field is necessary for calibration of magnetometers. Homogeneous field area generating by Helmholtz coils is confined to a small volume in the center of the coils.

As a result of work describes coils system to generating a homogeneous magnetic field with increased volume in comparison Helmholtz coils. Parameters of eight coils system to generating a homogeneous magnetic field are calculated. Modeling results of eight coils system showed that the homogeneity of the field created by a system of eight coils of 110 mm diameter is not less than 0.1 % at distances up to 27.5 mm from the center of the system.

The obtained results show that such a system can be used to calibrate the magnetometers.

References

[1] Warburton F 1955 *NAVORD Report* pp 3768–3821
[2] Barker J R 1949 *Review of Scientific Instruments* 26 pp 213–215.
[3] McKeehan L W 1936 *Review of Scientific Instruments* 7 pp 150–153.
[4] Morgan P, Conolly S, Mazovski A 1997 In: 5th Meeting ISMRM. Toronto, Canada p 1447
[5] On-Axis Field of an Ideal Helmholtz Coil http://www.netdenizen.com/emagnet/helmholtz/idealhelmholtz.htm
[6] Wang J, She S, Zhang S 2002 *Review of Scientific Instruments* **73** p 2175
[7] Hosoya M, Goto E 1991 *Review of Scientific Instruments* **62** pp 2472–2475

Features of mathematical model for ignition process of a metallized condensed substance by a local energy source

D Glushkov[1] and M Dmitrienko[1]

Tomsk Polytechnic University, Tomsk, Russia

E-mail: dmitriyog@tpu.ru

Abstract. Comparison of analysis results for regularities of a metallized condensed substance ignition by a single small-sized particle heated to high temperatures and a hot massive plate was carried out. Ignition delay time was defined for metallized condensed substance. Influence of local energy source heat content on integrated characteristics of process was established. It was shown that use of heat transfer mathematical model with perfect thermal contact conditions on surface of a condensed substance is impossible in case of its ignition by a local energy source with limited heat content.

Keywords: metallized condensed substance, particle, plate, heat transfer, ignition, mathematical modeling.

1. Introduction

Numerical and experimental research results of homogeneous solid [1–3] and liquid [4–6] condensed substances ignition at it interaction with single small-sized hot particle and besides massive energy source allowed to establish essential differences of the main integrated characteristic of studied process – ignition delay time for gunpowder, wood, polymethyl methacrylate, gasoline, kerosene, diesel fuel, etc. It is published [7,8] not so much theoretical research results of metallized condensed substance ignition conditions within the models considering structural heterogeneity of substance caused by aluminum particles. Mathematical models [7,8] rather difficult. It is expedient the assessment of it simplification possibility within approaches [9,10] for decrease calculation volume at the numerical researches of ignition processes.

The purpose of the present work is the numerical analysis of macroscopic regularities for metallized condensed substance ignition by local and massive energy sources.

2. Problem Statement

Heat transfer process at solid-phase ignition in a system "local energy source – metallized condensed substance – air" was considered (Figure 1). It was supposed that small-sized energy source (particle warmed to high temperature) in parallelepiped shape is besieged inertial on a surface of the metallized condensed substance [11] with known thermophysical and thermochemical characteristics. Warming of condensed substance near-surface layer happened due to heat accumulated in a hot particle. During the induction period the part of energy of a hot particle was taken away continuously from lateral sides in air. Such heat transfer considered at two-dimensional statement can have essential impact on integrated characteristics of condensed substance ignition by a small-sized hot particle in comparison with opportunity of a massive energy source when temperature on a condensed substance heated surface is maintained by a constant during the induction period.

Near the bottom side of an energy source ($y=y_2$) the rate of exothermic reaction increases and ignition happens as a result of metallized condensed substance near-surface layer warming.

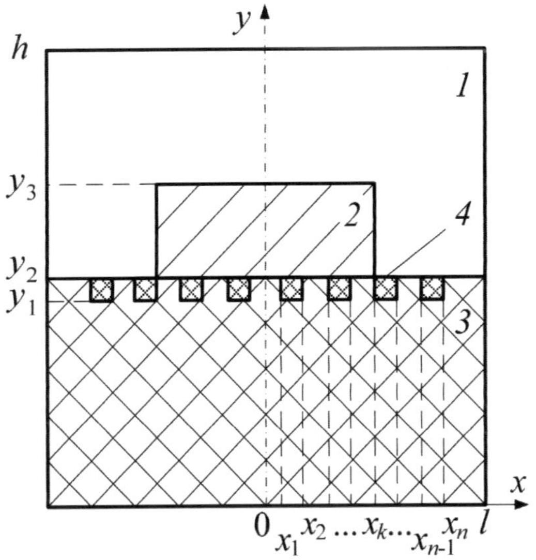

Figure 1. A scheme of the solution domain:
1 – air, 2 – hot particle, 3 – condensed substance, 4 – aluminum particle

At the condensed substance heterogeneous structure modeling (Figure 1) the sites corresponding to composite propellant (butyl rubber and ammonium perchlorate) and the sites corresponding to metal particles (aluminum) were allocated.

A single small-sized ($l_p=x_k=2.5\cdot10^{-3}$ m, $h_p=y_3-y_2=2.5\cdot10^{-3}$ m) heated to high temperatures particle in the parallelepiped shape was considered as an ignition source. In condensed substance and air the areas ($l=10\cdot10^{-3}$ m, $h=10\cdot10^{-3}$ m) were allocated that significantly exceeds a size of a local energy source. Sizes of aluminum particles was $l_g=x_2-x_1=0.1\cdot10^{-3}$ m, $h_g=y_2-y_1=0.1\cdot10^{-3}$ m.

Ignition conditions were accepted [9]:

1. The energy emitted as a result of exothermic reaction in a near-surface layer of a condensed substance is more than the heat quantity released from energy source to ignition zone.
2. Substance temperature in exothermic reaction zone above the initial temperature of a hot particle.

3. Mathematical Model and Solution Methods

Heat transfer processes in a system "local energy source – metallized condensed substance – air" were described by the system of nonlinear non-stationary differential equations in private derivatives.

Thermal conduction equation for air:

$$x_k < x < l,\ y_2 < y < y_3;\ 0 < x < l,\ y_3 < y < h,\ \rho_1 C_1 \frac{\partial T_1}{\partial t} = \lambda_1 \left(\frac{\partial^2 T_1}{\partial x^2} + \frac{\partial^2 T_1}{\partial y^2} \right). \quad (1)$$

Thermal conduction equation for hot steel particle:

$$0<x<x_k\,,\ y_2<y<y_3\,,\ \rho_2 C_2 \frac{\partial T_2}{\partial t} = \lambda_2 \left(\frac{\partial^2 T_2}{\partial x^2} + \frac{\partial^2 T_2}{\partial y^2} \right). \tag{2}$$

Thermal conduction equation for composite propellant:

$0<x<l$, $0<y<y_1$; $0<x<x_1$, $x_2<x<x_3$,..., $x_n<x<l$, $y_1<y<y_2$,

$$\rho_3 C_3 \frac{\partial T_3}{\partial t} = \lambda_3 \left(\frac{\partial^2 T_3}{\partial x^2} + \frac{\partial^2 T_3}{\partial y^2} \right) + Q^2 W_3\,, \tag{3}$$

where $W_3 = \rho_3 k_3^0 \exp\left(-\frac{E_3}{RT_3} \right)$ is mass rate of exothermic reaction [10].

Thermal conduction equation for aluminum particle:

$$x_1<x<x_2\,,...,\ x_{n-1}<x<x_n\,,\ y_1<y<y_2\,,\ \rho_4 C_4 \frac{\partial T_4}{\partial t} = \lambda_4 \left(\frac{\partial^2 T_4}{\partial x^2} + \frac{\partial^2 T_4}{\partial y^2} \right). \tag{4}$$

The nomenclature was accepted: ρ_i is density, kg/m^3; λ_i is thermal conductivity, W/(m·K); C_i is specific heat, J/(kg·K); T_i is temperature, K; k_3^0 is preexponential factor, s^{-1}; E_3 is activation energy of exothermic reaction, J/mole; Q_3 is thermal effect of exothermic reaction, J/kg; R is absolute gas constant, J/(mole·K); subscripts 1, 2, 3, 4 correspond to air, hot steel particle, composite propellant, aluminum particles.

The initial conditions at t=0:

$T_1=T_3=T_4=T_0$, $0<x<l$, $0<y<y_2$; $x_k<x<l$, $y_2<y<y_3$; $0<x<l$, $y_3<y<h$.

$T_2=T_p$, $0<x<x_k$, $y_2<y<y_3$.

The boundary conditions at $0<t<t_d$:

1. Conditions of equal to zero gradients of corresponding functions were accepted for all equations on symmetry axis and external borders:

$\frac{\partial T_i}{\partial x}=0$, i=1, 2, 3, x=0, $0<y<h$; $x=l$, $0<y<h$;

$\frac{\partial T_i}{\partial y}=0$, i=1, 3, y=0, $0<x<l$; $y=h$, $0<x<l$.

2. Perfect thermal contact conditions were accepted for all equations on borders between system components (Figure 1):

$x=x_1$, $x=x_3$,…,$x=x_{n-1}$, $y_1<y<y_2$: $T_3=T_4$, $-\lambda_3 \frac{\partial T_3}{\partial x} = -\lambda_4 \frac{\partial T_4}{\partial x}$;

$x=x_2$, $x=x_4$,…,$x=x_n$, $y_1<y<y_2$: $T_4=T_3$, $-\lambda_4 \frac{\partial T_4}{\partial x} = -\lambda_3 \frac{\partial T_3}{\partial x}$;

$x=x_k$, $y_2<y<y_3$, $T_2=T_1$: $-\lambda_2 \frac{\partial T_2}{\partial x} = -\lambda_1 \frac{\partial T_1}{\partial x} - q_r$;

$y=y_1$, $x_1<x<x_2$, $x_3<x<x_4$,…$x_{n-1}<x<x_n$: $T_3=T_4$, $-\lambda_3 \frac{\partial T_3}{\partial y} = -\lambda_4 \frac{\partial T_4}{\partial y}$;

$y=y_2$, $0<x<x_1$, $x_2<x<x_3$,…$x_{k-1}<x<x_k$: $T_3=T_2$, $-\lambda_3 \frac{\partial T_3}{\partial y} = -\lambda_2 \frac{\partial T_2}{\partial y}$;

$y=y_2$, $x_1<x<x_2$, $x_3<x<x_4$,…$x_{k-2}<x<x_{k-1}$: $T_4=T_2$, $-\lambda_4 \frac{\partial T_4}{\partial y} = -\lambda_2 \frac{\partial T_2}{\partial y}$;

$y=y_2$, $x_{k+1}<x<x_{k+2}$,…,$x_n<x<l$: $T_3=T_1$, $-\lambda_3 \frac{\partial T_3}{\partial y} = -\lambda_1 \frac{\partial T_1}{\partial y}$;

$y=y_2$, $x_k<x<x_{k+1},\ldots,x_{n-1}<x<x_n$: $T_4=T_1$, $-\lambda_4\dfrac{\partial T_4}{\partial y}=-\lambda_1\dfrac{\partial T_1}{\partial y}$;

$y=y_3$, $0<x<x_k$, $T_2=T_1$: $-\lambda_2\dfrac{\partial T_2}{\partial y}=-\lambda_1\dfrac{\partial T_1}{\partial y}-q_r$.

Thermal flow from a local energy source according to radiant heat exchange was calculated on a formula: $q_r=\sigma\varepsilon T^4$,
where σ is Stefan-Boltzmann constant, W/(m^2·K^4); ε is blackness degree of a particle.

Mathematical problem statement is similar given above at the heating of metallized condensed substance by hot steel plate. However perfect thermal contact conditions were accepted for thermal conduction equations on border "hot steel plate – metallized condensed substance" in case of a massive energy source.

The system of differential equations (1) – (4) with corresponding initial and boundary conditions was solved by the method of finite differences. The local one-dimensional method was applied to solve different analogues of differential equations. The sweep method using the implicit four-dot difference scheme was applied to solve one-dimensional equations in difference form.

4. Results and Discussion

Numerical investigations were carried out at the following values of process parameters [11]: initial temperature of air and metallized condensed substance T_0=300 K; initial temperature of energy source T_p=800÷1700 K. Blackness degree of a particle ε=0.6; Stefan-Boltzmann constant $\sigma=5.67\cdot10^{-8}$ W/(m^2·K^4); universal gas constant R=8.31 J/(mole·K). Kinetic parameters of ignition process $E_3=50\cdot10^3$ J/mole, $Q_h k_3^0=1.9\cdot10^9$ J/(kg·s).

Thermophysical characteristics of components (Figure 1):
λ_1=0.026 W/(m·K); ρ_1=1.161 kg/m^3; C_1=1190 J/(kg·K);
λ_2=36 W/(m·K); ρ_2=8100 kg/m^3; C_2=545 J/(kg·K);
λ_3=0.472 W/(m·K); ρ_3=1776 kg/m^3; C_3=1500 J/(kg·K);
λ_4=343 W/(m·K); ρ_4=2700 kg/m^3; C_4=930 J/(kg·K).

At the numerical research the dependence of metallized condensed substance ignition delay time on initial temperature of local energy source – t_{d1} and massive hot plate – t_{d2} was established (Table 1). The results obtained at use of mathematical model for a condensed substance ignition by an energy source with limited heat content (Figure 1) exceed more than for 100% the results calculated at use of mathematical model with perfect thermal contact conditions on a heated surface in all range of initial source temperature changing. At comparing t_{d1} with experimental values of ignition delay time t_{d3} [11] was established good correlation of research results. Obtained regularities are caused by specifics of studied process. Change of hot particle temperature happens during the induction period at local heating of metallized condensed substance. This result can be explained follows. At the heat transfer on horizontal coordinate quantity of energy of hot particle expenditure for air warming increases. Another part of energy is spent for warming near-surface layer of a metallized condensed substance. In case of a massive energy source such heat transfer on horizontal coordinate is absent and all accumulated energy is spent for warming of condensed substance.

Table 1. The dependence of metallized condensed substance ignition delay time on initial temperature of energy source at $l_p=h_p$=2 mm

T_p, K	1700	1600	1500	1400	1300	1200	1100	1000	900	800
t_{d1}, s	0.057	0.069	0.088	0.118	0.171	0.266	0.446	0.789	1.609	no ignition
t_{d2}, s	0.019	0.021	0.023	0.025	0.028	0.034	0.047	0.082	0.179	0.486
t_{d3}, s	0.064	0.091	0.132	0.198	0.305	0.487	0.809	1.413	2.614	no ignition

t_{d1} is metallized condensed substance ignition delay time at it interaction with hot particle (numerical research);

t_{d2} is metallized condensed substance ignition delay time at it interaction with hot plate (numerical research);

t_{d3} is metallized condensed substance ignition delay time at it interaction with hot particle (experimental research [11]).

The established specifics of studied process are illustrated a temperature field (Figure 2) in solution area at the ignition moment. Figure 2 shows that temperature change of a local energy source during the induction period makes more than 150 K (at T_p=1300 K). Also considerable change of air temperature near the hot particle was observed. It testifies to the additional heat sink in environment.

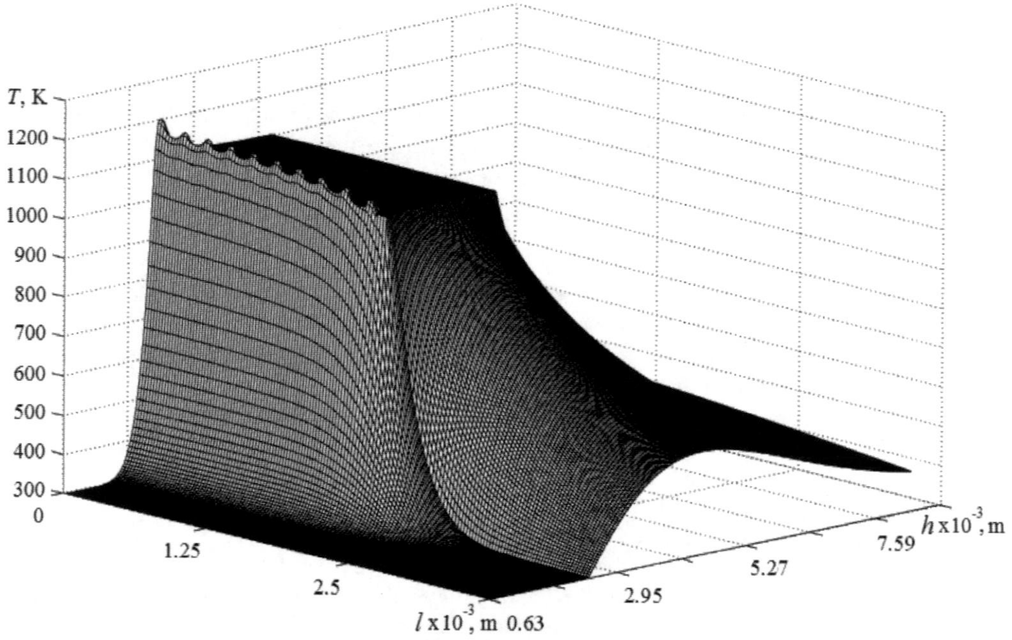

Figure 2. Temperature field in a system "local energy source – metallized condensed substance – air" at ignition moment (t_d=0.171 s) at T_p=1300 K

At the engineering calculations, as a rule, the great interest is represented to the limit conditions of ignition process in system (Figure 1). It was established that T=900 K is minimum initial temperature of a local energy source with sizes l_p=h_p=2 mm when ignition is realized.

5. Conclusion

As a result of numerical researches restrictions on application of mathematical model with perfect thermal contact conditions on a heated surface for analysis of ignition characteristics of a metallized condensed substance by energy source with limited heat content were set. Also minimum value of initial hot particle temperature T_p=900 K was defined at which in system "local energy source – metallized condensed substance – air" ignition happens.

Acknowledgments

The reported study is supported by the Grant of the President of the Russian Federation (project MK-2391.2014.8).

References

[1] Burkina R and Mikova E 2009 High-temperature ignition of a reactive material by a hot inert particle with a finite heat reserve *Combustion, Explosion and Shock Waves* **45** 144–150

[2] Kuznetsov G, Mamontov G, Taratushkina G 2004 Ignition of condensed matter by the "hot" particle *Russian Journal of Physical Chemistry* B **23** 67–72

[3] Khaneft A and Duginov E 2012 Effect of melting on the critical ignition energy of condensed explosives by a short laser pulse *Combustion, Explosion and Shock Waves* **48** 699–704

[4] Glushkov D, Kuznetsov G, Strizhak P 2014 Research of macroscopic regularities of heat and mass transfer at the ignition condition of a liquid high-energy material by an immersed source with a limited energy capacity *Advances in Mechanical Engineering* **2014** 9 pages, Article ID 764537

[5] Dagaut P and Cathonnet M 2006 The ignition, oxidation, and combustion of kerosene: A review of experimental and kinetic modeling *Progress in Energy and Combustion Science* **32** 48–92

[6] Kuznetsov G and Strizhak P 2010 Transient heat and mass transfer at the ignition of vapor and gas mixture by a moving hot particle *International Journal of Heat and Mass Transfer* **53** 923–930

[7] Glushkov D, Kuznetsov G, Strizhak P 2013 Numerical study of ignition of a metallized condensed substance by a source embedded into the subsurface layer *Russian Journal of Physical Chemistry B* **7** 269–275

[8] Glushkov D, Kuznetsov G, Strizhak P 2014 Solid-phase ignition of a composite propellant by a hot particle under free-convection heat sink into the environment *Russian Journal of Physical Chemistry B* **8** 196–204

[9] Vilyunov V and Zarko V 1989 *Ignition of Solids* Elsevier Amsterdam

[10] Frank-Kamenetsky D 1969 *Diffusion and Heat Transfer in Chemical Kinetics* Plenum New York

[11] Zakharevich A, Kuznetsov G, Maksimov V, Kuznetsov V 2008 Ignition of model composite propellants by a single particle heated to high temperatures *Combustion, Explosion and Shock Waves* **44** 543–546

IOP Conf. Series: Materials Science and Engineering **66** (2014) 012011 doi:10.1088/1757-899X/66/1/012011

Vessel ellipticity and eccentricity effect on automatic balancing accuracy

Evgeny N. Pashkov[1], **Nikita V. Martyushev, Igor A. Masson**
Tomsk Polytechnic University, 30, Lenina av., Tomsk, 634050, Russia

E-mail: epashkov@tpu.ru, martjushev@tpu.ru, massoni@sibmail.com

Abstract. The article investigates rotor spinning with a liquid layer on the chamber wall during a visco-elastic shaft operation within the framework of two-dimensional model. The outcomes of the mathematical analysis and trial runs on the stated subject-matter are provided. The article considers the influence of external friction forces and external torque on rotor spinning with the liquid auto- balancing device, located around vertical axis. As a result of the study, the dependencies of shaft bending, system disbalance, torque at a preset shaft angular velocity on the ratio of the reduced mass and rotor mass and external friction forces have been specified. In addition, the dependence, allowing calculation of the angular velocity, when the shaft "sticking" is possible in case of insufficient engine power, has been obtained.

1. Introduction

This article investigates the rotor spinning with a liquid layer, being on the chamber wall, at a visco-elastic shaft operation within the framework of two-dimensional model. It is assumed that the border of the independent liquid layer represents the circle with its centre on the rotation axis, and the liquid itself is spinning along with the rotor as a solid body.

2. Materials and Methods

In the previous works [1, 2] a steady motion of the out-of-balance rotor with liquid at presence of external resistance was discussed. It appears from these works that at a stationary motion the liquid, placed in a correction chamber, spins along with the rotor as a solid body. But this article considers the influence of external friction forces and external torque on the rotor spinning with liquid auto-balancing device around the vertical axis.

3. Results

We specify the torque M, applied to the shaft from the side of the engine, from the equilibrium equation of all the torques relatively AB axis (see figure 1, 2).

$$M = O_2O_1 \cdot \chi V_{O_1} = \chi a^2 \omega = \chi \omega (x^2 + y^2) \tag{1}$$

[1] To whom any correspondence should be addressed.

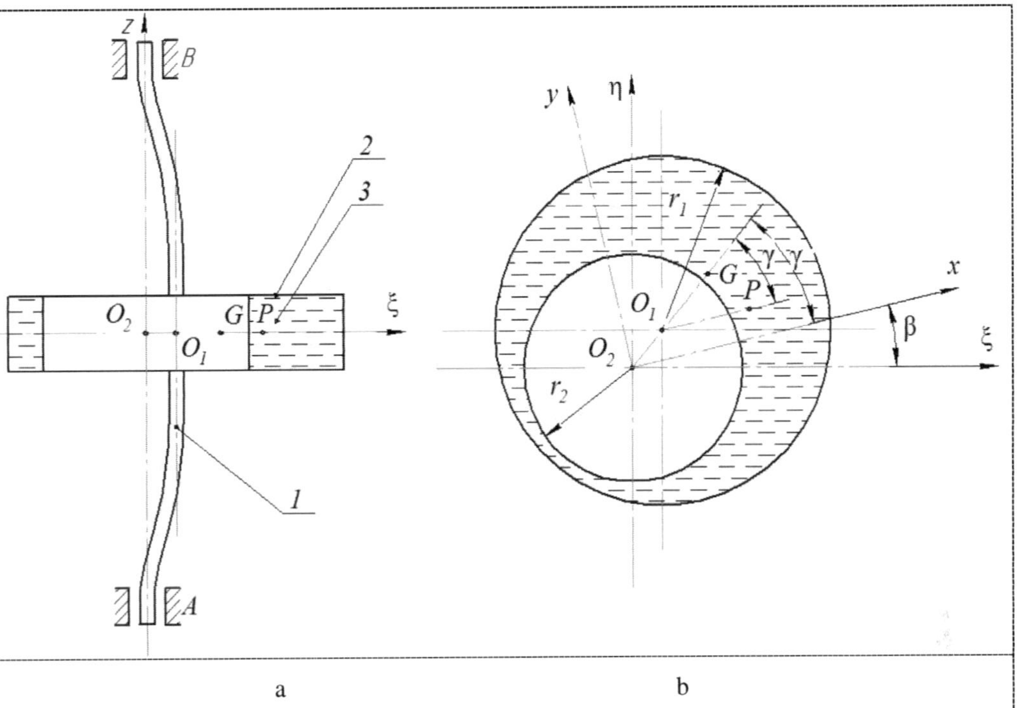

Figure 1. a – Rotor mounting diagram; b – Section of the rotor with a liquid

Complementing the calculations, carried out in [3-5], with equation 1, we can calculate the shaft bending $a=\sqrt{x^2+y^2}$; system disbalance $d=(m_1+m_2)\cdot r_c$; torque M in the following way:

$$a=\frac{ez}{\sqrt{D(\mu_1)}};\quad d=\frac{m_1 e\sqrt{1+nz}}{\sqrt{D(\mu_1)}};\quad M=\frac{e^2z^2c\sqrt{nz}}{D(\mu_1)};$$

$$D(\mu_1)=(1-\mu_1 z)^2+nz. \tag{2}$$

Here $\mu_1=m/m_1$ is the ratio of the system reduced mass to the rotor mass.

To compare the motion of the rotor with liquid (ABD) and without it, let us consider the following ratios:

$$\frac{a}{a_1}=\frac{d}{d_1}=\sqrt{\frac{D(1)}{D(\mu_1)}};\qquad \frac{M}{M_1}=\frac{D(1)}{D(\mu_1)}, \tag{3}$$

where a_1, d_1, M_1 are the shaft bending, disbalance, torque correspondingly, obtained from (2), when $\mu_1=1$, during rotor motion without balancing liquid.

Phase-shift angle of motion γ is specified by the formula

$$tg\gamma = y/x = -\frac{\chi\omega}{c - m\omega^2} = -\frac{\sqrt{nz}}{1-\mu_1 z} \tag{4}$$

In case of liquid absence, that is, when μ_1=1, this equation coincides with the similar equation in [6]. The ratio of shaft bendings a/a_1 , when n =0, coincides with the corresponding value, obtained for spinning without friction forces in [7].

From formulas (1–4) it follows that when $\omega\to\infty$ ($z\to\infty$), $a\to m_1e/m$, that is, the shaft bending becomes less than out-of-balance condition e, as $m_1/(m_1+rm_2)<1$; $r_c\to 0$, $\gamma\to\pi$, $x\to -m_1e/m$, $y\to 0$; the coordinates of rotor mass centres and the liquid layer obtain the following values: $x_p=rm_2e/m$, $y_p=0$ и $x_G=-rm_1e/m$, $y_G=0$; $d\to 0$; $M\to\infty$. Thus, in case of large ω values, the centre of system masses tends to take a position on rotation axis *AB*, and the system aligning occurs [8, 9].

Since $D(1) < D(\mu_1)$ when $z > 2/(1+\mu_1)$, according to (2), liquid ABD lessens the bending and system disbalance in comparison with the rotor without liqu id, having rotating frequencies, which are higher than $\omega = \sqrt{2/(1+\mu_1)\omega_0^2}$.

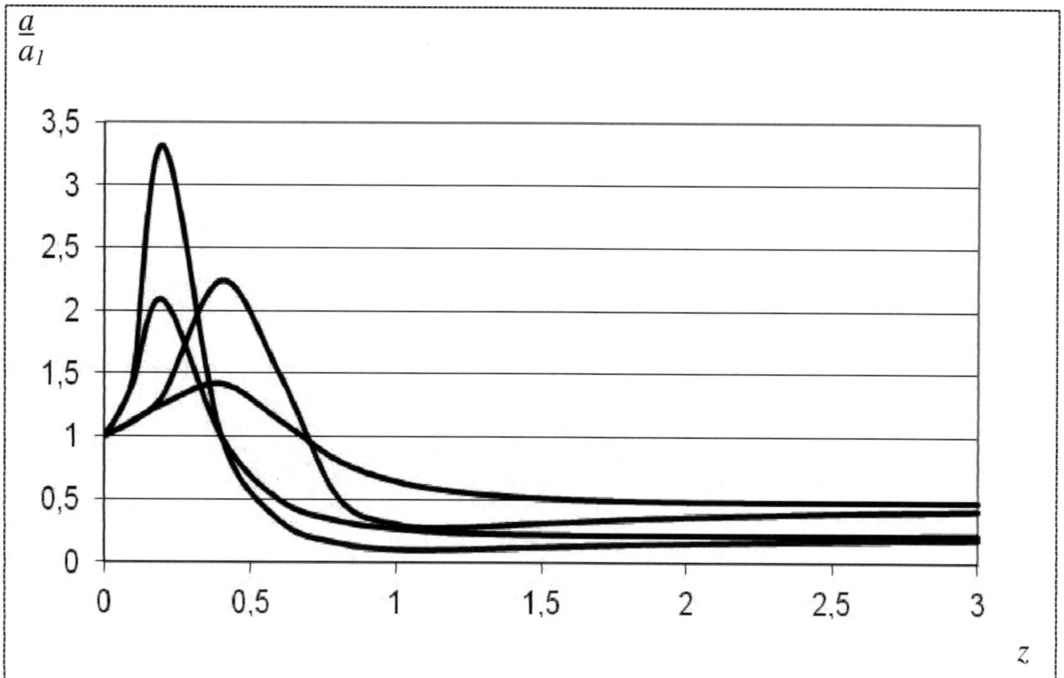

Figure 2. Dependence of the shaft bending on the angular velocity, when μ_1 and n have different values: 1 – μ_1=2, n=0.1; 2 – μ_1=2, n=0.7; 3 – μ_1=4, n=0.1; 4 – μ_1=4, n=0.7

Figure 2 represents the ratio of the rotor shaft bending with operating ABD to the bending without ABD, calculated for μ_1=2; 4 and n=0.1; 0.7. The figure shows that at the same μ value and at different n values the coincidence of the system shaft bendings occurs at the same z value. With the increase of μ parameter the critical rotating frequency decreases, and the extreme bending value decreases with the increase of n. The same curves describe the change in the ratios of disbalance and torque radicals.

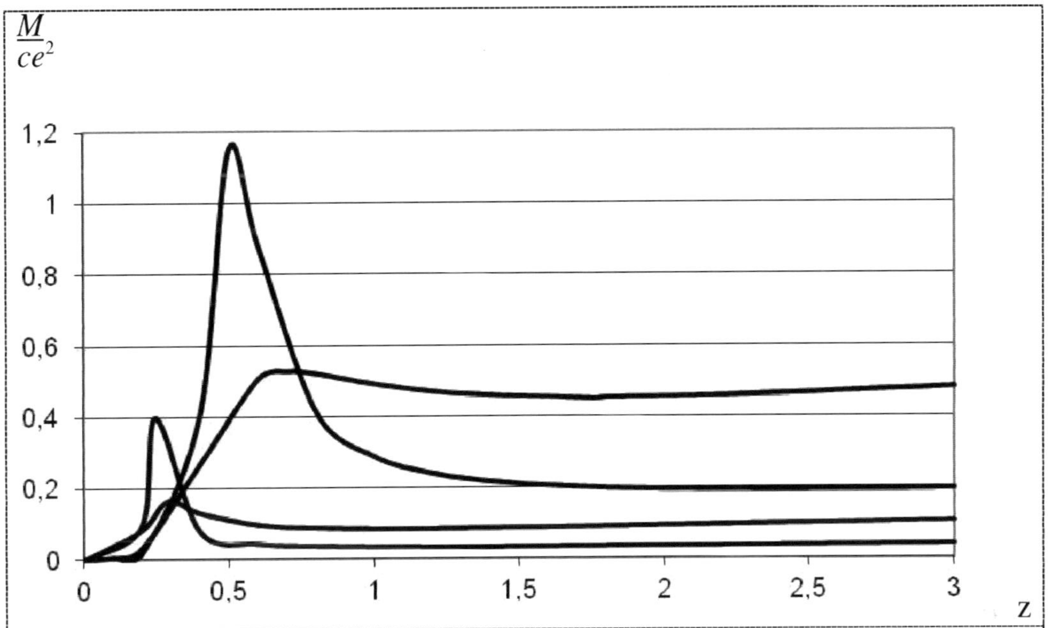

Figure 3. Dependence of the torque on the angular velocity at different μ_1 and n values: 1 – μ_1=2, n=0.1; 2 – μ_1=2, n=0.7; 3 – μ_1=4, n=0.1; 4 – μ_1=4, n=0.7

From (2) it follows that maximum shaft bending $a^{cr} = 2e/\sqrt{4\mu_1 n - n^2}$ occurs at critical angular velocity $z^{cr} = 2/(2\mu_1 - n)$. If liquid is absent from the correction chamber, $z_1^{cr} = 2/(2-n)$ and $a_1^{cr} = 2e/\sqrt{4n - n^2}$. Comparing these values, we can conclude: the liquid ABD decreases critical speed and maximum rotor deflection from the rotation axis. The alteration of the torque depending on the angular velocity is shown in figure 3. The calculations have shown that, when $\mu_1 > 2n$, these curves have two extreme rotating frequencies $z_{1,2} = (6\mu_1 - 3n \pm \sqrt{(6\mu_1 - 3n)^2 - 20\mu_1^2})/(2\mu_1^2)$. The first one (with a negative sign) corresponds to a maximum torque value, the second one corresponds to a minimum torque value. Hence, in case of low engine power when transferring through frequency z_1, shaft "sticking" is possible at this rotating frequency [10].

4. Conclusions

The dependencies of the shaft bending, system disbalance, torque at a preset angular shaft velocity on the ratio of the reduced mass and rotor mass and external friction forces have been established.

The equation for rotor rotating frequency, depending solely on the ratio of the system masses and rotor mass, above which the specified characteristics of motion of the system with ABD become less than for the rotor without liquid, has been obtained.

The dependence, which allows calculating the angular velocity, at which shaft "sticking" is possible in case of insufficient engine power, has been obtained. The results obtained should be taken into consideration when designing and applying liquid ABD for oscillation suppression of out-of-balance rotors, having vertical rotating axes.

5. Acknowledgments
The programme was realized and subsided within the framework of the Programme for enhancing Tomsk Polytechnic University's competitiveness.

6. References
[1] Pashkov E.N., Zijakaev G.R., Tsigankova M.V. 2013 Differential equations of processes for the hydropuls power mechanism of drill machines *Applied Mechanics and Materials.* **379** 91 - 94
[2] Sorokova S.N., Knyazeva A.G. 2010 Numerical study of the influence of the technological parameters on the composition and stressed-deformed state of a coating synthesized under electron-beam heating *Theoretical Foundations of Chemical Engineering.* **2** 172 - 185
[3] Martyushev N.V., Pashkov E.N. 2013 Tribotechnical properties lead bronzes *Applied Mechanics and Materials.* **379** 87 - 90
[4] Martyushev N.V., Pashkov E.N. 2013 Bronze sealing rings defects and ways of its elimination *Applied Mechanics and Materials.* **379** 82 - 86
[5] Nekrasova T.V., Melnikov A.G., Martyushev N.V. 2013 Creation of ceramic nanocomposite material on the basis of ZrO_2-Y_2O_3-Al_2O_3 with improved operational properties of the working surface *Applied Mechanics and Materials.* **379** 77 - 81
[6] Sorokova S.N., Knyazeva A.G. 2009 Estimation of the stresses in a coating growing on a flat plate under nonisothermal conditions *Russian Physics Journal.*. **12** 1309 - 1316
[7] Sorokova S.N., Knyazeva A.G. 2008 Simulation of coating phase structure formation in solid phase synthesis assisted by electron-beam treatment *Theoretical Foundations of Chemical Engineering.* **4** 443 - 451
[8] Chinakhov D.A., Vorobyov A.V., Tomchik A.A. 2013 Simulation of active shielding gas impact on heat distribution in the weld zone *Materials Science Forum.* **762** 717-721
[9] Gromov A.A., Pautova Y.I., Lider A.M., Korotkikh A.G., Teipel U., Chaplina E.V. 2011 Sigfusson T.I., Interaction of powdery Al, Zr and Ti with atmospheric nitrogen and subsequent nitride formation under the metal powder combustion in air *Powder Technology* . **214** 229 – 236
[10] Rubtsov V.E., Kolubaev A.V. 2004 Plastic deformation and quasi-periodic vibrations in a tribological system *Technical Physics.* **49** 1457 - 1463

MELTING OF CORROSION-RESISTING STEELS USING AIR IN BATH AGITATION AT THE END OF OXYGEN BLOWING

R.A. Gizatulin[1], D.V.Valuev[2], A.V. Valueva,[3] Ch.V.Yedesheva[4]
[1,2,3,4] National Research Tomsk Polytechnic University,Tomsk, Lenin Avenue 30, Russia

E-mail: valuev@tpu.ru

Abstract. A number of metallurgical plants employ ladle stirring with argon at the end of oxygen blowing during the melting process of corrosion-resisting steels [1, 2]. At the same time, the scarcity and relatively high cost of argon, its low pressure in a shop air pipeline restrain most plants from using argon for corrosion-resisting steel production. Compressed air was used instead of argon to intensify the process of decarbonizing when chromium-nickel stainless steels were made with a 40-ton electric arc furnace at the Kuznetsk Metallurgical Plant.

After the decrease of flame produced within the furnace, the bath was stirred with compressed air during the oxygen blow process. Compressed air was applied through refractory tubular lances of 3/4 inch in diameter and conveyed from the shop air pipeline at a pressure of 500 – 600 kPa. Lances were directed into the bath at a depth ranging from 300 to 500 mm and angle of 30 degrees to surface. The time of air blowing was from 3 to 5 min. Immediately before stirring the bath with air and after finishing the oxygen blowing, samples of material were taken to analyze an effect of the air stirring process on nature of carbon and chromium oxidation and to estimate a change in the nitrogen content in the metal. Two melts of X23H18 steel and eight melts of X18H10T steel in total were made using the air stirring process. Samples from 11 conventional melts of X18H10T steel were taken to test carbon, chromium and nitrogen in the final stage of oxygen blowing for comparison. The results on the experimental and comparison melts are given in Table 1.

Immediately after the start of air stirring, a visible repeated increase in flame was observed and boil in the bath was stronger. The earlier the air blowing started after the decrease of flame, the higher the rate of decarburization V_C was obtained in the final stage of oxygen blowing (see Figure 1). Despite the different levels of carbon concentration in the experimental and comparison melts at the time of taking samples, we can summarize that the decarburization process is intensified by bath agitation with air. An average rate of carbon oxidation in the melts with air stirring at the end of the blow process was $1{,}06 \cdot 10^{-2}$ %/min and an average rate of carbon oxidation in the melts without air stirring was $0{,}89 \cdot 10^{-2}$ %/min.

Table 1 – Results on experimental and comparison melts

Chemical composition, %			Blowing time, min	
C	Cr	N	Oxygen	Air
Melts with air stirring				
0,15* 0,07*	16,73* 16,73*	0,035* 0,037*	4,0*	1,6*
0,16* 0,06*	17,23* 16,44*	0,026* 0,024*	6,3*	4,8*
0,16 0,06	14,57 13,30	0,028 0,030	8,2	4,2
0,14 0,04	10,88 8,72	-	12,0	4,0

IOP Conf. Series: Materials Science and Engineering **66** (2014) 012012 doi:10.1088/1757-899X/66/1/012012

Table 1 Continuation

Chemical analysis, %			Blowing time, min	
C	Cr	N	oxygen	Air
Melts with air stirring				
0,05 / 0,03	10,31 / 9,30	0,018 / 0,019	6,0	5,0
0,12 / 0,07	10,60 / 9,63	0,020 / 0,022	5,3	5,3
0,11 / 0,06	11,03 / 10,13	0,020 / 0,020	5,0	4,0
0,16 / 0,06	10,74 / 9,32	-	8,0	4,0
0,18 / 0,09	10,93 / 9,58	-	10,0	5,0
0,11 / 0,06	9,68 / 8,92	-	8,0	5,3
0,07 / 0,04	9,13 / 8,42	0,021 / 0,021	6	-
0,07 / 0,04	12,54 / 11,69	0,021 / 0,021	4	-
0,18 / 0,08	10,01 / 9,02	-	11	-
0,20 / 0,06	11,26 / 9,62	-	17	-
0,26 / 0,04	11,57 / 9,28	-	18	-
0,11 / 0,05	11,88 / 10,19	-	8	-
0,14 / 0,04	12,40 / 10,88	-	9	-
0,22 / 0,09	13,25 / 10,92	-	23	-
0,24 / 0,05	11,84 / 9,18	-	14	-
0,21 / 0,06	12,48 / 10,31	-	19	-
0,18 / 0,06	12,91 / 11,70	-	12	-

Note: Numerators and denominators indicate the content of carbon, chrome and hydrogen at the beginning and the end of bath agitation with air respectively. Figures marked with an asterisk are given for melts of X23H18 steel; figures without asterisks are given for melts of X18H10T steel.

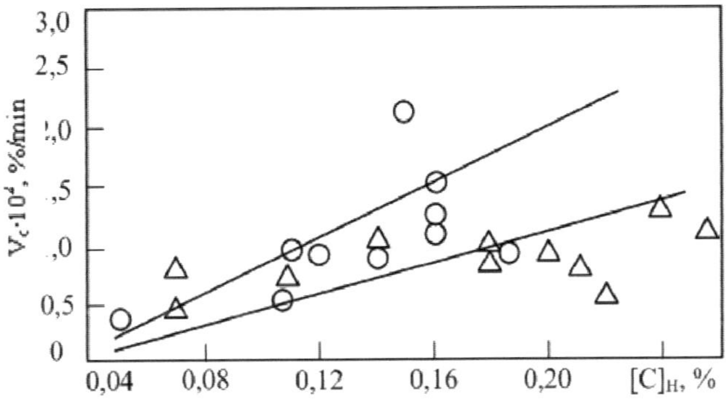

○ – Experimental melts in the case of air stirring;
△ – Comparison melts without air stirring

Figure 1 – Effect of air bath agitation on decarburization rates ($[C]_H$ – the carbon content at the beginning of air stirring)

The bath agitation in the case of stirring by air demonstrated a decreased volume of the melting loss of chromium in the final stage of oxygen blowing (see Figure 2). The mean melting loss of chromium in the final stage of decarburization in the melts with air stirring was 1.18% (absolute), in the conventional melts it was 1.64% (absolute).

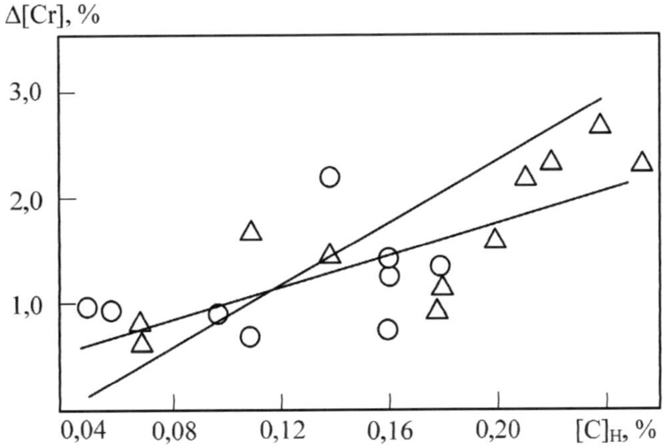

Figure 2 – Effect of bath agitation by air stirring on the melting loss of chromium in the final stage of oxygen blowing

In the context of the different concentration levels of carbon at the time of taking samples and respectively different values as to melting loss of chromium in the final stage of oxygen blowing, the case is more demonstrative when we compare values of melting loss of chromium in relation to 0,01 % of oxygenized carbon[3,4]. Figure 3 shows that in the case of bath agitation with air, the melting loss of chromium decreases relating to 0,01 % of oxygenized carbon.

Figures 1 – 3 illustrate that the efficiency of stirring can be increased if to start air blowing into a bath at the earliest possible time after the decrease of flame. The carbon content in the material is within 0,14 – 0,18 % at that moment. The time for stirring must exceed 5 min. When a bath is stirred with air at a time close to the end of oxygen blowing, this bath agitation causes lower intensification of carbon oxidation and decreased melting loss of chromium.

The use of air did not significantly increase the nitrogen content in the material by the end of decarburization. Absolute changes in nitrogen concentration in the final stage of oxygen blowing with the experimental melts did not exceed 0,002 %, and this value was within the accuracy of analysis on

IOP Conf. Series: Materials Science and Engineering 66 (2014) 012012 doi:10.1088/1757-899X/66/1/012012

nitrogen. (see above the given data). This result is explained by the following. When air is injected through lances, a rate of nitrogen diffusion into metal is very low because of the small contact surface of gas/metal phases and high degree of metal oxidation at the time of blowing[5, 6]. Additionally, the absorption of nitrogen from injected air is compensated by accelerating its removal in the form of carbon monoxide bubbles from the bath as a result of the increase in the rate of decarburization.

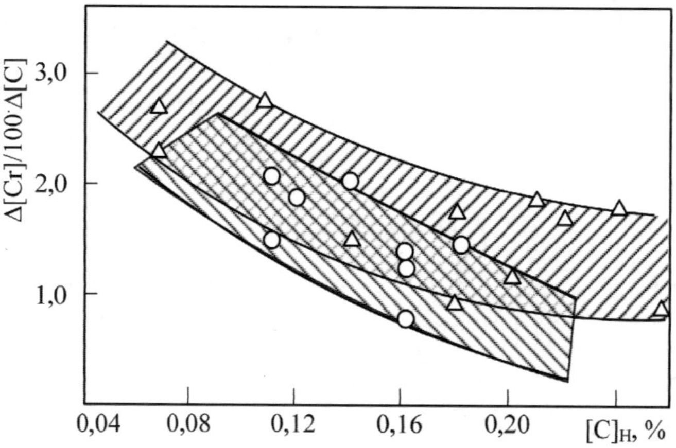

Figure 3 –Melting loss of chromium vs. oxygenized carbon dependency in the final stage of oxygen blowing

Conclusions

The use of compressed air in stirring stainless steels at the end of oxygen blowing intensifies the process of decarburization, reduces the melting loss of chromium and slightly increases the content of nitrogen in material. In order improve the efficiency of bath agitation, air blowing is required to start immediately after the decrease of flame, that corresponds to 0,14 – 0,18 % of carbon content. Compressed air for bath agitation can be used in any workshop making corrosion-resisting steels.

REFERENCES

[1] 1. Povolotskiy D.Ya., Gudim Yu.A., Katzman Tz.L. and colleges // Metallurgy and coke chemistry: Collection – Kiev: Tekhnika, 1980. - Iss. 69. P. 19 – 23. (In Russian)

[2] Povolotskiy D.Ya., Gudim Yu.A., Senyushkin L.I. // Production of electrical steels. 1980. No. 8. P. 54 – 57. (In Russian)

[3] Valuev D.V., Danilov V.I. Reasons for Negative Formation of Structures in Carbon Steel Processing of Pressure // 7th International Forum on Strategic Technology (IFOST - 2012): Proceedings: in 2 vol., Tomsk, September 18-21, 2012. - Tomsk: TPU Press, 2012 - Vol. 2 - p. 151-154

[4] Kuznetsov M.A. , Zernin E.A., Danilov V.I., Kartsev D.S. Application of nanostructured powders to control characteristic of electrode metal transfer and the process of weld structurization // Applied Mechanics and Materials. - 2013 - Vol. 379. - p. 199-203

[5] Ovcharenko V.E. , Mokhovikov A.A., Ignatjev A. S. Improving durability of cermets for metal cutting by generation of subsurface multilevel structures // Applied Mechanics and Materials. - 2013 - Vol. 379. - p. 131-138

[6] Chinakhov, D.A., Vorobyov, A.V., Tomchik, A.A.Simulation of active shielding gas impact on heat distribution in the weld zone // Materials Science Forum. – 2013 - Vol. 762. - , p. 717-721

Effect of protective release coatings on the basis of superdispersersed zirconium oxide powder on the formation of gas defects in bronze casting

Nikita V. Martyushev[1], **Nikolay A. Risto**
Tomsk Polytechnic University, 30, Lenina av., Tomsk, 634050, Russia

E-mail: martjushev@tpu.ru, ris_to555@list.ru

Abstract. This paper investigates the use of nanopowders in the composition of foundry coatings when casting leaded tin bronzes. Influence of the composition of the applied protective coating on surface finish is studied. The effects of the coatings of the following compositions are compared: non-stick coating (a mixture of low-dispersed chromium oxide powder and heat-treated vegetable oil); non-stick lubricant ASPF-2/RgU on the basis of low-dispersed graphite powder and heat-treated vegetable oil; patent #2297300 (a mixture of superdispersed zirconium dioxide powder with industrial oil). It is demonstrated that application of foundry coatings containing superdispersed metal oxide powders with low thermal conductivity makes it possible to significantly reduce irregularities and eliminate gas porosity on the surface of tin-leaded bronze castings.

1. Introduction

Nowadays, a considerable number of leaded tin bronze castings are obtained by casting into steel or cast iron chill molds. Chill casting of such alloys is often accompanied with the formation of numerous gas defects on the surface. The presence of surface defects is considered to be unacceptable for most parts produced this way including piston rings, gaskets, plain bearings, etc. There exist a few methods to prevent the formation of such defects and to ensure enhanced quality of the casting surface.

There are several techniques to improve surface quality when casting into metal molds. These are casting into forcibly cooled molds [1], and casting into pre-heated molds [2]. However, these techniques are rather complicated in terms of their implementation and often fail to fully eliminate the casting defects which are formed on the surface.

The most prospective method for achieving a high-quality casting surface is application of foundry coatings. It allows us to virtually eliminate the burn-on due to the appearance of an additional medium that separates the melt from the mold. Furthermore, these protective release coatings for molds are characterized by low thermal conductivity, which prolongs the period when the melt is in the liquid state and, thus, allows more efficient removal of gas from the melt. Such foundry coatings often contain organic binders, which, when being burnt, emit gases by themselves leading to the formation of pinholes on the surface [3, 4]. In addition, the most widespread coatings used when casting copper alloys contain relatively coarse material particles as fillers (graphite, chromium oxide), which can result in coating's layering.

[1] To whom any correspondence should be addressed.

Application of superdispersed powders as coating fillers is seen as a possible solution to the above mentioned problems [5, 6].

2. Materials and Methods

The researchers of the Department of Materials Science and Process Metallurgy of the Tomsk Polytechnic University investigated the effect of superdispersed powders used in the composition of a coating on castings' surface characteristics. When performing several contracts and casting piston and gasket rings for high-pressure compressors, it became clear that it was necessary to reduce the number of the workpieces rejected due to gas defects. The materials for rings were multicomponent bronzes. The parts were produced by using the technique of horizontal centrifugal casting in steel and cast iron molds. Casting was performed both with and without protective coating for comparison purposes. Prior to pouring, coatings of the following compositions were applied to molds preheated to a temperature of ~200-250°C:

- non-stick coating (a mixture of low-dispersed chromium oxide powder and heat-treated vegetable oil);
- non-stick material ASPF-2/RgU on the basis of graphite powder and heat-treated vegetable oil;
- mixture of superdispersed zirconium dioxide powder (~85 wt.%) with I-80 industrial oil (~15 wt.%) – patent # 2297300;
- mixture of superdispersed alumina powder (~85 wt.%) with I-80 industrial oil (~15 wt.%).

The particle size of the superdispersed alumina and zirconium dioxide powders were in the range from 0.1 to 0.5μm.

3. Results

When die casting without coatings, the rate of the finished castings rejection caused by gas defects reached 50% or even more. Steel molds with the 1:10 ratio of the casting mass to the mold mass were used. Mold temperature upon pouring was approximately 20°C. Increase in the cooling rate in this case did not lead to a decrease in the number of defective products. Conversely, numerous cold shuts and knots were formed on the castings' surface. Figure 1 shows the image of a casting obtained under the given conditions.

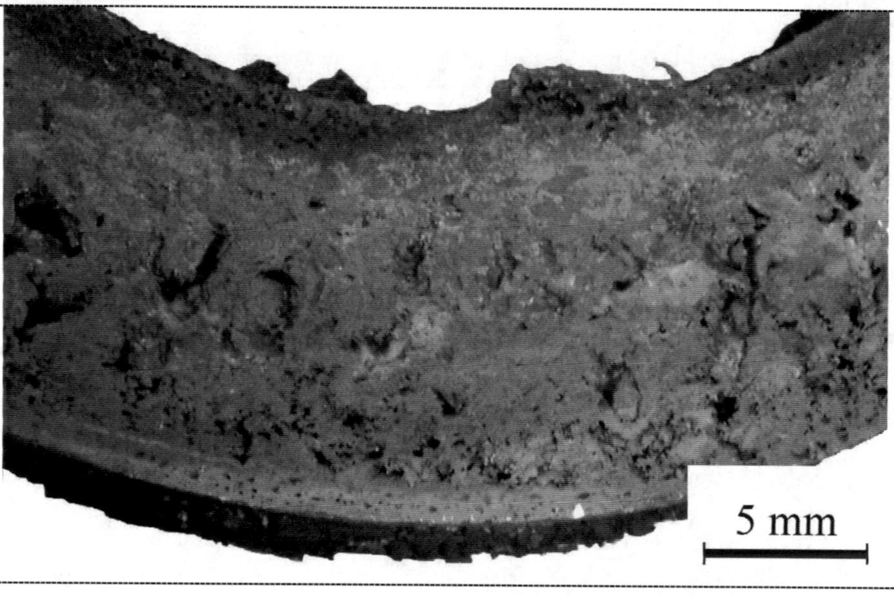

Figure 1. Appearance of a cast when casting at room temperature

Some improvement was observed when using non-stick coatings (a mixture of low-dispersed chromium oxide powder and heat-treated vegetable oil, and the anti-seize lubricant ASPF-2/RgU). The particle size of the powders composing these coatings was in the range from 0.3 to 0.06 mm. Prior to pouring, the mold was heated to a temperature of 250-300° C in order to improve the process of applying the coating and reduce the cooling rate. By using the described technique it became possible to reduce the rate of defective pieces from 50% to 40 % when using the coating on the basis of chromium oxide powder, and to 25%-30 % when using the ASPF-2/RgU anti-seize lubricant. Defective products, in this case, occurred due to the formation of pinhole porosity on the casting surface. Pinholes are known to be able to penetrate to a significant depth in a casting; the depth of penetration depends on the casting's thickness [7, 8]. In our case pinholes were found on the surface at a depth of about 5-6 mm (see figure 2).

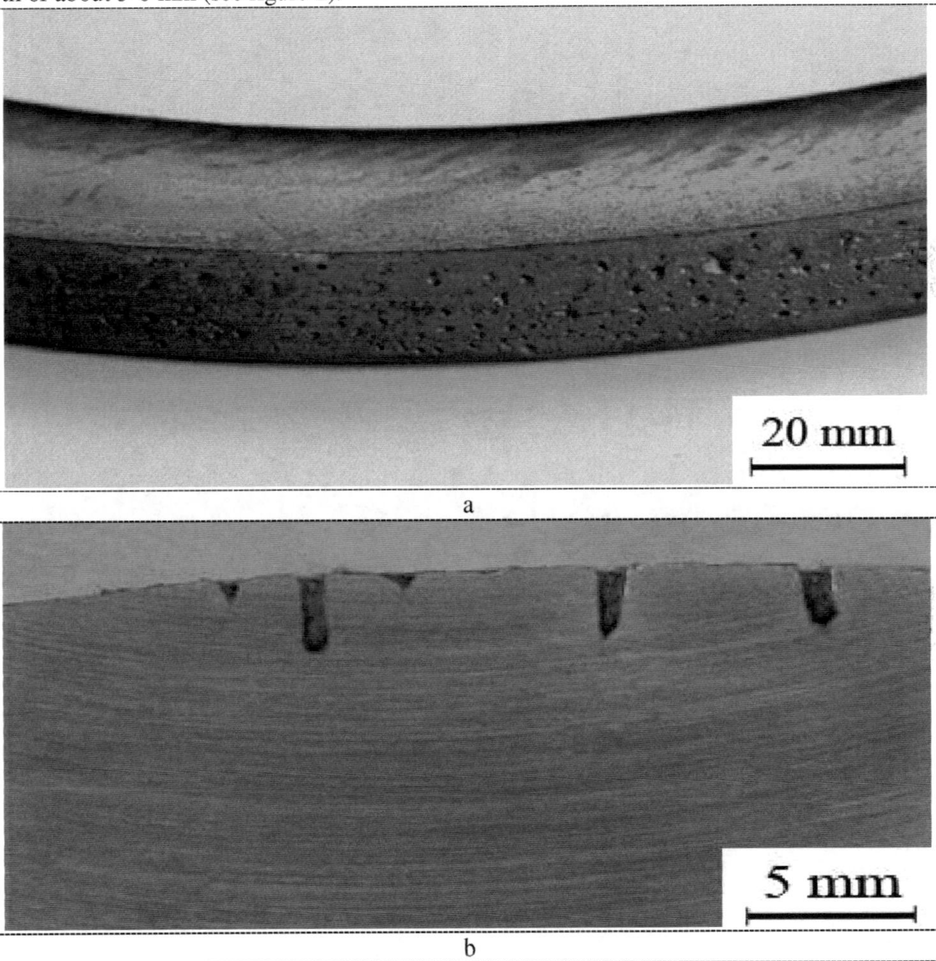

a

b

Figure 2. Pinholes in horizontal centrifugal casting from multicomponent bronze (with bronze, tin, lead, zinc, and nickel content): a – lateral face; b – lateral face after turning

The formation of pinhole porosity indicates that gas cannot go out in any other way than through the surface of the mold. The chromium powder particles in the coating of the first composition and graphite powder particles in the coating of the second composition are characterized by low porosity, which means that the gas flow through them is hindered. Application of superdispersed powders as coating fillers can be a solution to this problem. The following coating composition has been developed: patent #2297300 (a mixture of superdispersed zirconium dioxide powder with industrial oil). Application of the coating based on superdispersed zirconium dioxide powder leads to the elimination of both burn-on and pinhole porosity of a casting (see figure 3). Gas defects on the casting surface are still observed, but they are much less numerous and their shape is changed. When using conventional composition coatings the surface pores occur at a considerable depth and have sharp edges (see figure 2b), while when using superdispersed powders their shape tends to be conical (see figure 3). At the same time, the depth of the defects is not more than 1-1.5 mm, which does not extend beyond the lap.

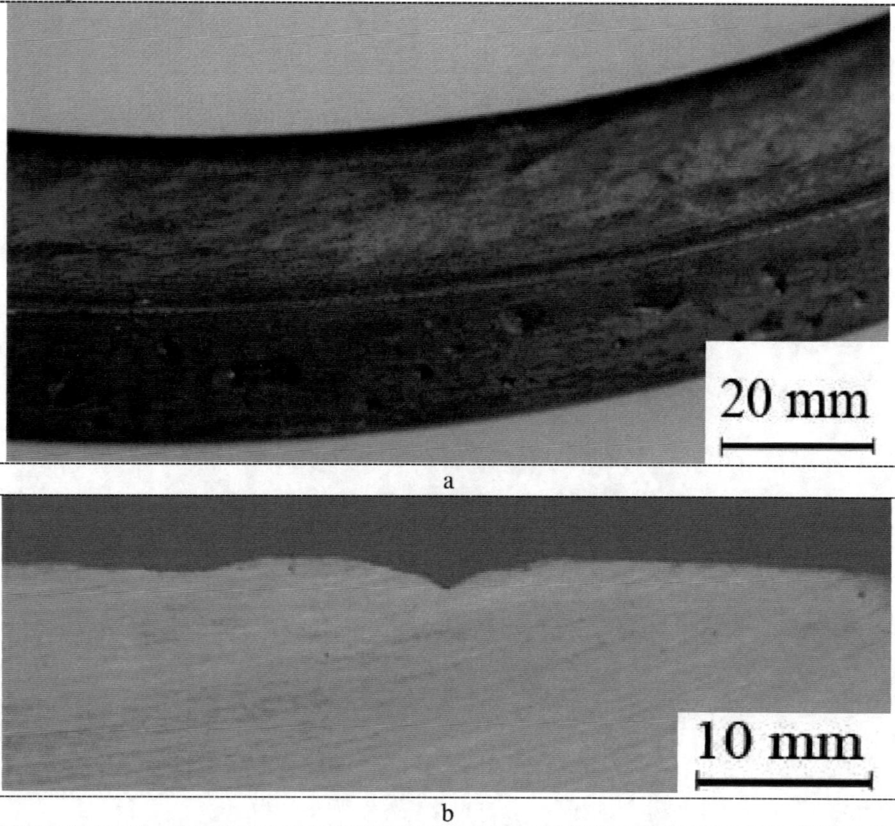

Figure 3. Surface of horizontal centrifugal casting from multicomponent bronze (with bronze, tin, lead, zinc, and nickel content): a – lateral face; b – lateral face after turning

During the first moments after pouring the melt into the mold the binder (in our case, industrial oil) is burned out, the evidence of which are the flames coming from the gating system. Gas gap occurs typically between the mold wall and the molten metal [9]. This resultant gas goes outwardly partially along the mold surface. In those sections where the melt having touched the mold solidifies, knots are formed. When the coating contains superdispersed powder particles in its composition, there appears a gas slurry consisting of nanopowder particles and the combustion products of industrial oil [6]. The nanopowder is uniformly distributed over the mold surface due to a small particle size (< 0.5 microns). Visual inspection of the mold after removal of the casting verifies the uniformity of nanoparticles distribution: the coated surface is covered with regular white film composed of the residual nanopowders. Thus, superdispersed powder particles prevent the contact of the molten metal with the mold surface and at the same time contribute to the removal of gases through themselves. Due to the extremely low thermal conductivity of the powder suspension the poured metal remains liquid for a longer period, which allows it to flow into the surfaces of the mold splitting with a thickness of 0.2-0.3 mm. The same phenomenon did not occur without the use of the developed coatings. As a result, a prolonged period of a metal being in the liquid state allows us to remove gases without the formation of pinholes, fill the entire mold space avoiding cold shuts and misruns, and produce castings with lower surface irregularity. By and large, when using the coating of the developed composition it has become possible to almost eliminate the presence of defective castings.

It is worth mentioning that the described coating composition allows us to achieve a casting surface roughness of Rz 30-40. Application of conventional coatings allows a surface roughness of only Rz 100-150 (on those surfaces where no gas porosity is formed). When using a coating based on superdispersed alumina powder a surface roughness of Rz 40-50 can be achieved.

4. Conclusions

Application of foundry coatings containing superdispersed metal oxide powders allows us to decrease castings' surface roughness by 2-4 times, eliminate gas porosity on the surface, and avoid cold shuts when casting leaded tin bronzes. This makes it possible to significantly reduce the machining allowance. This effect is achieved due to the small size of powder particles (< 1 mm): gases exuded from the mold surface and during the binder combustion pass between the powder particles as if through a sponge and are released in this way from the metal casting zone. The resulting layer between the mold walls and the poured metal consisting of the superdispersed powder coating and escaping gases will enable the melt to be in the liquid state for a longer period. As a result, molten metal occupies the entire mold cavity avoiding the formation of cold shuts and misruns.

5. Acknowledgments

The results presented herein were obtained with the assistance from Russian Presidential Grant MK-6661.2013.8.

6. References

[1] Martyushev N.V., Pashkov E.N. 2013 Tribotechnical properties lead bronzes *Applied Mechanics and Materials*. **379** 87 - 90

[2] Martyushev N.V., Pashkov E.N. 2013 Bronze sealing rings defects and ways of its elimination *Applied Mechanics and Materials*. **379** 82 - 86

[3] Pashkov E.N., Zijakaev G.R., Tsigankova M.V. 2013 Differential equations of processes for the hydropuls power mechanism of drill machines *Applied Mechanics and Materials*. **379** 91 - 94

[4] Sorokova S.N., Knyazeva A.G. 2010 Numerical study of the influence of the technological parameters on the composition and stressed-deformed state of a coating synthesized under electron-beam heating *Theoretical Foundations of Chemical Engineering*. **2** 172 - 185

[5] Nekrasova T.V., Melnikov A.G., Martyushev N.V. 2013 Creation of ceramic nanocomposite material on the basis of ZrO_2-Y_2O_3-Al_2O_3 with improved operational properties of the working surface *Applied Mechanics and Materials*. **379** 77 - 81

[6] Chinakhov D.A., Vorobyov A.V., Tomchik A.A. 2013 Simulation of active shielding gas impact on heat distribution in the weld zone *Materials Science Forum*. **762** 717-721

[7] Gromov A.A., Pautova Y.I., Lider A.M., Korotkikh A.G., Teipel U., Chaplina E.V. 2011 Sigfusson T.I., Interaction of powdery Al, Zr and Ti with atmospheric nitrogen and subsequent nitride formation under the metal powder combustion in air *Powder Technology* . **214** 229 – 236

[8] Rubtsov V.E., Kolubaev A.V. 2004 Plastic deformation and quasi-periodic vibrations in a tribological system *Technical Physics*. **49** 1457 - 1463

[9] Sorokova S.N., Knyazeva A.G. 2009 Estimation of the stresses in a coating growing on a flat plate under nonisothermal conditions *Russian Physics Journal*. **12** 1309 - 1316

An investigation into autobalancing devices with multi-reservoir system

Evgeny N. Pashkov[1], **Nikita V. Martyushev, Andrey V. Ponomarev**
Tomsk Polytechnic University, 30, Lenina av., Tomsk, 634050, Russia

E-mail: epashkov@tpu.ru, martjushev@tpu.ru, boom@mail.seversk.ru

Abstract. Behaviour of a liquid-type automatic balancing device is modeled in this paper. To perform mathematical research the authors use a rotor model which contains a ring functioning as a reservoir coupled to a rigid shaft being rotatable in bearings. An autobalancer with several reservoirs is used for the mathematical investigation. The article provides the layout of forces acting in a multi-reservoir balancing system. Data on the effect of various factors on balancing accuracy as well as the main calculation features of a multi-reservoir autobalancer are presented. The obtained modeling results indicate that autobalancing efficiency is enhanced by increasing the number of reservoirs. An increase in the number of reservoirs causes a decrease in a critical rotor speed.

1. Introduction

Known autobalancing devices are quite effective in reducing the mode change of rotor unbalance. However, this performance is achieved by using a heavy liquid (mercury) as a correction weight (mass). When designing machines and devices, engineers aim to avoid the use of highly toxic substances. At the same time, available nontoxic liquids are characterized by a density of no more than 2 g/cm^3. Therefore, using them instead of mercury as a correction weight in liquid-type autobalancers results in the reduced efficiency of automatic rotor balancing [1, 2]. It is due to relatively low efficiency of liquid-type autobalancers caused by a low density of such liquids, that the factors affecting the balancing accuracy have been understudied. At the Department of Theoretical and Applied Mechanics of the Tomsk Polytechnic University a device has been designed [3], which allows a multiple increase in the efficiency of the automatic rotor balancing by using liquid autobalancing devices. Hence, there appears the necessity to investigate various factors that affect the accuracy of balancing.

2. Materials and Methods

As can be seen in Fig. 1, only a part of the liquid of the reservoir's inner surface and the cylindrical surface touching the reservoir's surface and having an axis coincident with the rotor axis is involved in the balancing process in a liquid-type autobalancing device. The weight of this liquid satisfies the

[1] To whom any correspondence should be addressed.

boundary value under the condition of its sufficiency, i.e. at a specified shaft deflection value the free surface of a liquid touches the inner surface of a reservoir.

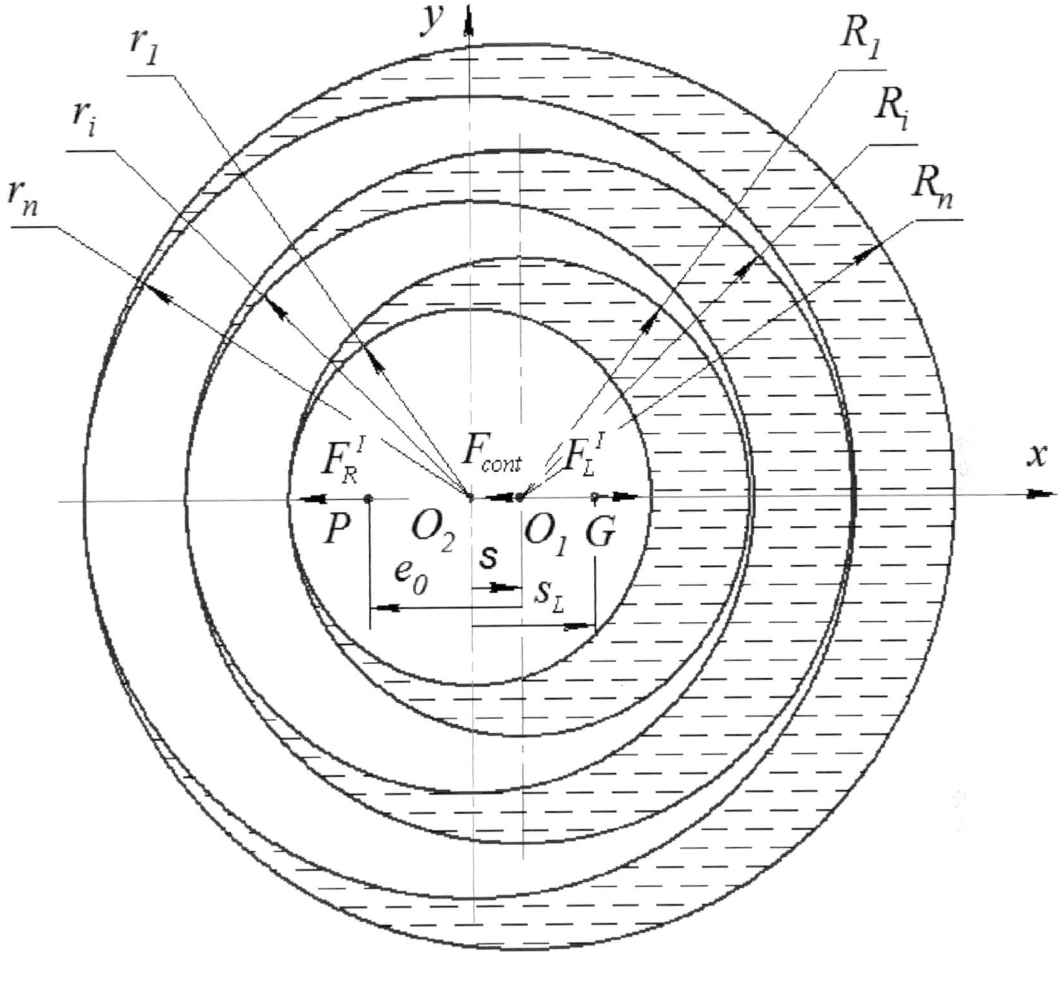

Figure 1. Diagram of forces in a multi-reservoir system.

The other part of the liquid is not involved in the balancing process. Based on this fact, an automatic balancing device with several concentric reservoirs, the axes of the internal cylindrical surfaces of which coincide with the rotor axis, has been proposed. This device will be further called a multi-reservoir device [5, 6]. The multi-reservoir autobalancing device operates like a liquid-type autobalancer with a single reservoir. Let us analyze its performance.

3. Results

On the assumption that the rotor has static unbalance (Fig. 1), i.e. the rotor's mass center (point P) is displaced with respect to the rotor axis (point O_1 in projection) by the value of e_0. As the rotor rotates at a constant speed ω that exceeds the critical value, the deflection of the rotor shaft occurs in such a way that the following equation of the forces acting on the system is satisfied:

$$\sum \overline{F} = \overline{F}_{cont} + \overline{F}^{I}_{R} + \overline{F}^{I}_{1} + \ldots + \overline{F}^{I}_{i} + \ldots + \overline{F}^{I}_{n} = 0$$

or in projections onto the x-axis:

$$-F_{cont} - F^{I}_{R} + F^{I}_{1} + \ldots + F^{I}_{i} + \ldots + F^{I}_{n} = 0, \tag{1}$$

where: $F^{I}_{i} = m_i \omega^2 s_i$ is the fluid inertia force in the i-reservoir applied to its mass centre;

$m_i = \rho \pi h (R_i^2 - r_i^2)$ is the fluid mass in the i-reservoir according to the condition of its sufficiency;

$s_i = s \dfrac{R_i^2}{(R_i^2 - r_i^2)}$ is the distance from the rotor's rotation axis to the fluid's mass centre for the i-reservoir;

R_i, r_i are the radius of the reservoir's inner surface and the radius of the fluid's free surface for the i-reservoir;

n – is the number of reservoirs.

The condition of the liquid's sufficiency in a multi-reservoir system differs from a similar condition in liquid autobalancing devices with a single reservoir. In a multi-reservoir device, the condition of the liquid's sufficiency is satisfied if the free surface of a liquid in the i-reservoir does not intersect with the reservoir's inner cylindrical surfaces formed by the cylindrical partitions [7-9]. Therefore, the condition of the liquid's sufficiency in a multi-reservoir device is as follows:

$$R_i - r_i \geq s \; ; \; r_i - (R_{i+1} + \delta) \geq s \, , \tag{2}$$

where δ is the thickness of the cylindrical partition's wall.

Taking account of the condition of the liquid's sufficiency (2), we define the rotor's vibration amplitude when balancing it with the use of a multi-reservoir device:

$$s = \frac{e_0 m \omega^2}{(m + \rho \pi h \sum_{i=1}^{n} R_i^2)\omega^2 - c} \, . \tag{3}$$

It is seen that in contrast to single-reservoir liquid autobalancers, in multi-reservoir devices the amplitude of rotor vibrations decreases with an increase in the number of reservoirs and in the radii of their internal surfaces. The vibration amplitude does not depend on the mass of the liquid in the chamber if the condition of its sufficiency is satisfied (2).

Using the expression (3) for a multi-reservoir autobalancing device, we can obtain the expressions for calculating the critical rotor speed and automatic balancing efficiency:

$$\omega_k = \sqrt{\frac{c}{m + k\rho \pi h \sum_{i=1}^{n} R_i^2}} \, ; \tag{4}$$

$$E_e = \frac{m + \rho \pi h \sum_{i=1}^{n} R_i^2}{m} = 1 + \frac{\rho \pi h \sum_{i=1}^{n} R_i^2}{m} \, . \tag{5}$$

4. Conclusions

The obtained results indicate that automatic balancing efficiency is enhanced with an increase in the number of reservoirs. An increase in the number of reservoirs causes a decrease in the critical speed of a rotor.

5. Acknowledgments
The programme was realized and subsided within the framework of the Programme for enhancing Tomsk Polytechnic University's competitiveness.

6. References
[1] Pashkov E.N., Zijakaev G.R., Tsigankova M.V. 2013 Differential equations of processes for the hydropuls power mechanism of drill machines *Applied Mechanics and Materials.* **379** 91 - 94
[2] Martyushev N.V., Petrenko Y.N. 2014 Effects of crystallization conditions on lead tin bronze properties *Advanced Materials Research.* **880** 174 - 178
[3] Martyushev N.V., Pashkov E.N. 2013 Tribotechnical properties lead bronzes *Applied Mechanics and Materials.* **379** 87 - 90
[4] Martyushev N.V., Pashkov E.N. 2013 Bronze sealing rings defects and ways of its elimination *Applied Mechanics and Materials.* **379** 82 - 86
[5] Nekrasova T.V., Melnikov A.G., Martyushev N.V. 2013 Creation of ceramic nanocomposite material on the basis of ZrO_2-Y_2O_3-Al_2O_3 with improved operational properties of the working surface *Applied Mechanics and Materials.* **379** 77 - 81
[6] Sorokova S.N., Knyazeva A.G. 2009 Estimation of the stresses in a coating growing on a flat plate under nonisothermal conditions *Russian Physics Journal.* **12** 1309 - 1316
[7] Sorokova S.N., Knyazeva A.G. 2008 Simulation of coating phase structure formation in solid phase synthesis assisted by electron-beam treatment *Theoretical Foundations of Chemical Engineering.* **4** 443 - 451
[8] Chinakhov D.A., Vorobyov A.V., Tomchik A.A. 2013 Simulation of active shielding gas impact on heat distribution in the weld zone *Materials Science Forum.* **762** 717-721
[9] Rubtsov V.E., Kolubaev A.V. 2004 Plastic deformation and quasi-periodic vibrations in a tribological system *Technical Physics.* **49** 1457 - 1463

Fault tolerant vector control of induction motor drive

G Odnokopylov[1, a], **A Bragin**[1, b]

[1] Tomsk Polytechnic University, Tomsk, Russia
[a]OGIz@yandex.ru, [b]lflenylol@gmail.com

Abstract. For electric composed of technical objects hazardous industries, such as nuclear, military, chemical, etc. an urgent task is to increase their resiliency and survivability. The construction principle of vector control system fault-tolerant asynchronous electric. Displaying recovery efficiency three-phase induction motor drive in emergency mode using two-phase vector control system. The process of formation of a simulation model of the asynchronous electric unbalance in emergency mode. When modeling used coordinate transformation, providing emergency operation electric unbalance work. The results of modeling transient phase loss motor stator. During a power failure phase induction motor cannot save circular rotating field in the air gap of the motor and ensure the restoration of its efficiency at rated torque and speed.

1. Introduction

The use of vector-controlled induction motor drives system is closely connected with increasing their resiliency and survivability in hazardous industries (nuclear, military, chemical, etc.). Induction motor vector control drives has a wide control range (up to 1:10,000), an extremely high speed.

Typically this drives are controlled by scalar control system in two phase operate mode [1, 2]. The purpose of this paper is development of vector-controlled induction motor drives.

2. Formation of failure-free vector control algorithm

Vector control of induction motor drive is based on a constant orientation of the coordinate system in the direction of a vector that allows the projection of this vector equate to another axis Cartesian coordinate system to zero. The control system produces control signals in a two-phase rotating coordinate system then this signals converted into three-phase coordinate system for controlling the frequency converter. Consider in detail the process for coordinate transformation in two-phase operate mode.

To derive the required expressions use the notion of a generalized space vector [3]. Denote the vectors arbitrarily shifted relative to the real axis coordinate angles λ, μ, ν, in the three-phase stator the coils A, B, C as IA, IB, IC and sum them, write the expression for the current space vector as:

$$\boldsymbol{I} = \frac{2}{3}(\boldsymbol{I}_A + \boldsymbol{I} \quad + \boldsymbol{I}_C) = \frac{2}{3}(I_A e^{j\lambda} + I_B e^{j\mu} + I_C e^{j\nu}),$$

$\boldsymbol{I}_A = I_A e^{j\lambda}$, $\boldsymbol{I}_B^B = I_B e^{j\mu}$, $\boldsymbol{I}_C = I \; e^{j\nu}$ vectors of the phase currents, I_A, I_B, I_C - their instantaneous values.

Denoting the real and imaginary axis as *a* and *b*, write the expression for the generalized current vector:

$$\boldsymbol{I} = i_a \quad ji \; .$$

Taking into account space shift of the currents angles λ, μ, ν and separating the real and imaginary parts, we obtain:

$$i_a = \mathrm{Re}(\boldsymbol{I}) = \frac{2}{3}(I_A \cos\lambda + I_B \cos\mu + I_C \cos\nu);$$

$$i_b = \mathrm{Im}(\boldsymbol{I}) = \frac{2}{3}(I_A \sin\lambda + I_B \sin\mu + I_C \sin\nu).$$

Denoting the component of zero sequence current as $i_0 = \frac{1}{3}(I_A + I_B + I_C)$ write expressions for the currents in the matrix form:

$$\begin{bmatrix} i_a \\ i_b \\ i_0 \end{bmatrix} = \frac{2}{3} \begin{bmatrix} \cos\lambda & \cos\mu & \cos\nu \\ \sin\lambda & \sin\mu & \sin\nu \\ \frac{1}{2} & \frac{1}{2} & \frac{1}{2} \end{bmatrix} \cdot \begin{bmatrix} I_A \\ I_B \\ I_C \end{bmatrix}.$$

The inverse transform of the two-phase stationary coordinate system a, b, 0 in the three-phase A, B, C will be as follows:

$$\begin{bmatrix} I_A \\ I_B \\ I_C \end{bmatrix} = \frac{3}{2} \begin{bmatrix} \cos\lambda & \cos\mu & \cos\nu \\ \sin\lambda & \sin\mu & \sin\nu \\ \frac{1}{2} & \frac{1}{2} & \frac{1}{2} \end{bmatrix}^{-1} \cdot \begin{bmatrix} i_a \\ i_b \\ i_0 \end{bmatrix}.$$

In general, if the drive is running in a symmetrical three-phase mode angles λ, μ, ν are 0 °, 120 ° and -120 °, respectively, and the transformation can be written in the form [3]:

$$\begin{bmatrix} I_A \\ I_B \\ I_C \end{bmatrix} = \begin{bmatrix} 1 & 0 & 1 \\ -\frac{1}{2} & \frac{\sqrt{3}}{2} & 1 \\ -\frac{1}{2} & -\frac{\sqrt{3}}{2} & 1 \end{bmatrix} \cdot \begin{bmatrix} i_a \\ i_b \\ i_0 \end{bmatrix}.$$

Then the transformation from the fixed coordinate system a, b to a three-phase coordinate system A, B, C given the absence of the neutral wire is carried by the expressions [4]:

$$I_A = i_a,$$

$$I_B = -\frac{i_a}{2} + \frac{\sqrt{3}}{2} i_b, \quad (1)$$

$$I_C = -\frac{i_a}{2} - \frac{\sqrt{3}}{2} i_b,$$

where I_A, I_B, I_C - three-phase stator currents in the fixed coordinate system, i_a, i_b - a two-phase stator currents in the fixed coordinate system.

Expressions (1) are valid for a three-phase mode of operation when the phase currents are symmetrical and are offset from each other by 120 °, but in case of emergency situations such as phase loss their use is not justified, because in case the motor circuit with decoupled phases appears in the neutral wire current i0.

Condition for the existence of a circular rotating field in an electric machine with two arbitrarily shifted in space windings is knew [5]:

$$\delta+\beta=\pi;\ F_{AM}=F_{CM}, \quad (2)$$

F_{AM}, F_{CM} - the amplitude values of magnetomotive forces, δ - angle spatial shift coils, β - angle timeshift winding currents asymmetric biphasic electrical machine. Spatial angle shift winding is constant and equal to $\delta=2\pi/3$ in the 3-phase electric machine, the time phase shift can be specified as $\beta=\pi/3$. In this case, the amplitude values of magnetomotive force of asymmetric two-phase motor are equal $F_{AM}=F_{CM}=F_M$ and circular rotating field is formed in the air gap of the motor.

Consider the algorithm of current vectors shift in two-phase operate mode. Fix the current vector I_A, taking the angle $\lambda = 0$. Current vector will be displaced by an angle-π / 3 and π / 3, respectively, upon the occurrence of an emergency phase B or C. Current vector IC is shifted to meet IB current vector by an angle π / 3 at phase loss A. Work of considered algorithms current vectors shift is shown in Fig. 1.

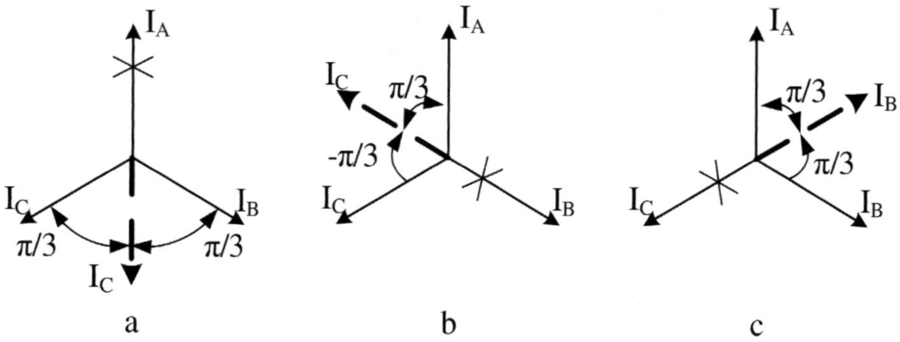

Figure 1. Current vectors shifts in an emergency two-phase mode: a - accident in phase A; b - an accident in phase B; in - an accident in the C phase

Having regard to the above conditions for the existence of a circular rotating field, we can write a general expression for coordinate transformation that takes into account the drive operate mode:

$$\begin{bmatrix}\bar{a} & 0 & 0\\ 0 & \bar{b} & 0\\ 0 & 0 & \bar{c}\end{bmatrix}\cdot\begin{bmatrix}I_A\\ I_B\\ I_C\end{bmatrix}=\frac{3}{2}\begin{bmatrix}1 & \cos(\frac{2\pi}{3}+c\cdot(-\frac{\pi}{3})+a\cdot\frac{\pi}{6}) & \cos(-\frac{2\pi}{3}+c\cdot\frac{\pi}{3}+a\cdot(-\frac{\pi}{6}))\\ 0 & \sin(\frac{2\pi}{3}+c\cdot(-\frac{\pi}{3})+a\cdot\frac{\pi}{6}) & \sin(\frac{2\pi}{3}+c\cdot(-\frac{\pi}{3})+a\cdot\frac{\pi}{6})\\ \frac{1}{2} & \frac{1}{2} & \frac{1}{2}\end{bmatrix}^{-1}\cdot\begin{bmatrix}i_a\\ i_b\\ i_0\end{bmatrix},$$

a, b, c - bit of failure phase A, B, C respectively $\bar{a}$, $\bar{b}$, $\bar{c}$ - inverted bit of failure.

Vector control system is shown in Fig. 2. It comprises a flux, speed and three current regulators. The reference signal on the current from the speed and the flux regulator flows in a coordinate converter which converts from the rotating coordinate system x, y to the fixed coordinate system a, b. Further, depending on the operating mode is converted according to the expression (3).

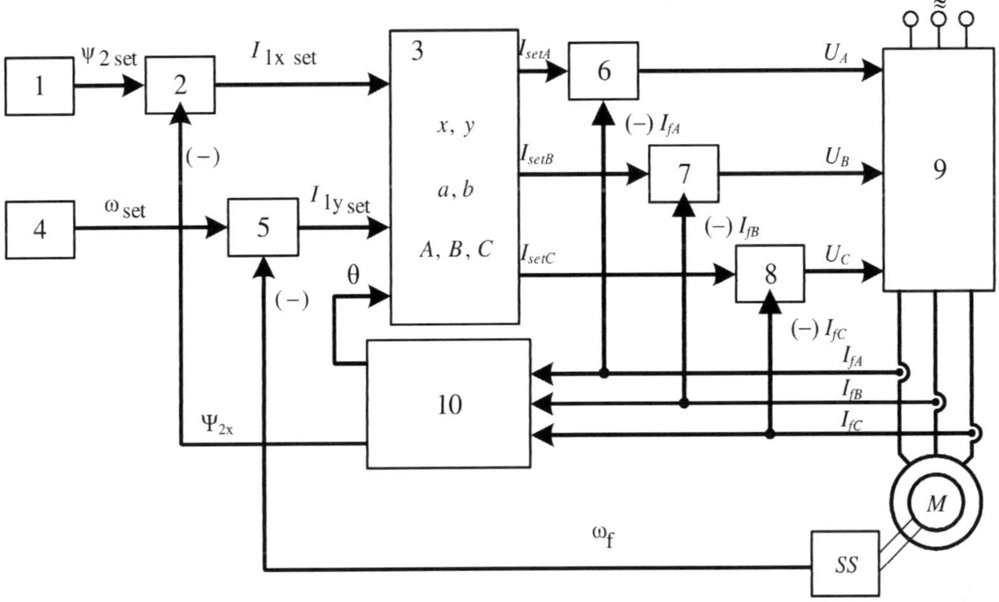

Figure 2. Vector control system of fault-tolerant induction motor drive: 1 - dial rotor flux linkage; 2 - the rotor flux linkage regulator; 3 - coordinate converter; 4 - speed dial; 5 - speed regulator; 6, 7, 8 - stator current regulators; 9 - frequency converter; 10 - calculating unit flux.

Induction motor AIR63A2 was chosen for the simulation. Loops were optimized by the method presented in [6]. Simulation was conducted among Matlab Simulink. Presented transient process of current, speed and torque occurring in the motor in the event of an accident with use of the fault tolerant vector control algorithm and without it fig. 3-4.

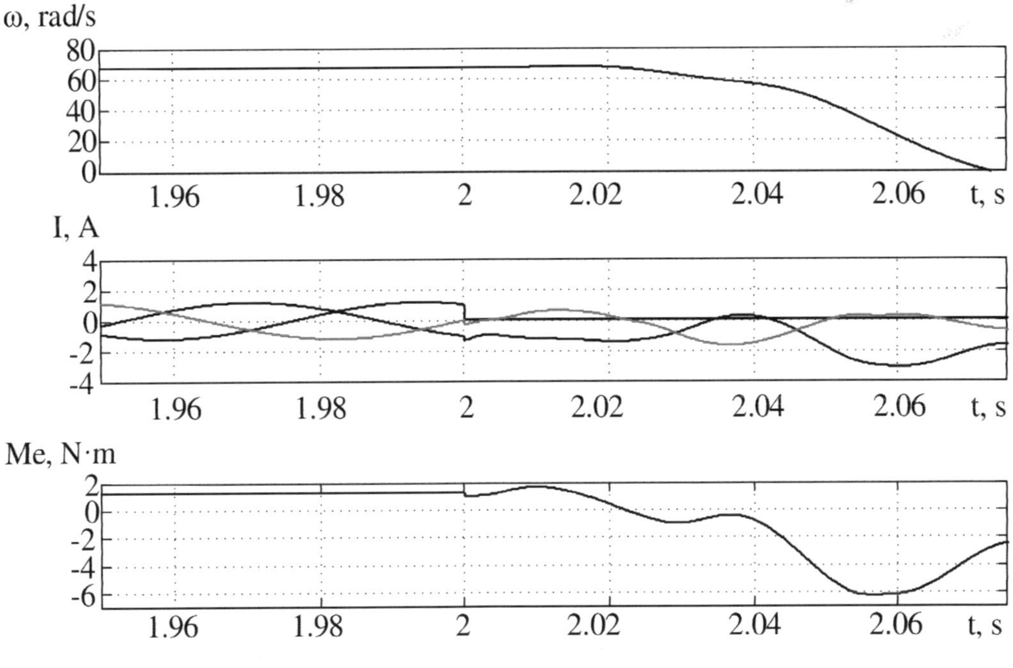

Figure 3. Transients process in the drive with vector control in stator phase loss case

Figure 4 shows the transient process stator B phase loss using recovery algorithm, represented by the expression (3).

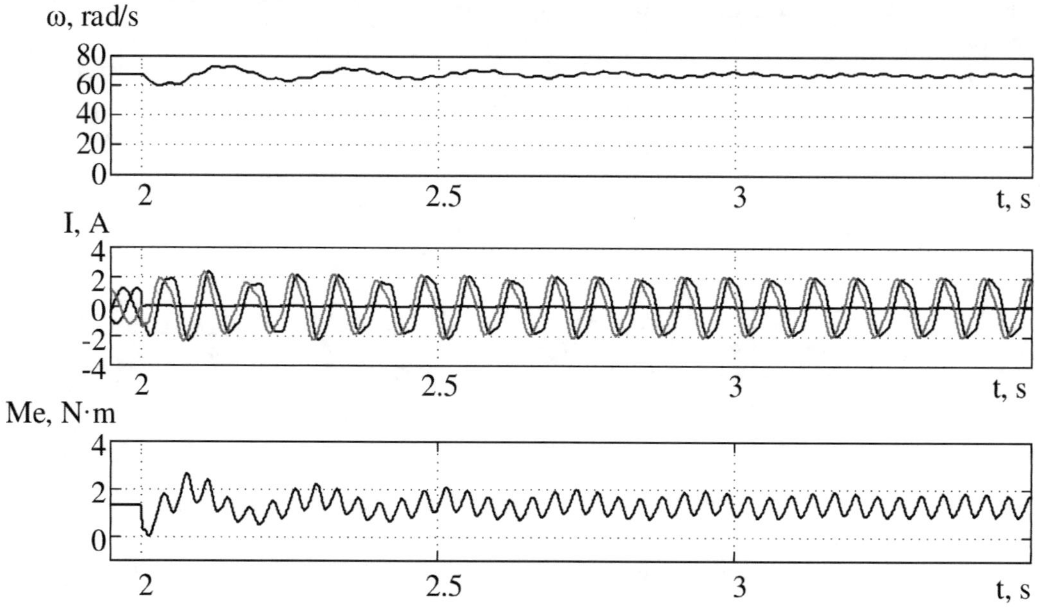

Figure 4. Transients process in the drive with vector control in stator phase loss case with using of recovery algorithm

The transients process (Fig. 3) shows that in the case of an emergency there is an increase of stator currents, and stalling of the motor. The electromagnetic torque changes its sign and becomes inhibitory. Elliptic field arises in the induction motor air gap. This field provides additional inhibitory torque and reduces the overload capacity of the motor. Upon transition to a two-phase operate mode when the motor circuit with decoupled sum of phase stator currents becomes zero, which leads to incorrect coordinate transformation coordinate in the forward converter. All of this features lead to inoperable of the electric drive in two-phase operation mode.

Amplitude of current increase when using recovery algorithm (Fig. 4), because of the need to compensate the lack of power has occurred due to loss of phase stator. Speed is oscillatory and the amplitude of the oscillations does not exceed 2.9% of the set speed, with the failure of the frequency of rotation of the transient time reaches 15.5% of the set frequency.

3. Conclusions:

1. A method of coordinate transformation that takes into account the shift of current vectors and the reconstruction algorithm allows to advance the vector control system. It allows you to keep a circular rotating field in the air gap and provide disaster recovery motor rated torque and speed after on of the motor phase falling.
2. A simulation model of three-phase induction motor drive with vector control system is developed in Matlab Simulink. It allows to investigate the drive in an emergency two-phase operation mode.
3. In a two-phase operating mode speed fluctuations in the steady state does not exceed 2.9% of the set speed when using fault tolerant vector control algorithm.

References

[1] Odnokopylov G I, Odnokopylov I G 2005 Providing of fault tolerance for AC motors: Monograph. *Publishing of Tomsk Polytechnic University* **187**

IOP Conf. Series: Materials Science and Engineering **66** (2014) 012015 doi:10.1088/1757-899X/66/1/012015

[2] Castellan S, Favot M, Menis R, Tessarolo A 2008 *Power Electronics, Electrical Drives, Automation and Motion: International Symposium on* **1175** 10140849
[3] Usolcev A A 2006 Frequency control of induction motors *St. Petersburg State University, versity of Information Technologies, Mechanics and optics* **94**
[4] Chernyshev A Ju, Dementiev Ju N, Chernyshev I A 2011 AC Drives: Tutorial *Publishing of Tomsk Polytechnic University* **213**
[5] Juferov F M 1988 Electric motors for automatic devices *High school* **479**
[6] Malceva O P, Udut L S, Kojain N V 2007 Electric drives control system: a tutorial *Publishing of Tomsk Polytechnic University* **152**

IOP Conf. Series: Materials Science and Engineering **66** (2014) 012016 doi:10.1088/1757-899X/66/1/012016

Microprocessor control unit of thyristor regulator of microhydroelectric power station ballast load

Yu Nomokonova [1] and E Bogdanov [1]

[1] Tomsk Polytechnic University Tomsk, Russia
E-mail: epbogdanov@mail.ru

Keywords: microprocessing unit, thyristor, switch, microhydroelectric power station, Fourier analysis.

Abstract. The operational principle of microhydroelectric power station ballast load is presented. The comparative overview of the mathematical modeling methods is performed. The ranges of thyristors optimal work are shown as a result of the regulator regimes analysis. Shows the necessity of regulation the ballast load in microhydroelectric power station with help of developed algorithm of the program for microprocessor control.

1. Regulation of autoballast

The autoballast regulation principle consists in parallel additional connecting (ballast) to the load of the same size (B) to station payload the through the regulator (fig. 1.). The payload (AL) change entails the change in a ballast size on account of the regulator and the total generator load remains unchanged [1]. The ballast load is a heating element. Take into account that a large amount of electrical energy is used to heat, then the energy dissipated by ballasts can be used effectively in such a way.

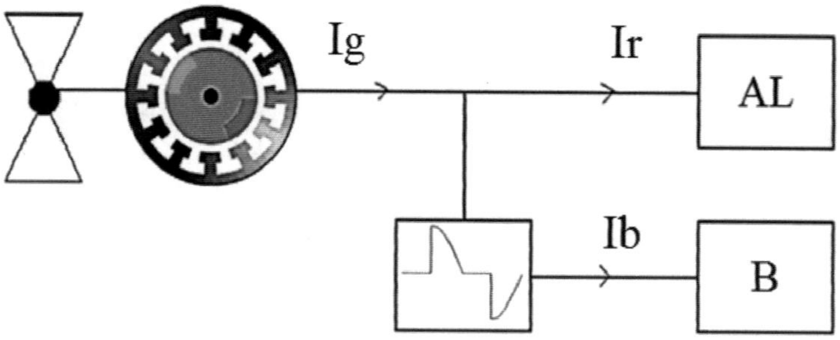

Fig. 1. Regulation ballast load of autonomous microhydroelectric power station

As a regulator, thyristor switches are used. Generally they are classified into the single-phase and three-phase in AC circuits. Three-phase thyristor switches are performed in symmetric bipolar thyristor cells and can be installed both in four-wire AC (Fig. 2,a), and in three-wire AC (Fig. 2,b) [2]. The analysis of the thyristor cell work with work neutral does not differ from the analysis of work in single-phase version. Therefore, it's using simplifies the mathematical model of the regulator.

The principle of any electrical energy converter operation is based on the nonlinear characteristics of thyristors and other nonlinear elements such as saturable inductors and transformers. Due to the nonlinearity of the elements that make up the converter devices, the analysis and calculation of electromagnetic processes in such devices involve significant mathematical difficulties [4].

IOP Conf. Series: Materials Science and Engineering **66** (2014) 012016 doi:10.1088/1757-899X/66/1/012016

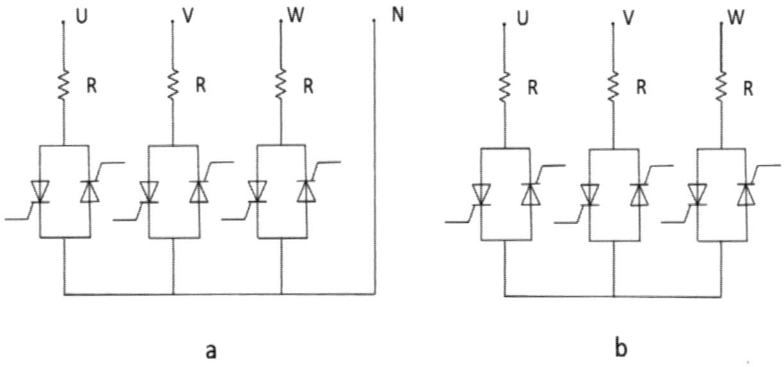

Fig. 2. Three-phase thyristor switches

2. Mathematical simulation

The analysis of the thyristor regulator work is carried out by using the Fourier analysis method. This method is used to investigate the periodic modes of semiconductor converters. The solution can be found in the form of full Fourier series coefficients that are determined in the calculation process. Approximation of the current-voltage characteristics of the thyristors accepted as piecewise linear. The current and voltage thyristor are submitted by Fourier series in a complex form [5].

The advantage of the Fourier analysis method consists in its allowing you to obtain the data about the harmonic composition of currents and voltages and avoid time-consuming determination of the roots of the characteristic equation. The latter is particularly important in this case, if the processes in the semiconductor circuit are described by the differential equations of higher orders.

The disadvantage is a large amount of computation that while analyzing the complex circuits if leads to the necessity to create equations complex systems.

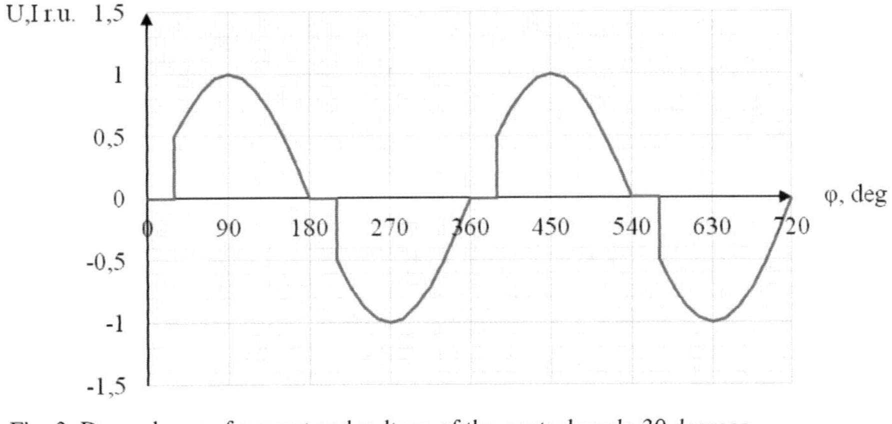

Fig. 3. Dependence of current and voltage of the control angle 30 degrees

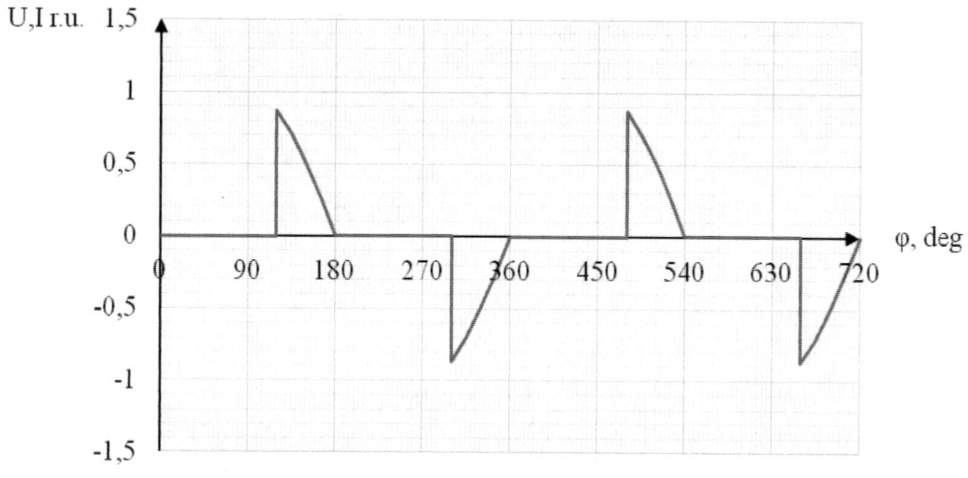

Fig. 4. Dependence of current and voltage of the control angle 120 degrees

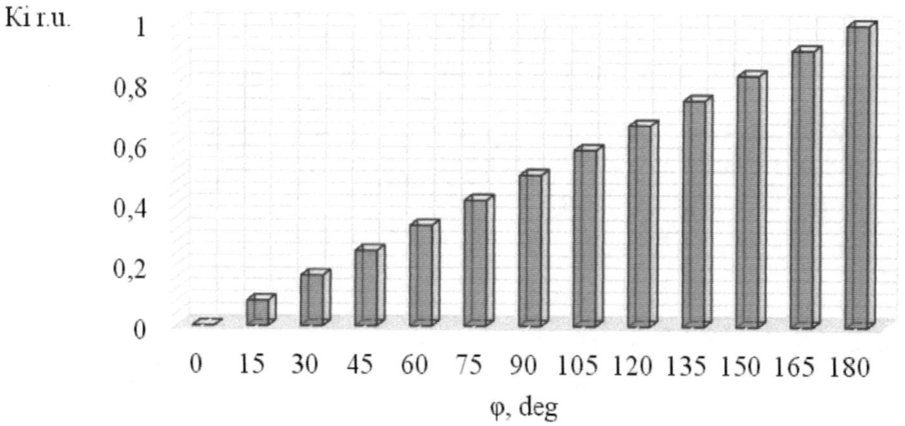

Fig. 5. Dependence of distortion factor of the control angle

The distortion of the output electrical parameters while increasing the control angle thyristor rise (distortion factor can reach 100%) [3]. Changing the control angle of the thyristors is recommended in the range of 30 to 90 degrees to obtain the optimal values of the output parameters. It is expedient to avoid output parameters regulation in a wide range to avoid exceeding the distorting standardized values.

3. Programming the microprocessor

To control the thyristor regulator ballast load a microprocessor device is used that changes the thyristor control angle. The algorithm work is programmed in a microprocessor by means of programming the environment Code Composer Studio (CCS).

CCS is designed to create software running on a DSP-processors [6]. CCS includes the tools for typing and editing the program, and the machine code is typed by using the C language, it's loading into the processor, running and debugging programs in real time. CCS works with a real device and simulator mode of the same model.

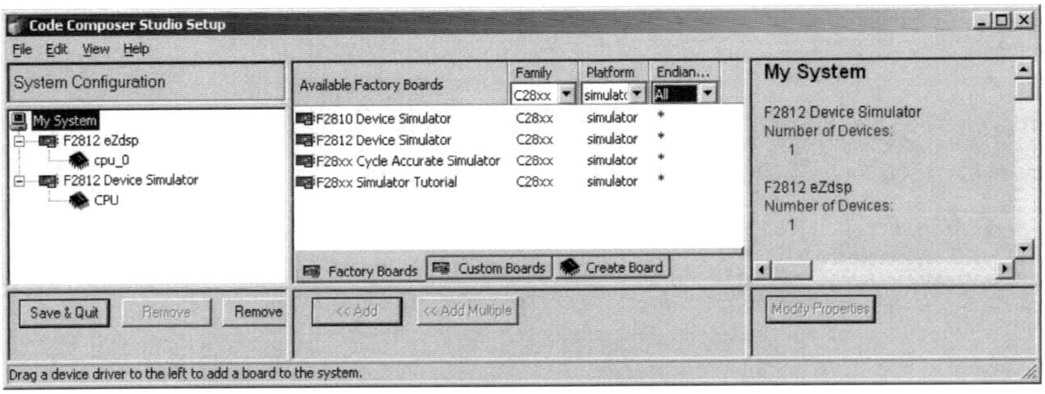

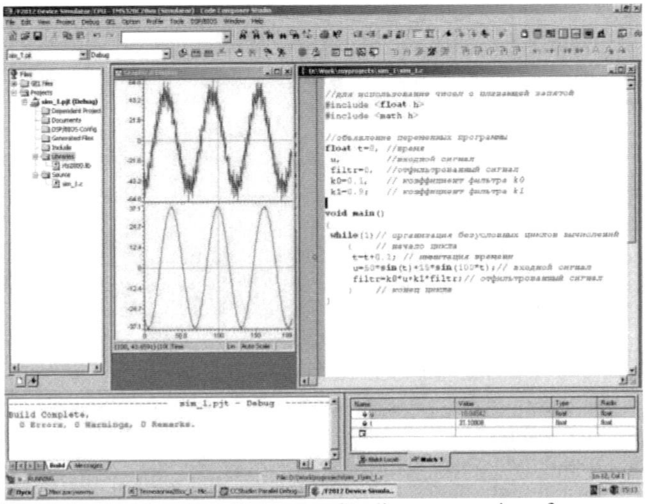

Fig. 6. The CodeComposerStudio program interface

4. Practical significance

Using the microprocessor will positively contribute to the management of the microhydroelectric power station by introducing a feedback system. Herewith control system will not be executed by the hardware implementation. Realization of control system based on software implementation instead of hardware gives possibility to improve the power system characteristics.

References

[1] Lukutin B V, Obukhov S G and Shandarova E B 2001 *Individual electricity from microhydroelectric power station* (Tomsk: STT) p 120

[2] Lukutin B V, Obukhov S G 1990 *Selection of control with microhydroelectric power station with autoballasting load* (Hydraulic Engineering) **7** pp 33–35

[3] Lukutin B V, Obukhov S G and Shandarova E B 2000 *Ways to improve the output voltage of microhydroelectric power station with thyristor autoballast* (Industrial Energy) **8** pp 49–52

[4] Partnun A and Antonien S 1985 *Microhydroelectric power station* (Hessler Erwin: Electrical, Switzerland) **36 12** pp 49–51

[5] Chernyshev A Y and Chernyshev I A 2005 *The Calculation of AC drives. Induction motor* (Tomsk Univ. Publ) **1** p 136

[6] Karakulov A S, Aksenov D S, Areschenko B V and Saidov V S 2007 *Development of software for control of electric motors* (Teaching aid. Tomsk Univ TUSUR) p 261

Oscillatory system of the jet electromagnetic vibrator

E Bogdanov [1], S Rikkonen [2] and Yu Nomokonova [1]
[1] Tomsk Polytechnic University Tomsk, Russia
[2] Tomsk State University Tomsk, Russia
E-mail: epbogdanov@mail.ru

Keywords: electromagnet, winding, frequency, current, voltage, oil.

Abstract. The article describes the system of nonlinear differential equations of the jet electromagnetic vibrator, designed to activate Newtonian and non-Newtonian fluids. The authors formulated the principles of designing electrical and hydro-mechanical components of the device based on the viscosity and density of the process liquid. The article describes experimental frequency characteristics of the system for liquids with the different rheological properties.

1. Introduction

The problem of the alternate/reciprocal motion of the confuser in liquid continuous technological environments is of great scientific and practical interest. The technological processes of preparation of polymer fluids, cement and drilling fluids, oil-water emulsions, paints, use turbulent stream regimes of submerged jets to reduce the viscosity of extra-heavy oil and oil products. The creation of submerged jets with high shear rates results in high quality mixing, and a high degree of dispersion, which significantly reduces the time needed for certain technological processes. Sometimes, these devices may contribute to creation of a completely new consumer properties of the process fluid. All these objects are non-Newtonian and Newtonian fluids. Newtonian and non-Newtonian fluids have absolutely different behaviours, especially in dynamics [1, 2, 4].

The jet electromagnetic vibrator is a device that produces submerged jets with high shear rates (Fig. 1). Such vibrators are starting to be used in liquids preparation processes, but currently there are no methods of calculation and design of oscillatory systems with active elements making alternate/reciprocal motion in a continuous flow medium [2].

The aim of this article is to analyze the jet electromagnetic vibrator as a single oscillating system and experimental studies of the jet electromagnetic vibrator with the definition of the dependence of the resonance frequencies from the properties of process fluids.

2. Theoretical background

The electromagnet in the device is powered by half-wave current (through the diode), and creates a traction force which overcomes the elastic force and the force of the hydraulic resistance. The electromagnet armature is made as a massive anchor with nozzles through which the unidirectional submerged jets are formed [2].

The behavior of the oscillating system is described with three differential equations: the equation of the voltage balance of electrical parts of the jet electromagnetic vibrator (1); the equation of the oscillatory system motion (2); the equation of mass liquid flow through the nozzle of the vibrating activator confuser (3).

Equations (1) and (3) are not difficult to formulate [3]. The equation of activator motion (2) has a complex component, the hydromechanical force of local resistances $F_{cop}(t)$, which varies depending on the fluid velocity and the direction of activator motion.

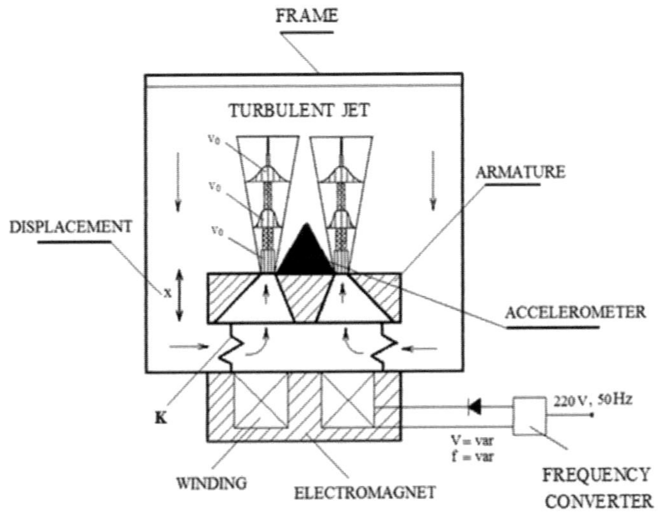

Fig. 1. Layout of the laboratory jet elecromagnetic vibrator

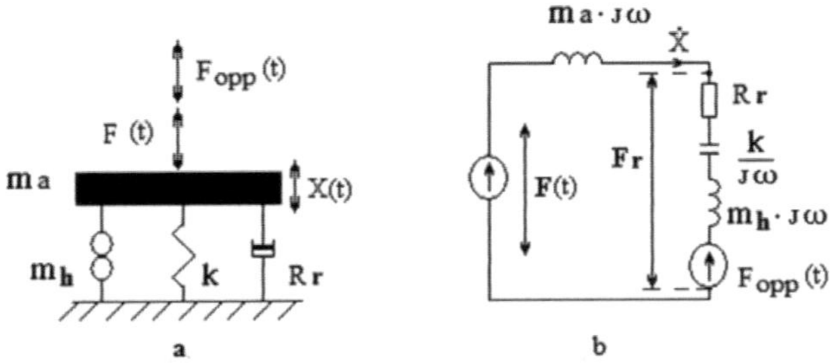

Fig. 2. Design of the Oscillatory system of the jet electromagnetic vibrator

$$U(t) = L_e(t)\frac{dI}{dt} + IR_e + I\frac{dX}{dt}\frac{dL_e}{dx} \quad (1)$$

$$F(t) = (m_a + m_h)\ddot{X} + R_r\dot{X} + K_cX + F_{cop}(t) \quad (2)$$

$$P(t) = L_g\frac{d^2m}{dt^2}\rho + R_g\rho\frac{dm}{dt} + \frac{\xi}{2}\rho\left(\frac{1}{S2\rho}\right)^2\left(\frac{S1}{S2}\right)^2\left(\frac{dm}{dt}\right)^2 \quad (3)$$

Formula (1) is the equation of the voltage balance of electrical parts of the jet electromagnetic vibrator, where $U(t)$ is the supply sine voltage, V; I is the magnet coil current, A; $L_e(t)$ is the magnet inductance, H; R_e is the magnet armature resistance, Ω.

Formula (2) is the equation of the oscillatory system motion, where $F(t)$ is the electromagnet force, N; $X(t)$ is the activator vibration displacement, m; m_a is the anchor mass, kg; $m_h = \frac{4}{3}\rho R_e^3$ is the medium added mass, kg; R_r is the active loss factor of the oscillatory system (losses in the spring, friction losses etc.), kg/s; K_c is the mechanical spring stiffness, N/m; $F_{cop}(t)$ is the hydromechanic activator motion resistance force, N.

The force of local hydraulic resistances to the activator motion will significantly depend on the geometry and dimensions of the channels through which fluids flow. The geometry and form of the channels vary depending on the changing gap between the electromagnet and the activator. The hydromechanical resistance to motion will also significantly depend on the direction of the activator movement [2].

The total hydraulic force of resistance to the forward stroke of the activator is the sum of resistance forces in the gap $F_{zaz}(\delta)$, in confuser F_{sop}, drag force F_{lob}, resistance force of the abrupt constriction of the channel $F_{s(1-2)}(\delta)$, resistance force of the abrupt enlargement of the channel $F_{r(2-3)}(\delta)$, resistance force of smooth constriction of the channel $F_{k(3)}$:

$$\mathbf{F_{opp1}(\delta) = F_{zaz}(\delta) + F_{sop} + F_{lob} + F_{s(1-2)}(\delta) + F_{r(2-3)}(\delta) + F_{k(3)} + F_{r(3-4)}}, \mathrm{N} \quad (4)$$

The total hydraulic force of resistance to the back stroke motion of the activator is the sum of resistance forces:

$$F_{opp2}(\delta) = F_{zaz}(\delta) + F_{sop} + F_{lob2} + F_{s(3-2)}(\delta) + F_{r(2-1)}(\delta) + F_{s(4-3)}(\delta) + F_{r(3)}, N \quad (5)$$

Formula (3) is the equation of mass liquid flow depending on the pressure under the activator and geometry of the channel, where *P(t)* is the fluid pressure under the activator, Pa; *m* is the mass of the fluid exiting the nozzle, kg; ρ is the fluid density, kg/m^3; $R_g = {128\mu L1}/{\pi d^4 l \rho^2}$, $L_g = {Ll^2}/{m_\Sigma}$, where Ll и $d^4 l$ are the equivalent length and diameter of the jet core, μ is the dynamic viscosity, Pa·s; $m_\Sigma = m_y + m_z$ is the total mass of the jet core fluid and the mass of the moving fluid in the nozzle.

The local hydraulic resistance factor depends on the nozzle design:

$$\zeta = \left(\frac{\lambda}{8\sin\alpha^l}\right)\left(\frac{n^2-1}{n^2}\right), n = \frac{S2}{S1},$$

where α is the jet taper angle which depends on the confuser (nozzle) shape.

All three equations (1, 2, 3) are nonlinear, and it is necessary to conduct a numerical or a full-scale experiment to understand the work parametres of the jet electromagnetic vibrator. According to Fig.1, the laboratory equipment was used to record frequency characteristics of the jet electromagnetic vibrator working in different technological liquids. Vibratory systems have maximum power transfer coefficients at resonance frequencies, and depending on the characteristics of the liquid these frequencies change [2, 3]. Fig. 3 shows temporal properties of the electromagnet current and vibratory acceleration of the activator at 49Hz resonance frequency when the activator is operated in water.

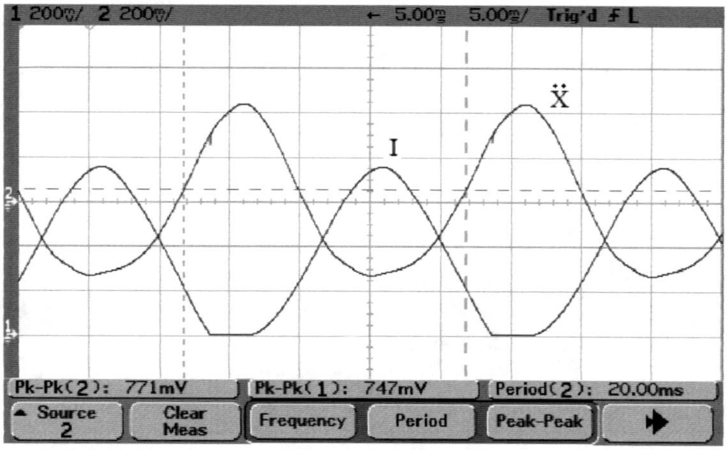

Fig. 3. Experimental temporal characteristics of the electromagnet current and vibratory acceleration operated in water. Um = 135V, Im = 3.7A, $\ddot{X}$ = 82.7 m/s^2, f = 49Hz

Fig.4 shows frequency characteristics of performance of the electromagnetic vibrator in water and in drilling agent. Fig.5 shows the frequency characteristics of the electromagnetic vibrator in cement slurries. The oscillatory systems have different resonance frequencies and different oscillation peaks, but we cannot compare these liquids because water is a Newtonian fluid, while drilling agent and cement slurry are non-Newtonian fluids and their viscosity substantially depends on the sheer speed of the fluid.

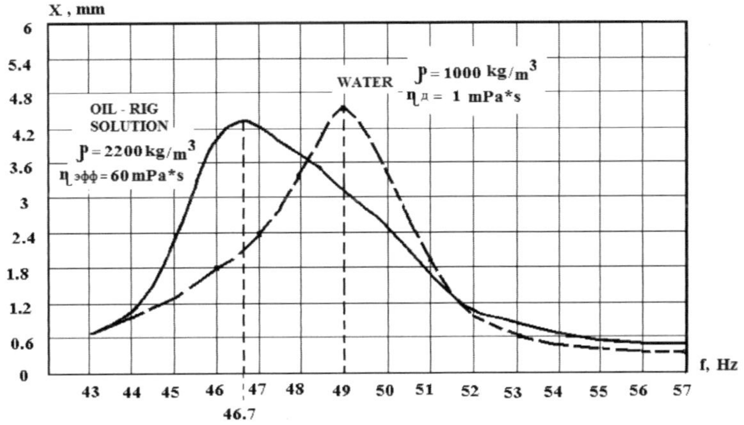

Fig. 4. Experimental frequency characteristics of performance of the activator operated in water and in drilling agent

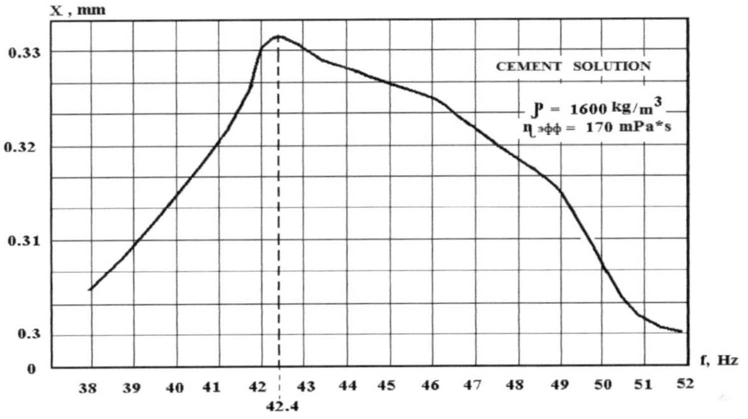

Fig. 5. Experimental frequency of the oscillatory system operated in cement slurries

3. Results and discussion

1. The oscillatory system of the jet electromagnetic vibrator is a nonlinear system, in its electrical, in its mechanical and in its hydromechanical components.
2. It is advisable to process the technological fluid in resonance frequencies of the oscillatory system.
3. The behavior of the jet electromagnetic vibrator, currents, resonance frequencies, vibration speed range and the mass flow depend on the flow properties of the technological fluid.
4. It is reasonable to design the oscillatory system and the electromagnetics of the jet electromagnetic vibrator for a specific type of technological fluids.

References

[1] Rikkonen S V 2006 *Energy development of viscous flow in continuous medium. Scientific and technological development of oil and gas industry. Reports of the fourth international Nadir conference* (National Engineering Academy of the Republic of Kazakhstan. Atyraus Institute of Oil and Gas. Almaty – Tomsk) pp 83–92

[2] Daneker V A, Rikkonen S V, Teplov A I and Khorkov A K 2000 *Jet vibrator processing of high-viscosity oil. Materials of NPF Geofit VNK Joint-Stock Company* (Tomsk University Publishing) p 131

[3] Koenig H E and Blackwell W A 1961 *Electromechanical System Theory* (New York: McGraw-Hill) p 504

[4] Chugaev R R 1982 *Hydraulics* (Leningrad: Energoizdat) p 401

Processing bronchial sonograms to detect respiratory cycle fragments

A Sh Bureev[1], D S Zhdanov[1], I Yu Zemlyakov[1], M V Svetlik[1]

[1] *Tomsk, 634055, Russia,* Diagnostika + Limited Liability Company

E-mail: D_S_Zhdanov@mail.ru

Abstract.

This article describes the authors' results of work on the development of a method for the automated assessment of the state of the human bronchopulmonary system based on acoustic data. In particular, the article covers the method of detecting breath sounds on bronchial sonograms obtained during the auscultation process.

Introduction

According to the data of the World Health Organization (2009), the global incidence of chronic obstructive lung diseases and bronchial asthma has become over 10 times higher since the beginning of 2000. The mortality rate for these conditions, including infectious diseases, takes the third place in Europe being inferior to cardiovascular and oncologic diseases. In Russia, the number of patients with bronchial asthma amounts approximately to 7 million people (2.2% of the population, the 18th place in the world); however, the country takes the second place in the world in terms of the number of lethal cases (28.6 per 100 thousand people). One of the key reasons for such a situation is late detection of the disease. The condition is diagnosed at the stages, when common therapeutic options cannot stop disease progression.

The simplest and the most common method to study the respiratory function is the auscultation (hearing) of breath sounds. The method was proposed by French physician R. Laennec in 1821. Despite the development of other diagnostic methods, this technique is still one of the most important patient examination methods. The key drawback of this technique is the subjectivity of a physicians' opinion during the analysis of these sounds – auscultation findings also depend on many hard-to-formalize factors, first of all, individual hearing parameters (musicality, hearing acuity, frequency range, physicians' age and gender). It is necessary to take into account the fact that a stethoscope might bring in inaccuracy, as it weakens the signals the frequency of which is higher than 120 Hz, and the human ear is characterized with poor sensitivity to low-frequency sounds [1-3].

Since the mid-1990s, scientists have been trying to make the methods of assessing normal and abnormal breath sounds more objective by means of bronchial sonography. The method's key idea implies recording and further analysis of patients' breath sounds and respiratory murmur. The most obvious solution is assessing the spectral density of sounds through time under the conditions of calm breathing and/ or forced expiration. In this connection, some authors [3-6] proposed new methods for acoustic lung diagnostics. They identified diagnostically significant objective signs: relations between the presence of sounds at the frequency of 0.2–1.2 kHz, 1.2–12.6 kHz and 1–12.6 kHz. The sensitivity of detecting bronchial obstruction with this method is 86.7%, the technique specificity is 86.5%.

Unfortunately, besides the statement of the fact of air obstruction in airways and quantitative assessment of this phenomenon, the models and methods described cannot provide any further information. According to many practicing physicians, one of the most important assessment parameters is the timbre nature of breath sounds and rattling from the viewpoint of their musicality: whether these sounds can be attributed to one of the common categories – rattling, sibilant rale, dry or moist rale, crepitation, pleural friction rubs, etc. [0, 0, 0]. It is exactly this identification of auscultative signs to be most subjective and often require a board of physicians to make a diagnosis.

Within the framework of State Contract No.2013-1.4-14-514-01 concluded between the Ministry of Education and Science of the Russian Federation and LLC Diagnostika + on the subject "Development and research on the methods of producing informational images of bronchial sonography signals and algorithms to detect pathological signs for more objective diagnostics of bronchopulmonary diseases in pediatrics", the authors developed an algorithm for the automated detection of respiratory cycles on patients' bronchial sonograms under study. Further analyses of the fragments detected allow building mathematic models to assess the state of the patients' bronchopulmonary system by means of specialized algorithms. The existing method including the algorithm for automated detection of respiratory cycles makes it possible to analyze patients' bronchopulmonary sounds in the automated mode most thoroughly, which allows significantly reducing the subjective component related to physicians' hearing peculiarities and personal experience when assessing patients' type of breathing. In addition, analyzing only significant fragments of the respiratory cycle makes it possible to detect bronchopulmonary diseases at the initial stages.

Theoretic rationale

The basis for the development of the algorithm to detect respiratory cycle fragments is bronchial sonograms recorded with a special model used to obtain auscultative signals from the patients' bronchopulmonary system.

The authors used a binaural pediatric stethoscope as an acoustic detector. The Panasonic WM-61 microphone was installed into the sonic locator of the device. In order to exclude noises produced by the frictions of the sound-amplifying membrane on the patients' skin, the authors used an open funnel of the stethoscope. The microphone was glued into the opening bored in the place of bifurcation of the sound channels, straight across their axis. Then a shielded lead was welded on the microphone, and the entire node was covered with a 5-8 mm layer of the silicone sealer. Such a structure of the acoustic detector allows reducing the level of external noises at the frequency of 50-1,000 Hz by 10-12 dB.

Bronchial sonograms were recorded at the therapeutic wards of pediatric and adult pulmonology departments, i.e., under the conditions that were maximally close to the actual operating settings of the authors' end product – software and hardware complex for invasive automated express diagnostics of human bronchopulmonary diseases. Patients were aged from 3 to 30 y. o. As a result, the bronchial sonograms contained a lot of noises (other than breath sounds), which made it hard to diagnose the state of the patients' respiratory system. In order to eliminate these noises, the authors used specialized digital filters that allowed detecting specifically breath sounds in the bronchial sonograms recorded. For this purpose, the authors used the Embarcadero RAD Studio XE4 development framework (C++ programming language) to create an application that allowed studying the frequency composition of the bronchial sonograms under consideration based on the fast Fourier transform (FFT) method. The bronchial sonograms under study were divided into sections, the length of which was equal to the size of the FFT window (1,024 sections of the bronchial sonograms under study), the sections underwent FFT, and the results were normalized to the sampling frequency of the section obtained according to the formula $\frac{i*f_s}{Size}$, where *i* is the number of the FFT window element; f_s- sampling frequency of the bronchial sonogram, *Size* – size of the FFT window. The research on the spectral composition of 100 bronchial sonograms obtained in patients aged 3-30 y. o. revealed that diagnostically significant breath sounds lie within the frequency range of 80-500 Hz. These values are true for the analysis of bronchial sonograms obtained under the conditions of patients' calm breathing. Therefore, it is necessary to design a digital filter that will allow singling out the above-mentioned frequency range [1-6].

Due to the peculiarities of the acoustic detector used to record patients' bronchial sonograms, they do not contain sounds at the frequency over 700 Hz. Therefore, the authors took a decision to use a high-frequency digital filter and noise suppression in the area of the rejection band at the level of 30 dB.

The authors used their own application (developed by means of the Embarcadero RAD Studio XE4 development framework, C++ programming language) to analyze and develop digital filters with finite impulse response (FIR) and infinite impulse responses (IIR). The key goal is to choose the most optimal implementation (in terms of fast action at the level of software) and highest-quality method to eliminate low-frequency noises hindering the correct detection of respiratory cycle fragments.

As for the FIR filter, the authors analyzed the probability to implement it by calculating window coefficients with the Hann window function (it provides noise suppression at the rejection band at the level of 44 dB). The filter was calculated in two stages. At the first stage, the authors set its order and calculated an ideal impulse response for the filter:

$$h(n) = \begin{cases} 1 - \frac{\omega_c}{\pi}; & n = M \\ -\frac{\sin(\omega_c(n-M))}{\pi(n-M)}; & n \neq M, \end{cases} \quad (1)$$

where $h(n)$ – element of the ideal impulse response of the filter, ω_c – cutoff frequency of the filter, n – number of the impulse response element of the filter, M – median element of the impulse response of the filter.

At the second stage, the authors calculated a window coefficient for the filter and corrected its ideal impulse response:

$$h(n) = h(n) * \left(0.5 * \left(1 - \cos\left(\frac{2\pi n}{N-1}\right)\right)\right), \quad (2)$$

where N – filter order. This filter type did not allow eliminating low-frequency noises hindering the detection of breath sounds in the bronchial sonograms under study. Moreover, in order to implement this filter, it is necessary to use a larger filter order (set of impulse response characteristics), which affects the speed of the algorithm used to detect breath fragments and express diagnostics of the bronchopulmonary system. The type 1 Chebyshev IIR filter would allow completely eliminating low-frequency noises. The coefficients of this filter in response characteristics are calculated according to the following formula:

$$H(z) = H_0 * (-1)^{N-M} * \frac{1}{\prod_{k=1}^{N}(1-p_k)} * (1 + z^{-1})^{N-M} * \frac{\prod_{k=1}^{M}(1-z^{-1})}{\prod_{k=1}^{N}\left(1-\frac{1+p_k}{1-p_k}z^{-1}\right)}, \quad (3)$$

where H_0 – constant calculated for the high-frequency type 1 Chebyshev filter, N – filter order, M – number of poles for the analogue prototype of the low-frequency filter, p_k – frequency of the pole. Following the elimination of noises, breath fragments that are significant assessing the state of the human bronchopulmonary system are singled out by means of the autocorrelation function. For this purpose, the autocorrelation function between the sections of the bronchial sonograms under study is calculated according to the following formula:

$$A(\tau) = \int f(t)f(t-\tau)dt, \quad (4)$$

where τ – offset value of the bronchial sonogram section; $f(t)$ – section of the bronchial sonogram; $f(t-\tau)$ – copy of the bronchial sonogram section shifted relatively to itself by τ.

Results and considerations

The authors developed specialized software able to automatically detect diagnostically significant sections of patients' bronchial sonograms. The theoretical data presented in the previous section and implemented in the software application developed were used to process 100 bronchial sonograms of patients aged 3-30 y. o. The method proposed is illustrated by the example of a 3-year-old girl's bronchial sonogram. A WAVE file with the sampling frequency of 10 kHz and duration of 9.5 seconds was chosen for processing. Figure 1 demonstrates the graphic representation of this record.

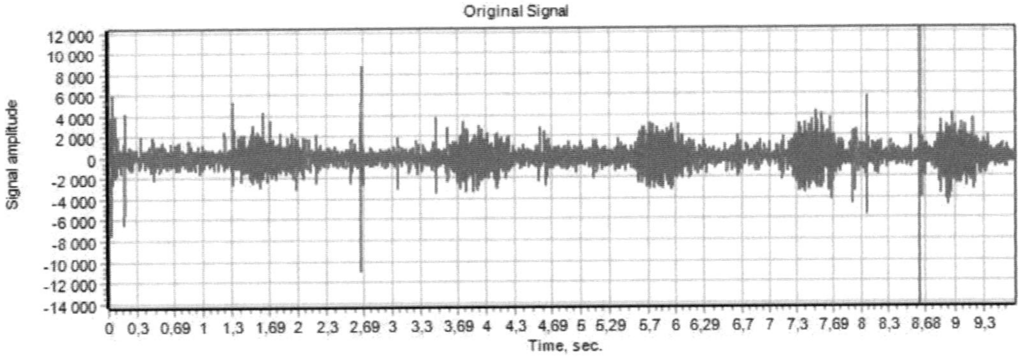

Figure 1. Graphic representation of the initial bronchial sonogram.

Then, in order to eliminate low-frequency ambient noises, the authors used the type 1 Chebyshev IIR high-frequency filter. Figure 2 demonstrates the results.

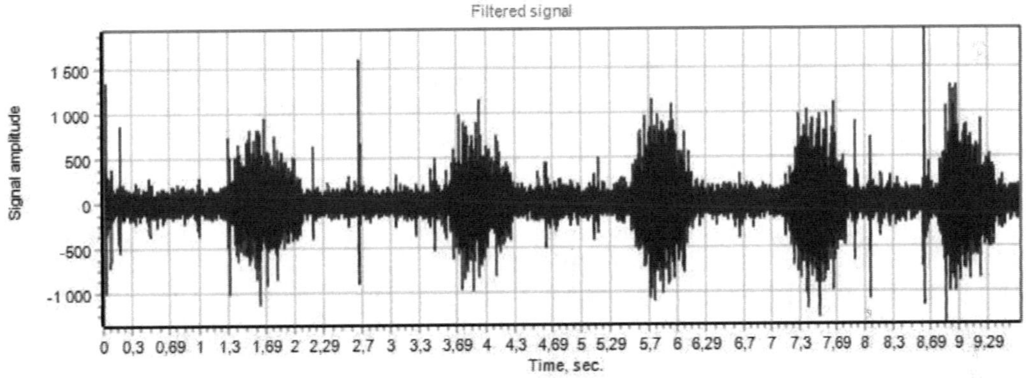

Figure 2. Graphic representation of the bronchial sonogram after eliminating low-frequency noises.

The graphic representation of the bronchial sonogram freed from low-frequency noises shown on Figure 2 allows one to see that noises hindering auscultation were eliminated. There are only clearly pronounced clicks caused by the contact between the acoustic detector and patients' skin. These clicks are not critical for the authors' method used to detect clinically significant breath sounds with the use of the autocorrelation function. The final automatically detected findings are represented on Figure 3.

The detected fragments of the respiratory cycle can be described as mathematic models for the purpose of further express diagnostics of bronchopulmonary diseases in patients of a varying age. Nowadays the authors are performing activities meant to describe various pathological conditions of the human bronchopulmonary system.

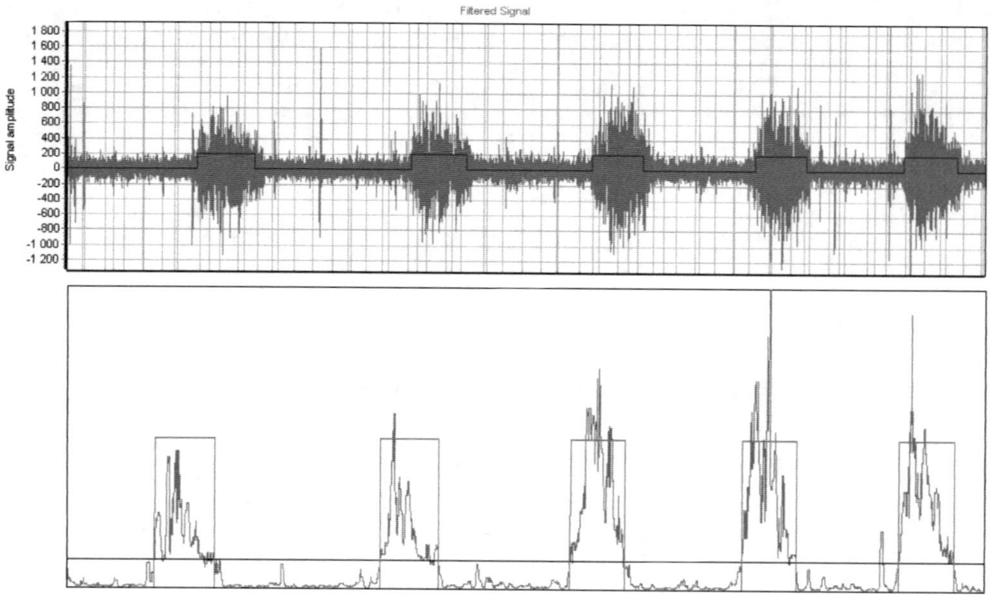

Figure 3. Graphic representation of detected clinically significant breath fragments.

References:

[1] Arati G, Scrafford C, Tielsch J, Levine O and Checkley W 2011 Computerized lung sound analysis as diagnostic aid for the detection of abnormal lung sounds: A systematic review and meta-analysis *Respiratory Medicine* vol. **105** pp 1396 – 1403.

[2] Arnal J, Wysocki M, Nafati C, Donati S, Granier I, Corno G and Durand-Gasselin J 2008 Automatic selection of breathing pattern using adaptive support ventilation. *Intensive Care Med* vol. **34** pp. 75-81.

[3] Guseinov A 2011 Bronhografiya: Acoustic criteria for the diagnosis of obstructive lung disease *Fundamental research* vol. **2** pp. 55-59.

[4] Jayant M and Praveen K 2013 Analysis of Lung Diseases and Detecting Deformities in Human Lung by Classifying Lung Sounds International *Journal of Advanced Research in Electrical, Electronics and Instrumentation Engineering* vol. **2** pp. 4528-4535.

[5] Sovijarvi A, Vanderschoot J. and Earis J. 2000 Standardization of computerized respiratory sound analysis *Eur. Respir. Rev.* vol. **10** pp. 585-588.

[6] Nagasaka Y 2012 Lung Sounds in Bronchial Asthma *Allergology International* vol. **61** pp. 353-363.

IOP Conf. Series: Materials Science and Engineering **66** (2014) 012019 doi:10.1088/1757-899X/66/1/012019

Simulation of electrochemical processes in cardiac tissue based on cellular automaton

S A Avdeev, N M Bogatov[1]
Kuban State University, 350040, Krasnodar, Stavropolskaya st., 149, Russia

E-mail: bogatov@phys.kubsu.ru

Abstract. A new class of cellular automata using special accumulative function for non-uniformity distribution is presented. Usage of this automata type for simulation of excitable media applied to electrochemical processes in human cardiac tissue is shown.

Keywords. cardiac tissue, cellular automata, computer simulation

1. Introduction

Currently the main method of diagnosis and analysis of cardiac arrhythmias is electrocardiography. Data analysis in electrocardiography is largely empirical. There is no complete understanding of relationship between graphical representation of electrocardiographic output and real processes occurring at the cellular level.

Some of the most common types of cardiovascular diseases are ventricular arrhythmias and fibrillation, the nature of which currently has not been completely studied.

Electrochemical processes during arrhythmias are nonlinear and non-stationary, self-sustaining autonomous waves underlying these phenomena are called autowaves. Methods of the theory of cellular automata are used for simulation of autowaves [1 - 4]. Advantages of such models over the models based on differential equations [5 - 7] are simple adaptation to the problem and low computational cost, which allows carrying out real-time simulation. Adaptive non-uniform automata are used to analyze and model various processes in biology, logistics, energetics, pattern recognition, cryptography.

Existing methods based on cellular automata have disadvantages that do not allow them to be applied to the problem of action potential propagation in excitable media or impose restrictions on the simulation. It is, for example, leak of isotropy [1], impossibility of modelling specific non-uniformity configurations of media, impossibility of modelling non-conservative forces amplifying or damping depending on distance from source and traversed path [1 - 4].

Simulation of autowave electrochemical processes in human heart with additional possibility of obtaining results in the form of simulated electrocardiograms allows making more informative diagnostics of heart diseases by electrocardiograms and suggesting methods of preventing life-threating arrhythmias.

Purpose of work. Creation of mathematical model and studying of electrochemical processes properties in human heart, leading to life-threating autowave arrhythmias: premature ventricular contractions of different grade, sustained and non-sustained ventricular tachycardia.

[1] N M Bogatov, Kuban State University, 350040, Krasnodar, Stavropolskaya st., 149, Russia, bogatov@phys.kubsu.ru.

IOP Conf. Series: Materials Science and Engineering **66** (2014) 012019 doi:10.1088/1757-899X/66/1/012019

2. Model of electrochemical processes in cardiac tissue

In this work model, which allows taking in account different action potential duration in different cells, was created. One of the most known effects of action potential's duration's differences in different areas of cardiac tissue is partially reverse direction of action potential propagation during repolarization. Solution of this problem involves the use of different transition functions δ in different cells of automaton.

The physical basis of the model is two types of electrochemical processes of intercellular interaction in cardiac tissue:

1. Amplification of K^+ ionic currents as a result of additional Ca^{2+} ions release from cellular depots.

First factor that can reduce duration of repolarization phase is the active membrane transport, related to transformation of cells internal energy, in the form of adenosine triphosphate (ATP), into electromotive force of ionic currents. Increase in Ca^{2+} concentration at the phase of early repolarization leads to additional Ca^{2+} ions release from sarcoplasmic reticulum, this phenomenon is called calcium-induced calcium release (CICR) [8]. Increase in Ca^{2+} concentration within the membrane as a result of this process leads to weakening of Ca^{2+} ionic current into the cell and amplification of K^+ ionic current as a result of additional K^+ ionic channels opening by Ca^{2+} ions. Together, this results in a more rapid membrane repolarization. Part of the ions is transferred to the neighboring cells, it leads to even higher intensity of release, and thus, significance of this process increases with increasing distance from the source of depolarization.

2. Impact of neighboring cells membrane potential on the duration of Ca^{2+} ion channels activity during repolarization.

2.1. Second factor that can reduce duration of repolarization phase is the presence of other membranes in a later phase of repolarization near the membrane. Electrical charge on the outer surface of such membranes is lower, that's why leaking current occurs between membranes through the gap junctions, and this current tends to equalize potential difference between them, as a result potential of membrane at earlier phase of repolarization decreases faster [9]. This in turn leads to faster closing of Ca^{2+} voltage dependent ionic channel in membrane, and as consequence, faster transition to rapid repolarization phase by K+ ionic current, together this leads to shortening of repolarization duration in such cells.

2.2. Other version of this process is repolarization near the depolarizing area. During the autowave process, for example, ventricular tachycardia, two adjacent areas can be in the state of depolarization and repolarization respectively at the same moment. In this case high potential of depolarizing region over the repolarizing one causes electrical current through gap junctions that will maintain high potential of repolarizing membrane during some time. This will increase the duration of Ca^{2+} ionic channels activity, and so, duration of repolarization phase for cells near depolarizing area will also increase.

Models with non-uniform action potential duration were developed by authors [10 – 13]. Different action potential duration in different cells in these models was obtained by adding special component T, which is function of action potential duration depending on cells location, to the cellular automaton, and using transition function that depends on T.

Adaptive non-uniform cellular automaton is non-uniform cellular automaton, non-uniformity distribution of which depends on parameters that change as a result of its functioning.

In the models of processes, which result depends on traversed path, adaptive non-uniform cellular automaton with transition function that depends on additional variable of non-uniformity can be used; value of additional variable must be recalculated for each iteration by adaptation's function depending on previous values of this variable for this cell.

Adaptive function of special form proposed, hereinafter referred to as accumulative function, it has next advantages over the memory functions of cellular automata [14, 15]:

1 – it introduces the ability to adaptively limit the upper limit of integration, depending on non-uniformity accumulative function value at the point (x,y) using delta function η;

2 – it takes in account not only the state of current cell, but also states of cells belonging to its neighborhood, also it uses memorizing with increment instead of simple memorizing.
Properties 1, 2 together guarantee spatial increment of function with gradient directed from the source that defined by function $g_0(x,y)$. Other important property of accumulative function is its above boundedness, which allows using its upper limit value, as variable t approaches infinity, in cellular automata as modelled values of forces, which are responsible for amplifying or damping. Examples of such forces are amplifying of Ca^{2+} ionic currents during mass repolarization and K^+ during mass depolarization of cardiac tissue, their effect on action potential duration at the cell of cellular automaton can be determined numerically by accumulative function upper limit value.
Improved adaptive non-uniform cellular automaton with modified adaptation equations $\gamma(\omega, g, q)$ and transition function $\delta(\omega, q, g, r_g)$ proposed for excitable media simulation taking in account phenomena of repolarization's duration variation, interaction of activation and repolarization and defined non-uniformity, related to structural damages of cardiac tissue:

$$M \equiv (B,\Omega,Q,G,q,q_0,g,g_0,\delta,\gamma,r_g,\varphi), \qquad (1)$$

$$\Omega \subset N,\ Q \subset N^2,\ q,q_0 \in Q,$$

$$q \equiv (u,v,t,z) \in Q \subset N^4,$$

$$\omega(q): Q^{|B|} \to \Omega \subset N^2,$$

$$g \equiv (\Delta,a,p,h) \in G \subset Z^4,\ g_0 \in G,$$

$$\delta(\omega,q,g,r_g): \Omega \times Q \times G \times Z \to Q,$$

$$\gamma(\omega,g,q): \Omega \times G^{|B|} \to G,$$

$$r_g(g): G^{|B|} \to Z,$$

B – set of neighbors, simple rectangular grid and Moore neighborhood are used. Moore neighborhood is 8 cells having at least one adjacent point with the center cell and 16 cells having at least one adjacent point with one of these 8 cells.
ω – function of input values, returns two values, ω_a – average level of depolarization excitation in cells neighborhood B and ω_b – average level of repolarization excitation, estimated by number of cells in depolarized and repolarized states respectively.
q – state vector, defined by four parameters u, v, t, z:
u – level of excitation,
v – level of refraction, responsible for duration of excitable media refraction phase,
t – delay timer that defines increasing of action potential duration ($t > 0$), takes in account potential difference between different areas of model, depends on non-uniformity parameter p,
z – current state indicator that shows which process, depolarization or repolarization occurs at present time;
φ – additional function for converting states of cellular automaton q to action potential values, range of function φ belongs to the set of real numbers;
g – non-uniformity vector, which consist of four parameters Δ, a, p, h:
Δ – parameter providing isotropy,

IOP Conf. Series: Materials Science and Engineering **66** (2014) 012019 doi:10.1088/1757-899X/66/1/012019

a – level of conduction abnormalities caused by structural damage expressed numerically, for example, delay of repolarization processes in this cell (abnormalities can be caused by various diseases, for example, myocardial infarction under conditions of ischemic heart disease); same as parameter Δ, this parameter does not changed by adaptive function between iterations, its value remains constant because processes of regeneration of damaged areas does not considered in this research.
p – parameter that models phenomena of repolarization's duration decreasing depending on distance from the source of depolarization, taking in account traversed path; it reflects the processes of type 1, increase of Ca^{2+} ionic concentration as a result of ions release from cellular ionic depots; it is an example of «non-conservative» phenomenon, because amplification during depolarization wave propagation passes along a curved path: through the right ventricular, skirting interventricular septum to the left ventricular, that's why it is necessary to use managed accumulative function for its adaptation.
h – parameter that models phenomena of repolarization's duration decreasing depending on distance from repolarization source, which means area of excitable media that repolarized first; it reflects electrochemical processes of type 2.1, impact of neighboring repolarizing cells potential on the duration of repolarization in this cell; it is also an example of «non-conservative» phenomenon, because wave propagation during repolarization can pass along curved path in the presence of barriers, such as excited cells in refractory state or areas with conduction abnormalities, that's why it is also necessary to use managed accumulative function for adaptation of this parameter.
$\gamma(\omega, \{g_j\}, q)$ – function of adaptation for non-uniformity variable, accumulative functions do not used for parameters Δ and α, function just passes values of parameters Δ and α to the next iteration:

$$\gamma(\omega,\{g_j\},(u,v,t,z)) = \left(\gamma_\Delta(\{g_j\}), \gamma_a(\{g_j\}), \gamma_p(\omega,\{g_j\},z), \gamma_h(\omega,\{g_j\},z)\right),$$

$$\gamma_\Delta(\Delta_j) = \Delta_{x,y},\ \gamma_a(a_j) = a_{x,y},$$

parameter p is adapted at each iteration using accumulative function:

$$\gamma_p(\omega,\{(\Delta,p)_j\},z) = p_{x,y} + u_p(\omega,\Delta)\eta(p_{x,y})\left(\frac{\sum_j p_{i,j}(t)}{\sum_j \rho(p_{i,j}(t)) + \beta} + \alpha_p\right), \tag{2}$$

function of adaptation for parameter h is constructed similarly to adaptation function for parameter p, because phenomena modelled by them, weakening of Ca^{2+} ionic currents as a result of ions release from cellular depots and as a result of intercellular currents, both are «non-conservative» phenomena, whose intensity depends on traversed path from the source:

$$\gamma_h(\omega,\{(\Delta,a,p,h)_j\},z) = h_{x,y} + u_h(\omega,\Delta)\eta(h_{x,y})\left(\frac{\sum_j h_{i,j}(t)}{\sum_j \rho(h_{i,j}(t)) + \beta} + \alpha_h\right), \tag{3}$$

cells, at which repolarization occurs first defined adaptively using managing function u_h:

$$u_{p,h}(\omega(t),\Delta) = \begin{cases} 1, \omega_b(t) > \Delta \\ 0, \omega_b(t) < \Delta \end{cases}, \tag{4}$$

this condition is similar to the condition of excitation transfer between model cells, managing function of this type allows constructing accumulative function for adaptation of non-uniformity parameter h taking in account traversed path from the source of repolarization.

$\Delta_{x,y}$, $a_{x,y}$, $p_{x,y}$ – values of non-uniformity parameters Δ, α, and p in current cell, before the adaptation, this values are contained in the set $\{g_j\}$, which contains values of non-uniformity variables for all neighboring cells from the set B.
$a_p > 0$, $a_h > 0$ – accumulation increments, they define increasing of values p and h, with increasing distance from the source of depolarization and repolarization respectively, different values of this parameters can be used for model calibration.
$r_g(g)$ – interaction function that models phenomena of repolarization duration increasing depending on distance from simultaneously depolarizing area, related to increase of Ca^{2+} ionic channels activity as a result of intercellular current between depolarizing and repolarizing area, this is electrochemical process of type 2.2.

$$r_p(g) = p_r \sum_j \rho\left(p_{i,j}(t)\right), \tag{5}$$

$p_r > 0$ – coefficient of interaction , that reflects intensity of ionic currents between cells in depolarizing and repolarizing states,
$\rho(x)$ – non-zero indicator function.

3. Simulation of reentry type arrhythmia initiation without structural damages

Results of idiopathic tachycardia modelling, initiated in cardiac tissue without structural damages, presented at figure 1, figure 2.

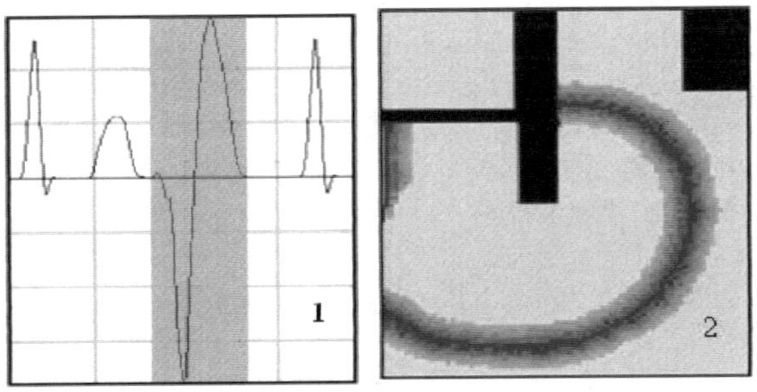

Figure 1. Premature ventricular contraction after repolarization without gradient ring break:
1 – simulated electrocardiogram (II standard lead), premature ventricular contraction marked blue;
2 – cellular automaton state graphical representation after repolarization, darkest green areas are repolarized last.

Natural gradient of repolarization's durations, marked green on figures 1.2, formed as a result of depolarization under conditions of Ca^{2+} ionic currents damping with the increase of distance from atrioventricular node, can became the unidirectional block if its break occurs. Such unidirectional conduction area marked red on figures 2.2. Break can occur as a result of repolarization duration alterations in a small area near the gradient ring, this area marked blue at the figures 2.2.
So, T-wave alterations, widely used for prediction of idiopathic arrhythmias, associated with the actual electrochemical processes inducing arrhythmia, spontaneous formation of unidirectional conduction block.
Therefore, idiopathic arrhythmia can be prevented by decreasing of media automatism coefficient, by prevention of premature ventricular contractions or by prevention of repolarization's duration gradient ring breaking.

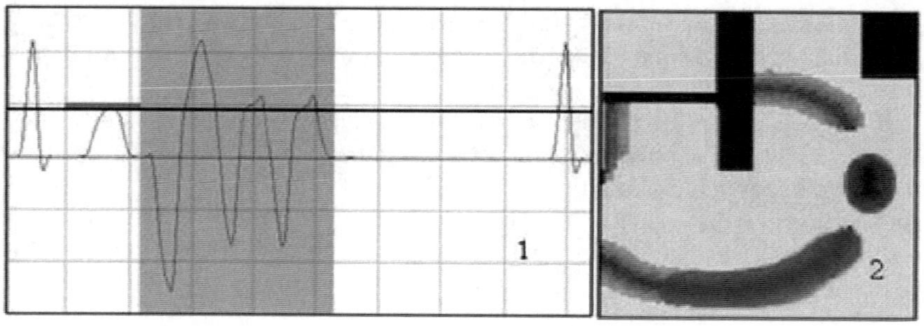

Figure 2. Result of premature ventricular contraction after repolarization with gradient ring break:
1 – simulated electrocardiogram (II standard lead), reentry complex marked red, T-wave alteration marked purple;
2 – cellular automaton state graphical representation after repolarization, darkest green and red areas are repolarized last.

4. Conclusion

New class of non-uniform cellular automata using special accumulative functions for non-uniformity distribution presented. This type of automaton used for modelling of excitable media in application to electrochemical processes in human cardiac tissue.
Model allows estimating of life-threating arrhythmia's risks depending on the size of damaged area of heart tissue, formed, for example, as a result of myocardial infarction.
Obtained results allow making more informative diagnostics of idiopathic arrhythmias by T-wave alterations using electrochemical interpretation of underlying processes of these phenomena.
Developed mathematical method could be used for real time simulation of large processes variety in different excitable media, for example, simulation of atrial arrhythmias, by the construction of the appropriate models.

5. References

[1] Moe G K, Rheinboldt W C, Abildskov J A 1964 *American Heart Journal* **67** 200
[2] Pourhasanzade F, Sabzpoushan S H 2010 *World Academy of Science, Engineering and Technology* **44** 917
[3] Gerhardt M, Schuster H, Tyson J J 1990 *Science* **247** 1563
[4] Markus M, Hess B 1990 *Nature* **347** 56
[5] Weimar J R, Tyson J J, Watson L T 1991 *Physica D* **55** 309
[6] FitzHug R 1955 *Bull. Math. Biophysics* **17** 257
[7] Aliev R R, Panfilov A V 1996 *Chaos, Solitons and Fractals* **7(3)** 293
[8] Venetucci L A, Trafford A W, O'Neill S C, Eisner D A 2008 *Cardiovascular Research* **77** 285
[9] Stephan R 2004 *Cardiovasc Res* **62** 309
[10] Tysler M, Turzova M, Svehlikova J 2003 *Measurement science review* **3** 37
[11] Farina D, Dossel O 2007 *Computers in cardiology* **34** 173
[12] Jiang Z, Mangharam R 2011 *Conf Proc IEEE Med. Biol. Soc.* 263
[13] Andreev S Yu *et al* 2010 *Izvestiya of Tomsk Polytechnic University* **317** 189
[14] Alonso-Sanz R, Martin M 2003 *Complex Systems* **14** 99
[15] Seck-Tuoh-Mora J C, Martinez G J, Alonso-Sanz R, Hernandz-Romero N 2012 *Information Sciences* **119** 125

IOP Conf. Series: Materials Science and Engineering **66** (2014) 012020 doi:10.1088/1757-899X/66/1/012020

Analysis of Steel Structures 5140 after Laser Treatment

O Lobankova[1,a], I Zykov[1,b] and A Melnikov[1,c]

[1]Tomsk Polytechnic University, Tomsk, Russia

E-mail: [a]lobankovaov@gmail.com, [b]zilju@yandex.ru, [c]melnikov_ag@tpu.ru

Keywords: steel grade 5140, laser irradiation, changing the steel structure.

Abstract. The main results of analysis of steel structure 5140 after laser treatment are presented. It is found that structural changes have occurred under the influence of heat radiation of the laser. This paper describes and demonstrates the changes. Method of studying the structure of steel, preparation of materials for the study and the method of laser processing are also described.

1. Introduction

At present time there is rapidly developing the production of embossed stamps (cliché). In particular, stamps fabrication by laser technology from hardened metal is very popular. This method of manufacturing avoids the additional processing of the stamp after laser processing of material. However, each material and each laser has its nuances in application to get a specific result. While manufacturing, laser radiation has a thermal effect on the residual material, often modifying its properties, making it difficult to optimize the process and, therefore, requires the study of the processes occurring in the metal.

2. Preparation for the research

It is necessary to choose the material for the work and the processing method before the research.

2.1. Material selection

Structural alloyed steel has a carbon content of 0.40 % and less than 1.5 % of chromium (steel 5140). This grade was chosen as a material for research because of demand in this area, price and wide spread.

2.2. Thermal treatment

Chamber electric furnaces were used for heating, quenching and tempering. Heating temperature for quenching was 850 C, which has allowed to achieve maximum hardness and durability of material. After shutter speed in a furnace the samples were cooled in water. Then there was the low-temperature tempering at a temperature of 200 C for 30 minutes, which made the sample more viscous.

2.3. Laser treatment

The laser complex sample equipped with an ytterbium fiber laser with a pulse repetition rate of 20 kHz was chosen for material processing. The velocity of the beam was 100 mm/s and an average output power of the laser 20 Watts. The number of passes of the laser beam and the pulse duration is varied in this case. [1]

This laser system called Minimarker2 - M20 was chosen due to widespread of the same complexes, a high coefficient of absorption of radiation with a wavelength of 1064 nanometers by metals and high efficiency of fiber lasers (about 25 %).

2.4. Metallographic analysis

The metallographic sections were prepared for carrying out microanalysis on samples by using grinding skins with different grain size of the abrasive. Final polishing was performed on cloth using special polishing pasta made in State Optical Institute. The structure was revealed by etching solution 4% HNO3 in ethanol.

Structure analysis was performed using an optical research microscope AxioObserver A1.m by Carl Zeiss (Germany) in a software product AxioVision v.4.6.

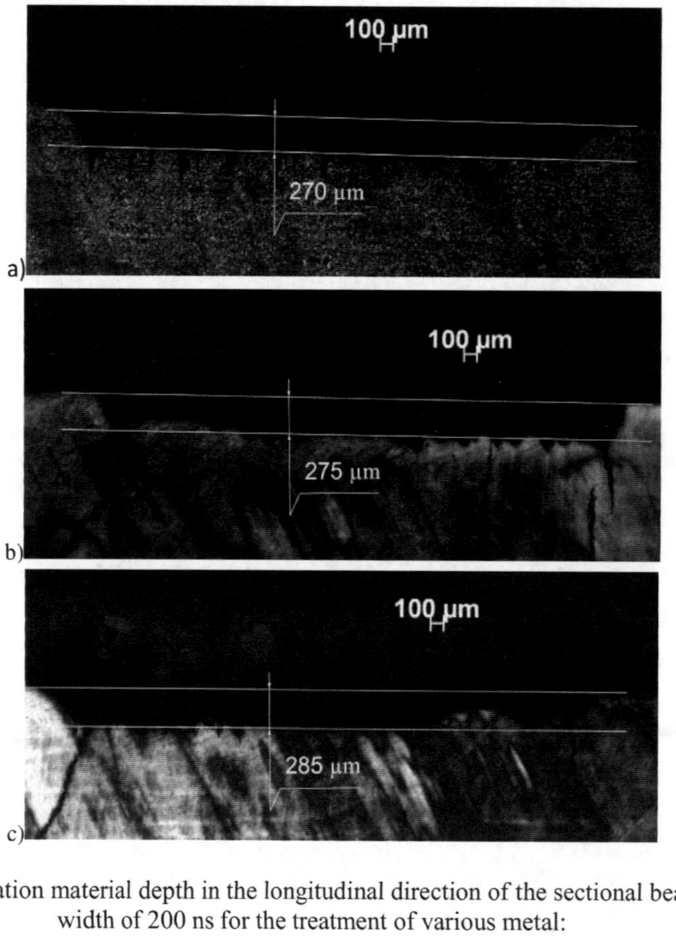

Figure 1. Evaporation material depth in the longitudinal direction of the sectional beam at a pulse width of 200 ns for the treatment of various metal:
1 - annealed 2 - hardened, 3 - hardened and tempered

2.5. *Hardness and microhardness analysis*

Hardness of the steel in the initial (annealed) condition was measured by the Brinell method, it was HB = 212-217. Hardness of the steel after quenching was 60-62 HRC, after the tempering - 51-52 HRC.

A diamond pyramid under a load of 50 g was pressed into material, and then the hardness was calculated in microscopically small volumes to determine the microhardness.

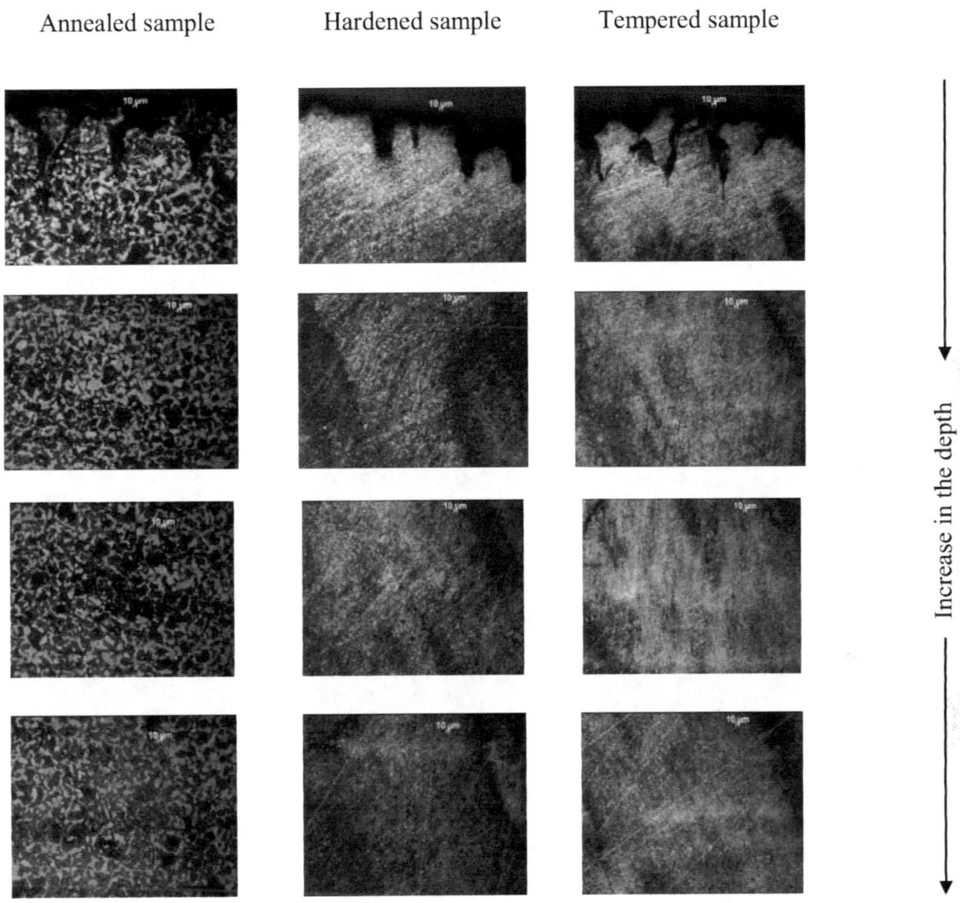

Figure 2. Steel structure at different depths below the zone of laser action

3. The research part

We consider the most profound holes formed during laser beam after 100 passes with the duration 200 nanoseconds in Figure 1. [2]

Thus leads us to believe that the quantity of removed material is independent of the pre-treatment of metal, or depends insignificantly.

Now examine the change in structure of the steel in the area of laser impact.

In Figure 2 it is clear to see that the steel structure doesn't change in all three samples under the influence of laser radiation. Laser processing complex Minimarker2-M20 has no effect on change in the structure and, hence, the properties of the material. It means, that after the metal sample have wells material within the groove should have the same hardness. [3] For the proof of this supposition microhardness was measured in the direction from the laser impact zone to non-exposed zone. Figure 3 illustrates that the microhardness of the material does not vary in the area of laser impact. [4]

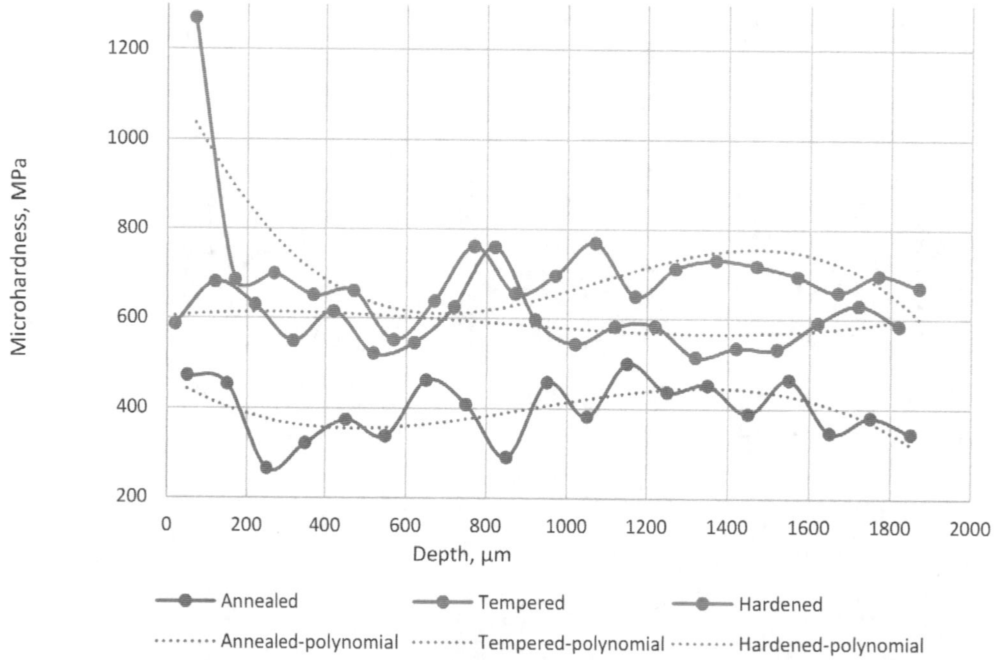

Figure 3. Microhardness of steel at different depths below the zone of laser action

4. Conclution

It is established that with the increase of pulse duration crater in the alloy also deepens stronger though emission power at pulse durations is less. It is assumed that this happens because of "screening" of radiation, i.e. the absorption of laser emission formed by plasma when exposed laser rays.

It is shown that the quantity of material evaporated by laser radiation is independent on the hardness (pretreating process) of the material.

We also revealed that the investigated steel 5140 changes neither its structure nor the properties after multiple sampling of material by laser. This simplifies the process of manufacturing a cliché in industrial areas without requiring additional research on changes in the characteristics of the samples. It should nevertheless be noted that this conclusion is correct for the steel grade is 5140 when processing its with impulse ytterbium fiber laser with an average output power of 20 watts. [5] In another research, the author of article concluded that the structure of another steel harden with fusion (with a liquid phase) under the influence of the same laser system Minimarker2 - M20.

References

[1] Babu P D, Buvanashekaran G, Balasubramanian K R 2012 Experimental studies on the microstructure and hardness of laser transformation hardening of low alloy steel *Transactions of the Canadian Society for Mechanical Engineering* **3** pp 241-257

[2] Smirnov N A, Misyurov A I 2012 Features of structures in laser treatment *Vestnik MSTU* pp

115-129

[3] Qiu F, Uusitalo J, Kujanpaa V 2012 Laser transformation hardening of carbon steel: microhardness analysis on microstructural phases *Surface Engineering* **29** pp 34-40

[4] Magin D Y, Kostromin S V 2013 The structure and properties of high-heat-resistant steels after volume heat treatment and laser surface hardening *Proceedings of the Nizhny Novgorod State Technical University* **101** pp 256-261

[5] Valiulin A, Mountain S, Grechko Y, Patra M, Yudin K, Jurevich V 2007 Laser marking of materials *Photonics* **3** pp 16-22

INVESTIGATION OF THE STRUCTURAL-PHASE STATE OF ULTRAFINE PLASMOCHEMICAL ZrO_2 (Y) POWDERS

S A Ghyngazov[1], I P Vasiliev[1], T S Frangulyan[1], V M Khaidukova[1], T S Mylnikova[1]
[1]Tomsk, 634050, Russia, National Research, Tomsk Polytechnic University
E-mail: ghyngazov@tpu.ru

Annotation. The methods of X-ray phase analysis, thermogravimetry and differential scanning calorimetry were used to study the structural and phase state of ultrafine powders (UFP) of ZrO_2(Y) obtained by plasmochemical synthesis. It is shown that these UFPs are in a highly nonequilibrium state due to the presence of adsorbates and bound water, and the distribution of the stabilizing impurity in the volume of the particles is nonuniform.

Keywords: zirconia, structural-phase state, X-ray diffraction, differential scanning calorimetry, ultrafine powders, plasmochemical synthesis.

1. Introduction

Advancement of science and technology is impossible without materials which possess a unique combination of physical and mechanical properties and improved performance characteristics. Over the last few years, great emphasis has been put into the problem of producing ceramics for structural and instrumental purposes with high mechanical strength and fracture strength.

There are two basic approaches to tackle the problem. One of them is to produce nanostructered ceramic materials. Currently, this problem can be solved due to a substantial progress in the technology of UFP production. Since the surface energy of UFP is high if compared to coarse powders, the powders can be sintered at lower temperatures, and this reduces the rate of grain growth due to intense recrystallization processes. The second approach is based on the effect of polymorphic transformation under external mechanical loads.

Both of the approaches can be realized in ceramic structures made on the base of the nano-size particles of stabilized zirconia [1–5].

In practice, to develop high-strength nanoceramics a wide range of additional factors that contribute to its physical and mechanical properties is to be considered. These factors include characteristics of the nanodispersed powder medium used (grain-size composition, morphology, degree of agglomeration, presence of stabilizing additives), which essentially depend on the technology of UFP production. There are currently more than 15 fundamentally different methods of oxide UFP synthesis. The efficiency of compacting and sintering of UFP depend on the method of their manufacturing. Therefore, ceramic materials may acquire a particular set of properties depending on the nature of the initial green powder.

An efficient plasmochemical technology to manufacture a wide range of oxide nanopowders, including UFP of stabilized zirconia, has been developed at the Siberian Chemical Combine, Tomsk, [6]. High-performance of this technology makes it attractive to be used in mass production.

To choose the technological regime of powder compaction and sintering it is necessary to obtain sufficiently full data on the structural-phase state of ultrafine zirconia plasmochemical powders, which can significantly impact the effectiveness of compression, the shrinkage kinetics and properties of the sintered ceramics. In [7], features of the structural-phase state of composite ultrafine plasmochemical powders of the $(ZrO_2\text{-}Y)\text{-}Al_2O_3$ system and phase transformations during thermal heating are investigated.

The present research addresses the study of thermally-stimulated structural-phase transformation in ultrafine plasmochemical powders of stabilized zirconia.

2. Experimental techniques

UFPs of the ZrO_2 – 3 mol. % Y_2O_3 solid solution were synthesized at the Siberian Chemical Combine by decomposition of aqueous solutions of zirconium nitrate salts and yttrium in HF discharge plasma. The change in mass and thermal effects which occurred during heating of the test nanopowders were recorded simultaneously using the analyzer STA 449 C Jupiter (Netzsch, Germany), the sensitivity of weights being 0.1 µg. For correct measurement of the baselines a reference crucible was loaded with the inert Al_2O_3 powder, its mass being equal to the mass of the test material. Heating was carried out in air. The X-ray phase analysis of the powders was carried with the ARL X 'tra powder diffractometer using monochromatic Cu K□.

3. Experimental Results

Figure 1 shows the TG and DSC curves for UFP of the partially stabilized zirconia ZrO_2 – 3mol. % Y_2O_3.

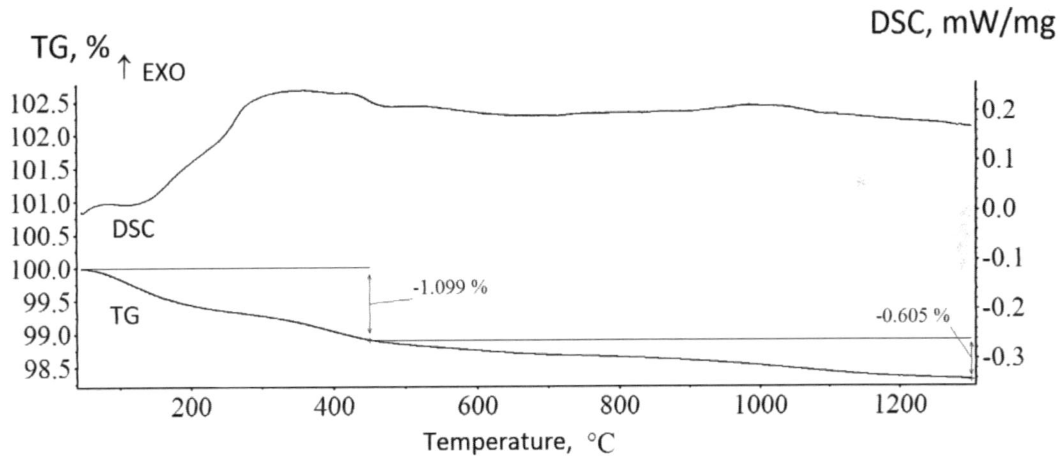

Fig.1. TG and DSC curves for zirconia powder

Noteworthy is the fact that during heating the mass of the powder reduces. When heated to T = 1300°C, the total mass loss is about 1.8 %.

The process of the powder mass reduction occurs in several stages. The first stage, which falls within the temperature range T = (100–200)°C, is due to the removal of the adsorbed moisture, and this is confirmed by the presence of the low-temperature endothermic peak in DSC at T = 110°C. The second stage, during which the powder mass reduces, covers the temperature range T = (300–500)°C, and within this stage an exothermic effect occurs. The exothermic peak is nonelementary.

The mass-spectrometric studies [8] showed that in this temperature interval, the bound water evolves from zirconia nanoparticles synthesized by plasmochemical method. This process is irreversible. The exothermic heat effect is due to the structural changes in the particles caused by changes in their chemical composition as a result of water removal. The subsequent stages of the process most probably are due to the removal of crystallization and constitution water.

The X-ray phase analysis showed that the test zirconia UFP consisted of a mechanical mixture of monoclinic and tetragonal phases in the amount of 36 and 64 mass%, respectively. At T□1000°C the monoclinic phase is known to be unstable, and it is to transform to a tetragonal modification. This transformation should be accompanied by an endothermic peak in the DSC curve. Our results indicate the absence of this peak in the DSC curve.

Figure 2 shows the diffraction patterns of ZrO_2-Y_2O_3 powder annealed at T = 300, 500, 700 and 1000°C for 30 minutes.

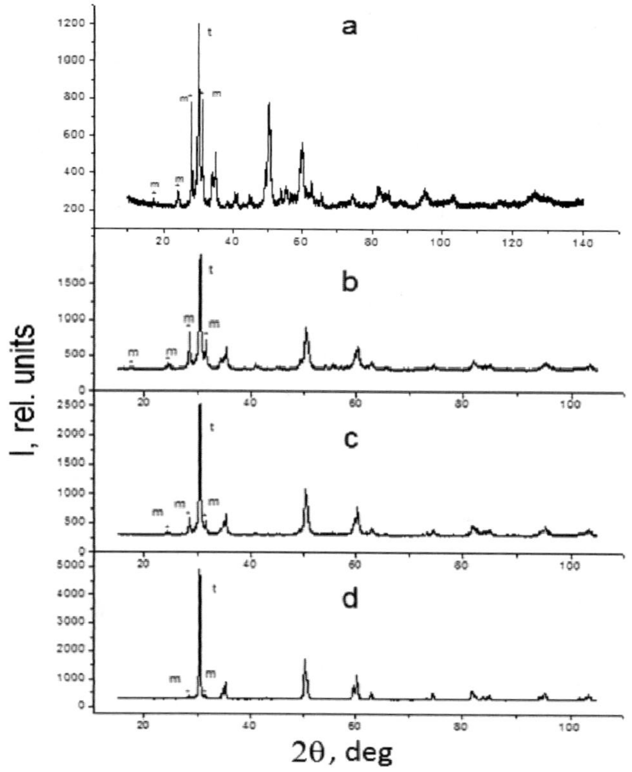

Fig. 2. Diffraction patterns for zirconia powder after annealing at different temperatures: *a is* unannealed and after annealing at T = 300 and 500 °C; *b, c,* and *d* are after annealing at T =700, 900 and 1000°C, respectively.

In our opinion, this fact can be consistently interpreted if we take into account the existence of the volume chemical heterogeneity of the synthesized plasmochemical powder of ZrO_2-Y_2O_3 [6]. It is well known that impurities in solid solutions tend to segregate in different interface regions including those on the surface of the particles and grains [9].

This means that there are a smaller number of yttrium ions in the particle volume than that on the particle surface. This heterogeneous chemical composition of the powder contributes to the increased content of the monoclinic phase. During sintering of the compact the diffusion of the impurity into the grain volume is stimulated, and this causes increase in the content of the tetragonal phase in ceramics. Since t-phase formation is limited by the impurity diffusion, the heat effects are observed as if prolonged over the time and within a wide temperature range.

This is what can account for a considerably wide minimum in the DSC curve which covers the temperature range T = (500–1000)°C, within which the monoclinic phase transforms into a tetragonal one. This assumption is confirmed by the results obtained in the study of the effect of stepped annealing in the temperature range T = (300–1000)°C on the structural-phase state of the test powders.

The summarized results of the analysis of the diffraction patterns are presented in Table 1. According to the data in Table 1, thermal annealing stimulates the transition of the monoclinic modification of zirconia in a tetragonal one.

Table 1. The phase composition and structural characteristics of $ZrO_2 – Y_2O_3$ powder

Annealing temperature T, °C	m-phase content mass%	t-phase content mass%	Lattice parameters of t-ZrO_2, Å	Lattice parameters of m-ZrO_2, Å	L,nm t-ZrO_2	L,nm m-ZrO_2	□d/d □ 10^{-3} t-ZrO_2	□d/d□ 10^{-3} m-ZrO_2
Unannealed	38	62	a=в=3.609; c=5.172	a=5.174; в=5.216; c=5.329	31	25	0.4	1.4
300	39	61	a=в=3.609; c=5.172	a=5.174; в=5.224; c=5.327	23	-	1.2	-
500	34	66	a=в=3.608; c=5.170	a=5.178; в=5.212; c=5.331	27	37	0.9	1.2
700	20.8	79.2	a=в=3.610; c=5.173	a=5.174; в=5.214; c=5.337	35	27	0.9	0.8
1000	4.0	96.0	a=в=3.608; c=5.174	-	53	-	-	-
1200	-	100	a=в=3.607; c=5.173	-	96		0.2	-
1400	-	100	a=в=3.607; c=5.174	-	73		0.2	-
1500	-	100	a=в=3.606; c=5.174	-	101		0.3	-
1600	traces	90	a=в=3.605; c=5.175	-	120		0.3	-

Note: L is the size of coherent-scattering region of X-rays; □d/d is the value of microstrains.

4. Conclusion

The studies conducted have revealed that UFP of the stabilized zirconia obtained by plasmochemical method are in a strongly nonequilibrium state. They contain a significant amount of adsorbates and bound water. It is shown that the test zirconia UFP consists of a mechanical mixture of the monoclinic and tetragonal phases in the amount of 38 and 62 mass%, respectively. This can be attributed to a nonuniform distribution of the stabilizing yttrium impurities in the volume of the powder particles. Under heating, the phase composition of the powders changes as a result of diffusion of the stabilizing additive from the particle surface into the particle volume, where it is incorporated into ZrO_2 lattice.

This research is supported by The Ministry of Education and Science of the Russian Federation in part of the "Science" program.

5. References

[1] Gusev A I 2007 Nanomaterials, nanostructures, nanotechnology (Moscow: Fizmatlit) 416 p

[2] Lukin E S, Makarov N A, Kozlov A I, Popova N A, Anufrieva E V, Vartanian M A, Kozlov I A, Safina M N, Lemeshev D O, Gorelik E I and Bakunov V S 2009 Nanopowders for a new generation of ceramic oxide *New refractories* **11** pp 29-34.

[3] Pliner S Y and Dabizha A A 1986 Hardening of zirconia ceramics by tetragonal-monoclinic transformation *Refractory* **3** pp 58-62

[4] Lange F F Transformation toughening. Part 3: Experimental observation in the $ZrO_2 - Y_2O_3$ system 1982 *Journal of Materials Science* vol **17** pp 240-246

[5] Bakunov V S, Balkevich V L, Vlasov A S, Guzman I Ja and Lukin E S 1977 *Ceramics from highly refractory oxides* (Moscow: Metallurgy) 304p

[6] Larin V K , KondakovV M, Malyi E N, Matyukha V A, Dedov N V et al 2003 Plasma-chemical method to obtain ultrafine (nano) powders of metal oxides and prospective lines of their development *Izvestiya Vuzov: Tsvetnaya metallurgiya* **5** pp 59-64

[7] Surzhikov A P, Franguljyan T S and Gyngazov S A 2014 A thermoanalysis of phase transformations and linear shrinkage kinetics of ceramics made from ultrafine plasmochemical $ZrO_2(Y)–Al_2O_3$ powders *Journal of Thermal Analysis and Calorimetry* vol **115** pp 1439-1445

[8] Gyngazov S A, Lysenko E N, Surzhikov A P, Frangulyan T S 2013 Control of the thermal effect on the structural-phase state of ultrafine plasma-chemical zirconia powders *Control. Diagnostics* **13** pp 122-126

[9] Theunissen G S A, Winnubst A J A, Burggraaf A J 1992 Surface and grain boundary analysis of doped zirconia ceramics by AES and XPS *Journal of Materials Science* vol **27** pp 5057-5066

Formation of wear-resistant chromium-vanadium cast iron coating using a continuous and pulsed electron beam

N V Tatarkina[1], A I Ziganshin[2], B V Dampilon[1,2], V G Durakov[2] and A M Tolstokulakov[1]

[1]Tomsk Polytechnic University, Tomsk, Russia
[2]Institute of Strength Physics and Materials Science, Tomsk, Russia

E-mail: dampilon@ispms.tsc.ru

Abstract. The paper reports the results of an investigation of structure formation in the eutectic chromium-vanadium cast iron coating deposited by electron-beam hardfacing (EBH) and subjected to multipoint pulsed electron-beam treatment (MPEBT) of the surface layers. The coating material structure in the zones modified by MPEBT changes considerably, showing the presence of a large amount of supersaturated austenite matrix and eutectic colonies in the form of fine isolated inclusions. The eutectic colonies are fan-like. The leading phase in the eutectic colonies is vanadium carbide. After aging within the temperature range from 900 to 1100°C the structure of the modified zones exhibits a large amount of fine secondary carbides Cr_7C_3 rejected from the matrix. The secondary carbides are uniformly distributed in the bulk of the modified zones. The matrix after aging undergoes $\gamma \rightarrow \alpha$ transformation and mainly has the martensitic structure. Aging increases significantly the hardness of the modified zones and abrasive wear resistance of the entire coating.

1. Introduction

Today, more and more structures, constructions and equipment used in various industries (power engineering, transport, oil and gas production, mining) work in the conditions of high contact loads accompanied by abrasive and impact-abrasive wear, which raises a widespread problem of machinery breakdown.

A primary task in this situation is to provide long and cost effective service life and reliability of working parts of machines, technological equipment and tools as these factors are directly related to the enhancement of wear resistance.

Currently, new methods of materials production have been intensively developed and applied. These are the technology of very rapid melt cooling during coating deposition or production of solid materials, dispersion, thermomechanical treatment, ultrasonic treatment, ultrafast deformation, synthesis, and so on. A combination of these technologies often allows new structural and functional materials to be designed [1-3].

The development of a method for producing materials with high structural homogeneity and fine strengthening phases from a highly nonequilibrium structural state using rapid melt crystallization and subsequent aging opens up new possibilities for hardening of materials. Rapid melt crystallization results in the formation of nonequilibrium structural states, supersaturated solid solutions, new metastable phases and structures. Subsequent aging leads to the rejection of a large amount of fine secondary strengthening phases from a supersaturated solid solution with the formation of a homogeneous material structure. The given method thus requires the study of the features and regularities of structure formation and phase transformations in such materials and coatings.

A promising method for a limited production of materials with high structural homogeneity and fine strengthening phases can be a complex electron-beam technology which includes electron-beam hardfacing in vacuum and multipoint pulsed electron-beam treatment [4] with subsequent heat treatment (artificial aging).

This paper reports the results of an investigation of structure formation in modified zones produced by MPEBT on the surface of chromium-vanadium cast iron coatings.

2. Materials and experimental procedure

Coatings of eutectic chromium-vanadium cast iron with rated chemical composition 2.5C, 17Cr, 5V, 1Si, 1Ni, 1Mn, Feres.(mass %) were deposited by electron-beam hardfacing in vacuum on low-carbon steel substrates measured 350×35×5 mm. The thickness of the deposited coatings was 3 mm. After polishing (down to 2 mm) the coating surface was locally treated at certain intervals (Fig. 1) by a pulsed, point-focused electron beam (∅1 mm). After the treatment the entire coating surface was covered by modified zones with discrete point distribution in a square packing. The diameter of each zone was about 1000 μm and the depth was 600 μm. Depending on the electron beam power and focusing, the zones can be hemispherical or conical in cross-section. Each modified zone was formed by one electron beam pulse of duration 15 ms and power density 56 kW/cm^2. Heat treatment of specimens by "aging" was carried out in a vacuum furnace at temperatures 900, 1000 and 1100°C within 30 minutes of exposure. The rate of heating up to the holding temperature and subsequent cooling down to 500°C was 15°C/min. X-ray phase analysis of the modified zones was performed using an X-ray diffractometer Shimadzu XRD 6000. X-ray spectral microanalysis was performed using a microscope Leo Evo 50. The hardness of the modified zones was measured on the Nanotest system.

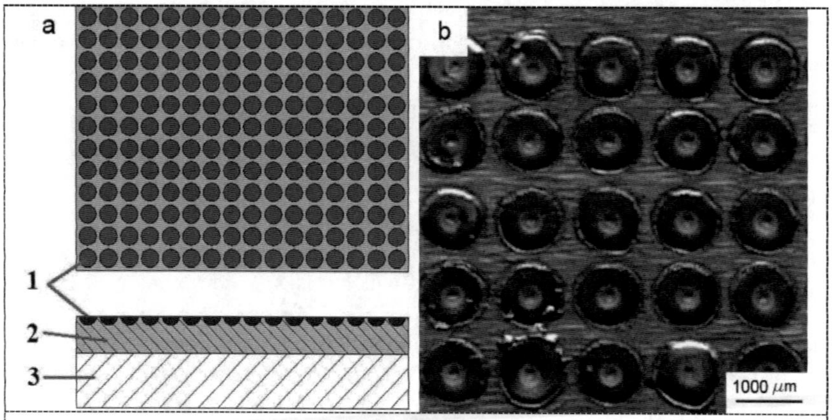

Figure 1. Schematic drawing (a) and fragment of the coating surface with modified zones produced by MPEBT (b): 1 — modified zone; 2 — deposited coating; 3 — steel substrate.

3. Results and discussion

The micrograph of the chromium-vanadium cast iron coating surface with modified zones produced by MPEBT is given in Fig. 1b. Between the modified zones there are coating regions not subjected to MPEBT which provide for stress relaxation.

Metallographic analysis has revealed considerable microstructural changes that occurred in the coating after MPEBT (Fig. 2).

IOP Conf. Series: Materials Science and Engineering **66** (2014) 012022 doi:10.1088/1757-899X/66/1/012022

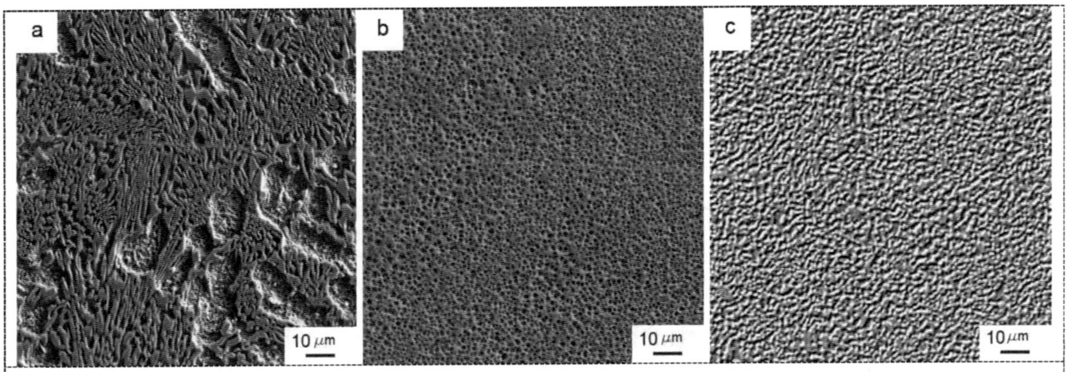

Figure 2. Microstructure of the eutectic chromium-vanadium cast iron after EBH (a), after MPEBT (b) and after aging at T = 1100□C (SEM)

As distinct from the initial state (after hardfacing) (Fig. 2a), the structure of the modified zones is highly homogeneous due to a uniform distribution of fine (up to 3 μm) inclusions of round shape (Fig. 2b) observed as fan-like eutectic colonies on the high-power micrograph in Fig. 3a. Vanadium carbides are the first precipitates that are formed from the melt and make up eutectic colonies for the short lifetime of the melt during the electron beam pulse, being the nucleation centers of the colonies (Fig. 3b).

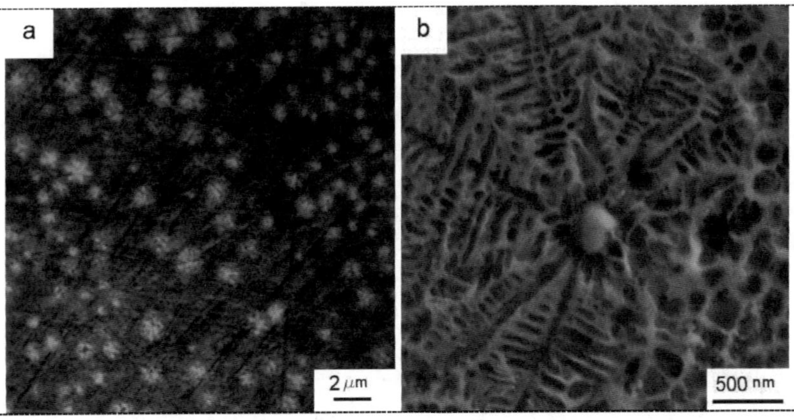

Figure 3. Eutectic colonies in the modified zone structure (a) and eutectic colony with vanadium carbide in the center (b).

X-ray structure analysis results (Fig. 4) demonstrate that the matrix of the deposited coating consists of austenite (25 %) and martensite (30 %), and the carbide phase is represented by carbides M_7C_3 (28 vol. %), VC (12 %) and V_2C (5 %). After MPEBT the material of the modified zones consists mainly of austenite (60 vol. %) and vanadium carbide VC (25 %) which is the leading phase in the eutectic colonies (Fig. 3b). Evidently, a liquid phase is formed in the coating material in the zone heated by the electron beam for pulse time 15 ms which is immediately crystallized when the electron beam is turned off. Vanadium carbides are the first precipitates that are formed from the melt and make up fine fan-like eutectic colonies for the lifetime of the liquid phase.

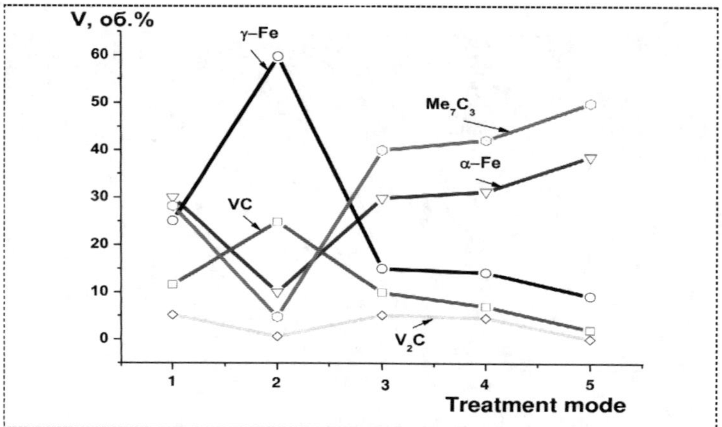

Figure 4. Volume fraction of structural components depending on the treatment mode: after electron-beam hardfacing (1); after MPEBT (2); heat treatment at 900 (3), 1000 (4) and 1100°C (5).

X-ray spectral microanalysis of the eutectic colonies shows that in the center of the colonies there are particles containing on average up to 60 mass % V, 15 % Cr and 7 % Fe. Although most of the colony falls within the zone of identification, the high vanadium concentration indicates that particles in the center of the colonies are vanadium carbides. The total volume fraction of carbides in the modified zones decreases from 45 to 30 vol. % (Fig. 5a) due to their partial dissolution in austenite. Correspondingly, the volume fraction of supersaturated austenite increases from 25 to 60 vol. %. According to X-ray spectral microanalysis data, the modified zone matrix contains on average 17.68Cr, 3.64V, 0.56Si, 0.62Mn, 1.47Ni, 76.03Fe (mass %), and the initial coating matrix contains 11.12Cr, 1.47V, 0.66Si, 0.65Mn, 1.7Ni, 84.39Fe (mass %) (Table 1). It is shown that the chromium and vanadium concentration in the austenite matrix increases 1.6 and 2.5 times, respectively, as compared to the initial coating matrix. Hence follows that MPEBT leads to the formation of modified zones with highly nonequilibrium structural-phase state. The formed matrix consists of a large amount of metastable supersaturated austenite, which determines a reduced hardness of the modified zone (1.2 GPa) as compared to the initial coating hardness (5.7 GPa) (Fig. 5).

Table 1. X-ray spectral microanalysis data for the initial coating and the modified zone produced by MPEBT*.

Chem. element	Matrix composition in the modified zone *, mass %				Matrix composition in the initial coating *, mass %		
	1	**2**	**3**	**4**	**5**	**6**	**7**
Si	0.55	0.56	0.52	0.6	0.64	0.66	0.68
V	3.92	3.72	3.63	3.29	1.47	1.59	1.36
Cr	19.22	17.95	17.66	15.90	11.45	11.55	10.37
Mn	0.69	0.6	0.64	0.56	0.61	0.62	0.72
Ni	1.35	1.62	1.50	1.41	1.94	1.62	1.55
Fe	74.27	75.55	76.05	78.24	83.89	83.96	85.32

*—carbon not taken into account.

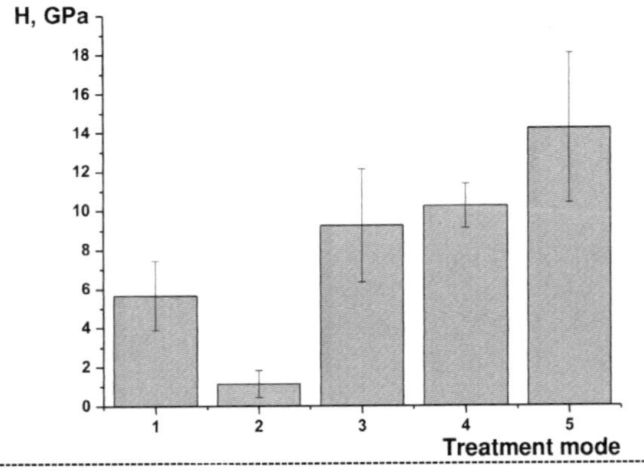

Figure 5. Hardness of the coating material depending on the treatment mode: after electron-beam hardfacing (1); after MPEBT (2); heat treatment at 900 (3), 1000 (4) and 1100°C (5).

In order to reject secondary carbides from the supersaturated austenite matrix of the modified zones, the specimens were subjected to aging in the temperature range from 900 to 1100°C. It was established experimentally that during aging the volume fraction of carbides and martensite increases significantly in the entire temperature range (Fig. 4). For example, after aging from 1100°C the total fraction of carbides increased from 30 to 52 vol. %, martensite—from 10 to 39 vol. %, and the austenite fraction decreased from 60 to 9 vol. % as compared to the material subjected to MPBET. Investigations showed that during aging a large amount of fine carbides Cr_7C_3 is rejected from supersaturated austenite, while austenite undergoes $\gamma \rightarrow \alpha$ transformation with martensite formation. In this case, the structure is characterized by the presence of a large amount of fine (1–2 μm) secondary carbides and by their highly homogeneous distribution in the bulk of the modified zones (Fig. 2c). It is found that the increase in the volume fraction of carbides Cr_7C_3 and martensite as well as the decrease in the residual austenite faction are accompanied by hardness growth (Fig. 5). The maximum hardness of 14 GPa corresponds to the modified zones with the highest fraction of carbides and martensite (aging at 1100°C). The decrease in the fraction of vanadium carbides after aging is evidently due to the fact that they became the nucleation centers of secondary carbides Cr_7C_3 formed on their surface.

The results of abrasive wear testing of the coatings (GOST 23.208-79 "Abrasive wear testing of materials in friction against non-fixed abrasive particles", primary standard: steel 45, Kw = 1) are the following: the coefficient of relative abrasive wear resistance of the coated specimens is Kw = 0±0.7, for the coatings subjected to MPEBT Kw = 7±1.2 and to heat treatment (aging at 1100°C) Kw = 25±1. The reduced wear resistance of the specimens after MPEBT is, first, due to a high fraction of metastable austenite (60 vol. %) in the structure of the modified zones and, second, due to a relatively low hardness of the modified zones on the whole. The maximum wear resistance of the coatings is achieved after aging at 1100°C, as a result of the growing volume fraction of carbides (52 vol. %) and martensite (39 vol. %) in the modified zones.

Conclusion

1. Multipoint pulsed electron-beam treatment of the surface of chromium-vanadium cast iron coating leads to the formation of modified zones consisting mainly of metastable supersaturated austenite and fine isolated eutectic colonies of round shape fanning out from the colony centers with the leading VC phase.
2. Heat treatment (aging) in the temperature range from 900 to 1100°C within 30 minutes of exposure causes intensive rejection of fine secondary carbides Cr_7C_3 from supersaturated austenite uniformly

IOP Conf. Series: Materials Science and Engineering **66** (2014) 012022 doi:10.1088/1757-899X/66/1/012022

distributed in the modified zones, austenitic-martensitic transformation and a significant increase in the hardness of the modified zones and abrasive wear resistance of the entire coating.

3. The complex technology including electron-beam hardfacing, multipoint pulsed electron-beam and subsequent heat treatment allows designing coatings with specified geometry and periodic distribution of modified zones on their surface.

References

[1] Gnyusov S and Tarasov S 2014 *J. Appl. Surf. Sci.* V **293**

[2] Proskurovsky D et al 1998 *J. Vac. Sci .Technol.* A **1694**

[3] Ozur G et al 2003 *J. Laser and Particle Beams.* **21**

[4] Dampilon B et al *RF Patent* RU 2494154 C1

Authors:

1. Natalia V. Tatarkina, graduate student of the Department of Materials Science in Mechanical Engineering of National Research Tomsk Polytechnic University

2. Artem I. Ziganshin, 2/4, postgraduate student at Institute of Strength Physics and Materials Science of Siberian Branch Russian Academy of Sciences

3. Bair V. Dampilon, PhD, associate professor of the Department of Materials Science in Mechanical Engineering of National Research Tomsk Polytechnic University, research fellow at Institute of Strength Physics and Materials Science of Siberian Branch Russian Academy of Sciences

4. Vasiliy G. Durakov, PhD, senior researcher at Institute of Strength Physics and Materials Science of Siberian Branch Russian Academy of Sciences

5. Aleksey M. Tolstokulakov, student of the Department of Materials Science in Mechanical Engineering of National Research Tomsk Polytechnic University.

IOP Conf. Series: Materials Science and Engineering **66** (2014) 012023 doi:10.1088/1757-899X/66/1/012023

The Effect of Substrate Temperature on the Structure and Magnetic Properties of Cobalt Films Deposited by CVD

R Hairullin[1], **S Dorovskikh**[2]

[1] Tomsk, 634050, Russia, National Research Tomsk Polytechnic University.

[2] Novosibirsk, 630090, Russia, Nikolaev Institute of Inorganic Chemistry SB RAS.

E-mail: hairullin@list.ru

Abstract. In this work the effect of substrate temperature on the structural parameters (sizes of coherent scattering region, values of microstresses), phase and chemical composition, surface morphology of Co films is revealed. Moreover, the correlation between structure, cobalt content and magnetic, electrical characteristics of Co films is presented. Co films were deposited on Si (100) substrates by chemical vapor deposition using the diiminate complex $Co(N'acN'ac)_2$ as a precursor. The sizes of coherent scattering region, values of microsresses and phase composition of Co films were determined by the X-ray diffraction analysis. The chemical composition was identified by the Energy-dispersive X-ray spectroscopy. The surface morphology of Co films was investigated by atomic-force microscope. Magnetic characteristics were measured by vibromagnetometer and the electrical resistivity was measured by four-probe dc method. It is found that the variation of deposition conditions allows widely changing structural parameters and chemical composition of Co films.

1. Introduction

Thin films of cobalt have been the subject of significant scientific research and have attracted great interest with respect to applications in composite magnets for micromechanics, data storage devices and sensors [1, 2].

There are currently many deposition techniques for metal films. However, chemical vapor deposition (CVD) has a special place among them. CVD is a technique which offers potential for producing films with high uniformity of thickness and composition, high purity, minimal substrate damage, high deposition rates and the possibility for selected area growth [3]. Moreover, CVD allows obtaining thin films with a different structure (different size and shape of the grains, texture) and effectively varying the thickness, phase and chemical composition of films in a wide range during their growth [4].

Properties of the films obtained by CVD, essentially depend on the deposition parameters (substrate temperature, vaporization temperature, duration of deposition process).The effect of CVD-deposition conditions on the properties of Co films has already been studied [5]. Nevertheless, very little empirical research has been carried out to investigate Co films deposited by CVD where a precursor was the diiminate complex $Co(N'acN'ac)_2$ that was characterized by a good volatility and high stability [6].

Therefore, to obtain films of Co with the specified operational characteristics by CVD using the diiminate complex as a precursor it is necessary to reveal the dependence of structure, phase and chemical composition on deposition conditions. The aim of this research is to investigate the effect of substrate temperature on the structural parameters (sizes of coherent scattering region, values of microstresses), phase and chemical composition, surface morphology of Co thin films deposited by CVD. Furthermore, the correlation between structure, cobalt content and magnetic, electrical characteristics of Co films is presented in this work. It is expected that the variation of deposition conditions allows widely changing structural parameters and chemical composition of Co films.

2. Experimental details

Co films were deposited on Si (100) substrates by CVD. Co(*N'acN'ac*)$_2$ was used as a precursor. The deposition conditions were as follows: gas-carrier flow rate (Ar) was 1 l/h, gas reagent flow rate (H_2) - 4 l/h, the operating pressure was atmospheric (~760 *Torr*), the deposition process was carried out during a period of 4h. The vaporization temperature (T_{vap}) was fixed and was equal to 130 °C, while the substrate temperatures (T_s) were varied in the range from 300 to 340 °C.

The X-ray diffraction (XRD) analysis of the samples was performed on a DRON-SEIFERT-RM4 diffractometer (Cu Kα radiation, λ = 1.54051 Å). The chemical composition was identified by the Energy-dispersive X-ray spectroscopy. The surface morphology of Co films was investigated by atomic-force microscope (AFM) Solver HV. Magnetic characteristics were measured by vibromagnetometer VM-23K in the easy direction of magnetization. The electrical resistivity was measured by four-probe dc method. All measurements were carried out at an atmospheric pressure and at room temperature.

3. Results and discussion

According to XRD data, the most intensive diffraction peak centered at around 44,2 – 44,7° in 2Θ is observed in the samples deposited at substrate temperatures T_s =300-340°C (figure 1). Detailed analysis of XRD pattern reveals that this diffraction peak is positioned between (111) of β-Co (face centered cubic) at 44,3° and the (002) reflection of α-Co (hexagonal close-packed) at 44,6°.

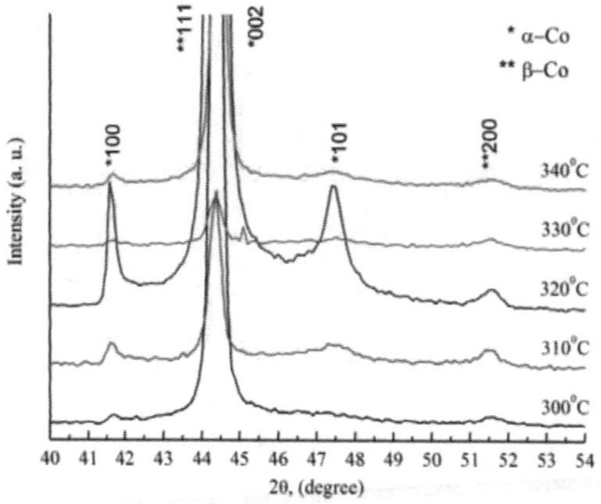

Figure 1. XRD patterns of the Co films deposited at different substrate temperatures.

Moreover, the XRD pattern for Co samples deposited at T_s = 300 – 340 °C contains additional peaks, which could be indexed to the α-Co (100) (2Θ = 41.7°), α-Co (101) (2Θ = 47.6°) and β-Co (200) (2Θ = 51.7°) (figure 1). The height of the diffraction peaks is governed by the substrate temperature. The XRD peak becomes higher with increasing substrate temperature from T_s = 300 to T_s = 320 °C. However, further growth of substrate temperature up to T_s = 330 and 340 °C results in a gradual weakening of the peaks. The Co film deposited at substrate temperature T_s = 320 °C is noted to be characterized by well-pronounced texture (the maximum height of the diffraction peaks).

The sizes of the coherent scattering region (CSR), microstress values and chemical composition of Co films under study are presented in table 1. As seen from the table 1, the CSR size of films weakly depends on the substrate temperature in the range of temperatures from 300 to 330 °C. However the CSR size starts to decrease after increasing temperature over 330 °C.

It should be noted that substrate temperature does not affect microstresses of Co films. Values of microstresses are kept constant within the measurement error (table 1).

As listed in the table 1, there are atoms of carbon, oxygen and nitrogen in Co films. The Energy-dispersive X-ray spectroscopy demonstrates that the chemical composition of Co films is characterized by the extreme dependence on the substrate temperature. Actually, the increase in the substrate temperature from 300 to 330 °C on the one hand is accompanied by a growth of the cobalt content from 84.8 to 93.5 at. % and on the other hand leads to a decline in the carbon content from 14.2 to 6 at. %. However, further rise in substrate temperature up to 340 °C causes both a decrease in the cobalt content down to 90.9 at. % and an increase in the carbon content up to 8.8 at. %.

Table 1. Chemical composition, CSR size, microstresses σ and electrical resistivity ρ of Co films deposited at different substrate T_s temperatures.

T_s (°C)	Chemical composition	CSR (nm)	σ (GPa)	ρ (μΩ*m)
300	Co 84.8%, C 14.2%, O 0.1%, N 0.9%	35	0,4	0,36
310	Co 86.5%, C 12. 6%, O 0.2%, N	26	0,5	0,34
320	Co 92.3%, C 7. 2%, O 0.5%	33	0,2	0,12
330	Co 93.5%, C 6.0%, O 0.5%	26	0,4	0,07
340	Co 90.9%, C 8.8%, O 0.3%	20	0,3	0,1

According to AFM investigations, Co films deposited at substrate temperature T_s = 300 – 330 °C are characterized by fine – grained surface relief (figure 2, a-d). However, further growth of temperature up to T_s = 340 °C leads to a disappearance of the film's grain structure. As seen in figure 2, e, the presence of grain structure is less apparent.

Structural parameters and chemical composition determine magnetic and electrical properties of Co films. A comparison of hysteresis loops of films deposited at different substrate temperatures indicates that the Co films, obtained at T_s = 330 °C, are characterized by higher residual magnetization and saturation magnetization compared to films deposited at 300 °C (figure 3, a, c). The latter is associated with an increase in cobalt content in films deposited at higher substrate temperatures (table 1). The electrical resistivity of Co films is also governed by the content of cobalt. Indeed, films deposited at T_s = 330 °C, are characterized by the maximum cobalt content (93.5%), and hence the minimum electrical resistivity (0.07 μΩ*m) (table 1). In contrast, the coercive force of Co films mainly depends on their degree of texture. Films, deposited at T_s = 320 °C, are characterized by higher value of the coercive force because these films were most textured (figure 3, b and figure 1).

4. Conclusion

The results of the performed investigations show that Co films deposited by chemical vapor deposition consist of α-Co and β-Co crystals. Varying the substrate and vaporization temperatures allows us to modify widely the microstructure and chemical composition of Co films. A rise in substrate temperature causes a decline in sizes of the coherent scattering region and results in the disappearance of the grain structure of Co films. The cobalt content and texture of films are characterized by the extreme dependence on the substrate temperature with a maximum at 320-330 °C. However, microstresses of films don't depend on substrate temperature and remain unchanged.

The selection of the optimal combination of deposition parameters allows obtaining films of cobalt with required magnetic and electrical properties. The residual magnetization and saturation magnetization of films rise and the electrical resistivity decreases with increasing cobalt content. An increasing the degree of texture of Co films raises their coercive force.

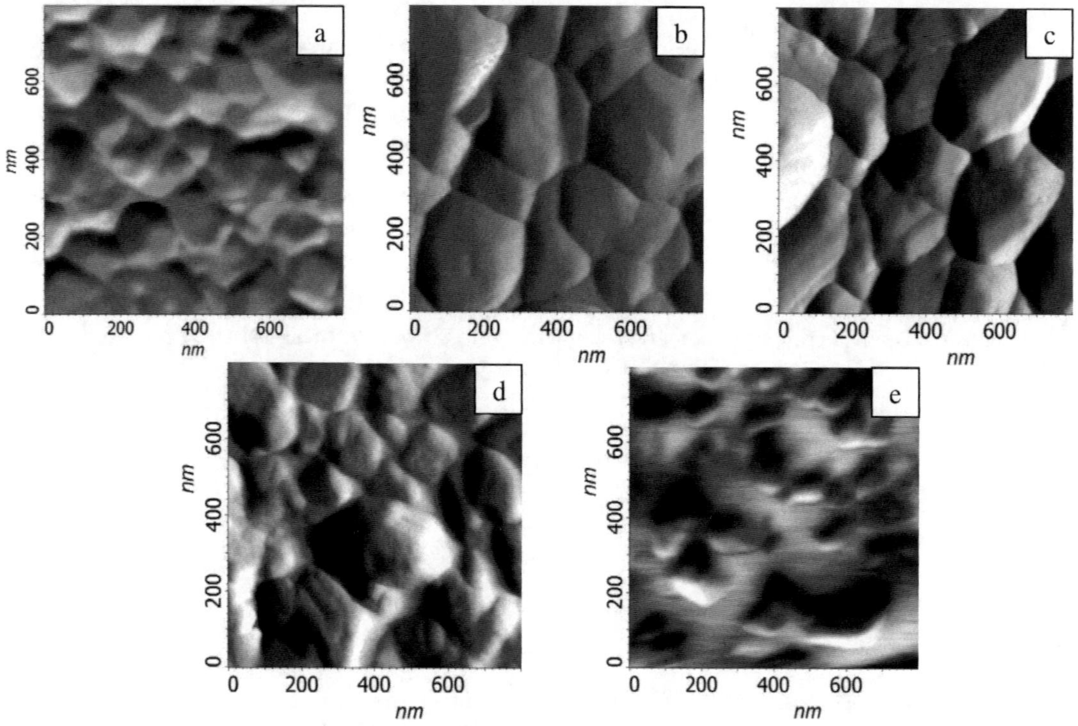

Figure 2. AFM-images of the Co films deposited at T_{vap} =130 °C and T_s = 300 (a), 310 (b), 320 (c), 330 (d) and 340 °C (e).

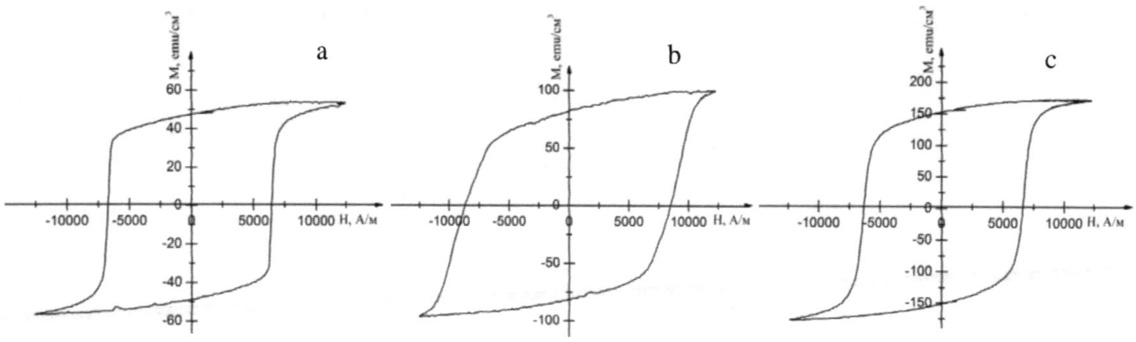

Figure 3. Hysteresis loops of the Co films deposited at T_{vap} =130 °C and T_s = 300 (a), 320 (b) and 330 °C (c).

References

[1] Chioncel M F and Haycock P W 2006 Cobalt thin films deposited by photoassisted MOCVD exhibiting inverted magnetic hysteresis *Ch. Vap. Dep.* **12** pp 670–678
[2] Kazakov V G 1997 Thin magnetic films *Sor. Obr. J.* **1** pp 107-114

[3] Chioncel M F and Nagaraja H S 2007 Domain structures of MOCVD cobalt thin films *J. of Mag. and Mag. Mat.* **313** pp 135–141
[4] Syrkin V G 2000 *CVD-method: chemical vapor deposition* (Moscow: Nauka) p 483
[5] Paranjape M A Mane A U and Raychaudhuri A K 2002 Metal–organic chemical vapor deposition of thin films of cobalt on different substrates: study of microstructure *Th. Sol. Fil.* **413** pp 8-15
[6] Morozova N B Stabnikov P A and Baidina I A 2007 Structure and thermal properties of volatile copper(II) complexes with β-diimine derivatives of acetylacetone and the structure of 2-(methylamino)-4-(methylimino)-pentene-2 crystals *J. of Str. Ch.* **48** pp 889-898

IOP Conf. Series: Materials Science and Engineering **66** (2014) 012024 doi:10.1088/1757-899X/66/1/012024

Investigation of various criteria for evaluation of aluminum thin foil "smart sensors" images

S V Panin[1, 2], A V Eremin[1], P S Lyubutin [1, 2], M V Burkov [1, 2]

[1] Institute of Strength Physics and Materials Science of SB RAS
SB RAS, Tomsk, Russia
[2] Tomsk Polytechnic University Tomsk , Russia

E-mail: svp@ispms.tsc.ru

Abstract. Various criteria for processing of aluminum foil "smart sensors" images for fatigue evaluation of carbon fiber reinforced polymer (CFRP) were analyzed. There are informative parameters used to assess image quality and surface relief and accordingly to characterize the fatigue damage state of CFRP. The sensitivity of all criteria to distortion influences, particularly, to Gaussian noise, blurring and JPEG compression was investigated. The main purpose of the research is related to the search of informative parameters for fatigue evaluation, which are the least sensitive to different distortions.

1. Introduction

CFRP are widely used in industry to produce structures and machine parts with improved mechanical properties. Moreover, the application of CFRPs in aerospace reduces the costs due to the possibility of production of large-size parts with minimum joints (e.g. the solid CFRP fuselage sections with no fasteners). The structures during operation are mainly loaded cyclically and the fatigue failure may occur. Therefore, the mechanical state of CFRP should be inspected throughout the operation. Recent papers on the subject of non-destructive testing are devoted to the Structural Health Monitoring (SHM) systems, which can provide information as the damage occurs and improve the safety.

One approach [1] of built-in SHM methods is based on optical image registration of aluminum foil (sensor) glued to the investigated structure. Strain-induced relief is formed on the surface of the sensor due to the cyclic deformation and registered via digital camera. Then the relief is analyzed using different parameters to estimate the mechanical state of CFRP.

In order to provide reliable results the informative parameters for image estimation should be unaffected by noise, blurring, illumination irregularities or defects on the captured image. In the previous work [2] the adapted bilateral filtration was used for image processing and the fractal dimension was used as an informative parameter to assess the mechanical state. The aim of the present work is to determine the influence of a) noise; b) blur; and c) compression; on informative parameters used to process the images of aluminum foil sensors during CFRP cyclic tests.

2. Materials and methods

Rectangular flat CFRP specimens with two edge U-shaped notches were tested cyclically using servohydraulic machine UTM Biss-00-201 (the test mode is uniaxial cyclic tension with stress ratio R=0.1). Polycrystalline aluminum foil was glued on the specimen's surface and then polished. The images of foil were captured during the test by DSLR camera Canon EOS 550D connected with microscope lenses. This optical setup provides images with 5184×3456 pixel size (12×8 mm^2). The areas of 1024x1024 pixel size were cut from initial images for numerical evaluation.

The illumination scheme used differs with the setups described in literature. In [3] the foil was removed using acetone and then imaged using microscope: the undeformed areas are bright and folds are dark. In [4] the images of foil were captured while the specimen was clamped in grips during the test, however the diffuse lighting provided similar images: bright undeformed areas and dark folds.

In the present work, the illumination scheme includes two light sources: halogen lamp and point LED (Fig. 1). The halogen lamp (background lighting) was mounted in xz-plane at an angle of 45° to the x-axis. The LED used to increase the contrast of deformation folds was located in xz-plane at an angle of 10° to the x-axis. The illumination setup provides dark image at the initial of the cyclic test, because the light reflects from the mirror-polished foil missing the camera lens. Then the relief forms and the light diffusely reflects from folds, thus deformed areas become bright.

In our previous research [4] the metal specimens were tested and aluminum foil was chemically etched. In the present paper the foil was polished in order to obtain mirror surface. This preparation technique improves repeatability and the comparison of different specimens becomes more correct.

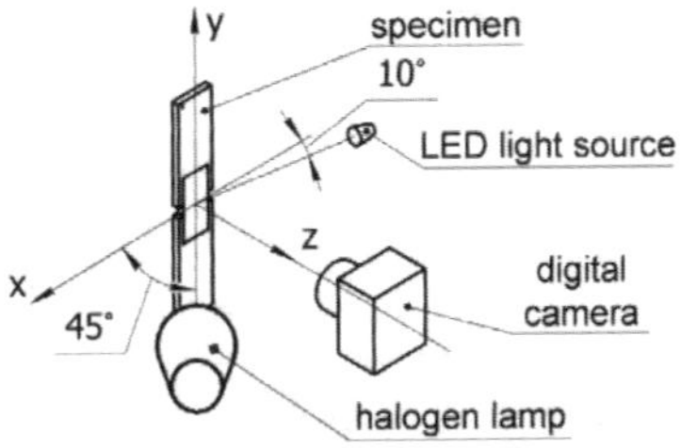

Fig. 1. The scheme of specimen illumination and image capturing

Fig. 2 presents the images captured at different number of cycles after the initial of the test. The first image (Fig. 2,a) captured at 100 cycles is homogeneously dark. At N=1600 cycles (Fig. 2,b) the folds are starting to appear. Further cyclic loading to N=6 400 is accompanied with the significant growth of deformation relief (Fig. 2,c). The relief continues to rise and, finally, after the N=29 000 (Fig. 2,d) the relief stay practically unchanged. The plot on the Fig. 3 (line **10**) is a dependence of the image mean brightness I_{mean} on the operating cycles.

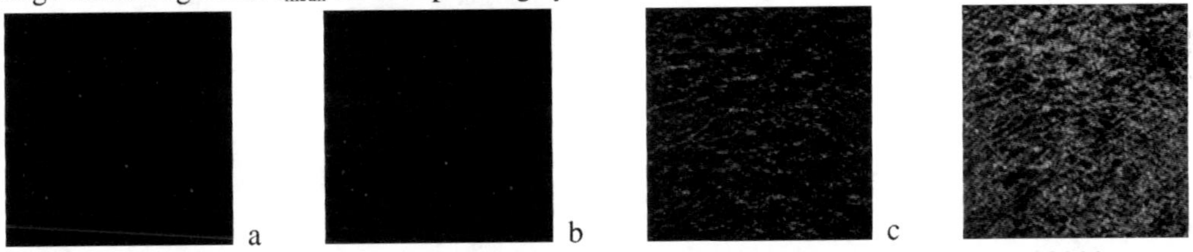

Fig. 2. The images of foil. Number of cycles to failure N: a) 100, b) 1600, c) 6400, d) 29000.

The images were analyzed using following informative parameters [5]: MSE (mean square error); H (Shannon entropy); F_d (fractal dimension); E_{FS} (Fourier energy spectrum); PSNR (peak signal-to-noise ratio); UIQ (universal image quality); VSNR (visual signal-to-noise ratio); Sq (white to black pixel ratio). Then the parameters were investigated for the sensitivity to various distortions.

Image blurring. The captured images of aluminum foil were blurred using of the Gaussian filter. Gaussian distribution in N dimensions is calculated as follows [6]:

$$G(x,y) = C \cdot e^{-\frac{x^2+y^2}{2\sigma_b^2}}, \qquad (1)$$

where σ_b – standard deviation for Gaussian distribution (it defines the degree of blurring), c – scale coefficient. The degree of blurring was varied from 1 to 10.

Gaussian noise. It is a statistical noise, which has a probability density function corresponding to a normal distribution. Probability density function P of Gaussian random variable z is specified as:

$$P_G(z) = \frac{1}{\sigma_n\sqrt{2\pi}} e^{-\frac{(z-\mu)^2}{2\sigma_n^2}}, \quad (2)$$

where z – gray level, μ – mean value, σ_n – standard deviation, which defines noise amplitude [7]. Noise level value was taken from the range 0.03-0.3.

JPEG image compression. Photo and video cameras are used as detectors of visual information, but the disadvantage is a data compression, which leads to information loss. The aim of this section is evaluate a dependence of informative parameter values on image compression rate. For this purpose JPEG compression was chosen as one of the most widely used. The XnView was used to compress the images with the following parameters: DKT – slow, smoothing coefficient – 0, discretization (default) – 2x2, 1x1, and 1x1. Compression degree was varied from 10 to 100 with a step of 10 units.

3. Materials and methods

Undistorted images comparison. In order to compare all informative parameters (quality assessment criteria) they were calculated for initial (without applying any distortion) image series of aluminum foil obtained at different number of operation cycles. All results were normalized and shown on Fig. 3. It is seen, that all parameters have three-stage behavior, but the duration of each stage and dynamic variation range of each parameter are different. The values of the almost all informative parameters within the range of N=1 000 cycles stay constant. Further (N>1 000 cycles) the significant increase of following parameters is observed: H (line 1), F_d (2), PSNR (5), UIQ (6), VSNR (7), H-MSE (9). For the rest parameters the second stage (where the parameter starts to rise) begins at higher operation cycles. The third stage has the largest duration wherein the most parameters don't change their values except H (1), MSE (4), PSNR (5), Sq (8) and H-MSE (9). The Sq value decreases on the third stage, which makes this parameter unsuitable as a criterion for surface relief evaluation. The mean brightness of an image (I_{mean}) has similar behavior with H, but I_{mean} has the less dynamic range. Such behavior of brightness related to the illumination setup. Apparently in the other illuminating conditions the mean brightness curve may differ.

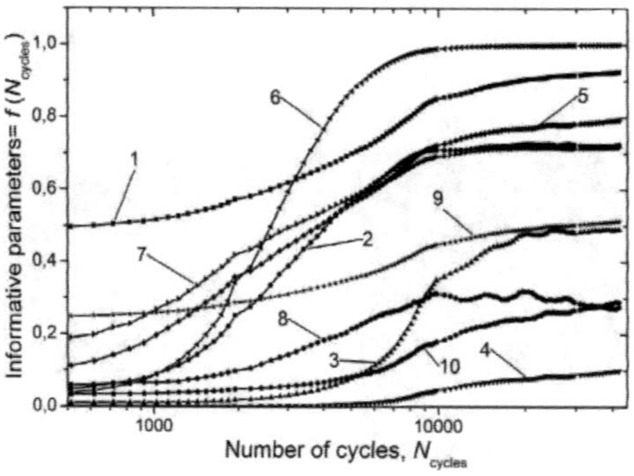

Fig. 3. Graphs of informative parameters for image quality assessment the quantity of cycles: 1) H; 2) F_d; 3) E_{FS}; 4) MSE; 5) PSNR; 6) UIQ; 7) VSNR; 8) Sq; 9) H-MSE; 10) I_{mean}

According to a maximum of dynamic range the appropriate criteria are UIQ (ΔUIQ=1,0), PSNR (ΔPSNR=0,8), F_d (ΔF_d=0,7), E_{FS} (ΔE_{FS}=0,5), VSNR (ΔVSNR=0,5). From the sensitivity viewpoint (the parameter to be sensitive to the minor changes of relief on the third stage) the following parameters are denoted: H (**1**), F_d (**2**), E_{FS} (**3**), PSNR (**5**), VSNR (**7**) and H-MSE (**9**).

The calculation and the analysis of informative parameters was performed and then the various distortions (Gaussian noise, Gaussian blurring and JPEG compression) were applied to the initial images in order to investigate the criteria's sensitivity to different influences. Two additional characteristics were calculated for numerical evaluation of distortion degree: root-mean-square error (RMSE – not the same with MSE) as a function of noise, blurring, compression ratio; and correlation coefficient (Kr) between parameter values calculated for initial and distorted images as a function of distortion ratio.

Consequently, the most appropriate parameters should have the minimal RMSE value and the Kr should be as close to 1.

Gaussian blurring. Fig. 4 shows the plots of RMSE and correlation coefficient Kr from blurring rate. For the most of the parameters in the whole range of σ_b from 1 to 10 (Fig. 4a) the RMSE is less than $5\cdot10^{-3}$, but there are three parameters where RMSE increases: F_d (**2**), E_{FS} (**3**) and Sq (**8**). On the other hand, the Kr (Fig. 4b) has the maximum values for H (**1**), E_{FS} (**3**), MSE (**4**) and H-MSE (**9**). Kr has a value in a range of 0.8-0.9 for F_d (**2**), PSNR (**5**), UIQ (**6**), VSNR (**7**). For the Sq parameter Kr has a dependence of a curve with a maximum at σ_b=6.

Joint analysis of the two graphs allows us to make following conclusions. The most insensitive to blurring is MSE parameter, which has Kr=0,975 in the whole range (σ_p=1-10) and minimum RMSE value. Then they are H (Kr=0.96) and E_{FS} (Cr=0.95-0.94), despite the E_{FS} is sensitive to blurring according to RMSE. Sufficiently high Kr values have VSNR (Kr=0.85-0.89), PSNR (Kr=0.84) and UIQ (Kr=0.8-0.83) accompanied with the minimal values of RMSE. F_d also have high Kr (0.85-0.81), but with the increase of blurring rate the RMSE for F_d significantly increases. Sq is the worst criterion where Kr changes within the range from 0.7 to 0.91 (maximum) and RMSE value is the highest. Finally, the H-MSE, based on the calculation of two parameters, has small RMSE and Kr=0.96.

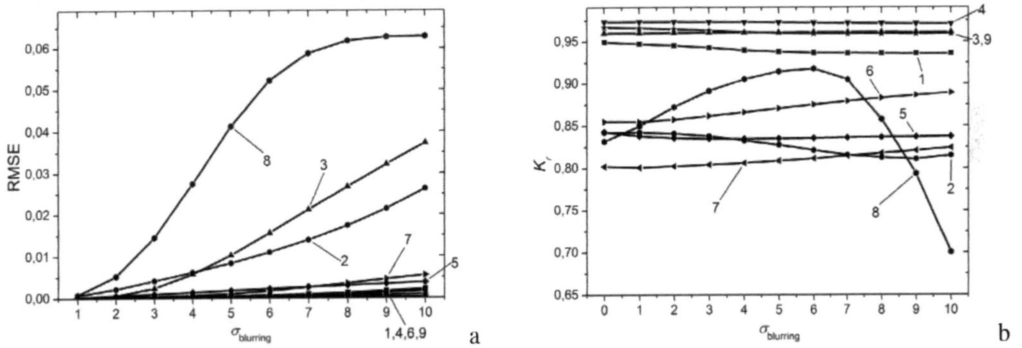

Fig. 4. The graphs for different blurring rates: a) RMSE; b) correlation coefficient Kr. 1) H; 2) F_d; 3) E_{FS}; 4) MSE; 5) PSNR; 6) UIQ; 7) VSNR; 8) Sq; 9) H-MSE

Gaussian noise. Fig. 5 shows RMSE and Kr dependencies on the degree of Gaussian noise, applied to the initial images for each informative parameter.

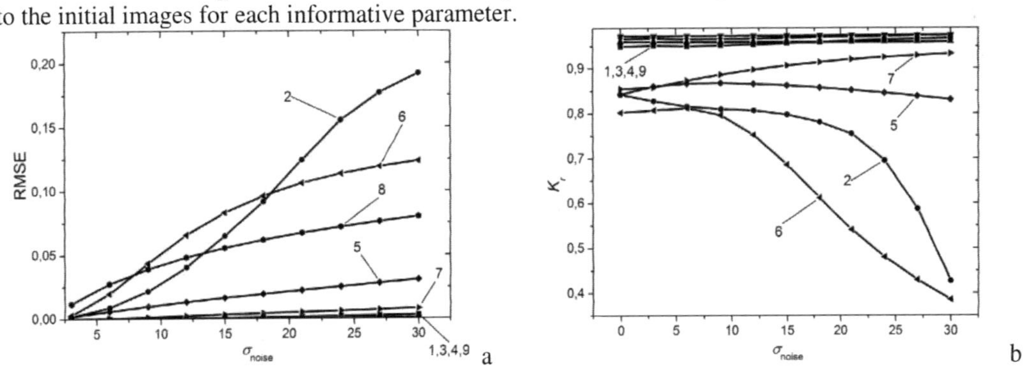

Fig. 5. The graphs for different Gaussian noise degrees: a) RMSE; b) correlation coefficient Kr. 1) H; 2) F_d; 3) E_{FS}; 4) MSE; 5) PSNR; 6) UIQ; 7) VSNR; 8) Sq; 9) H-MSE

It is seen that four parameters have constant RMSE in analyzed range of noise degree: H (**1**), E_{FS} (**3**), MSE (**4**), H-MSE (**9**). While the error increases for F_d (**2**), PSNR (**5**), VSNR (**7**), Sq (**8**) and UIQ (**6**). The similar, but an opposite trend is typical for UIQ (**6**) and F_d (**2**) which are shown on graph Kr=f(σ_n). The Kr value of VSNR parameter (**7**) increases and for PSNR (**5**) stays unchanged (Kr=0.85). The maximum Kr is typical for H (**1**), E_{FS} (**3**), MSE (**4**) and H-MSE (**9**).

MSE is the least sensitive to the noise with Kr=0.97 (RMSE is constant). In addition, the criteria quite insensitive to noise are H (Kr=0.95-0.965), E_{FS} (Kr=0.96-0.955) and less "stable" criteria are VSNR (Kr=0.85-0.94) and PSNR (Kr=0.84-0.86) with low values of RMSE. F_d and UIQ are characterized by slight decrease of Kr with the rise of noise level and sufficiently high RMSE. H-MSE is also insensitive to noise: it has small RMSE and high correlation coefficient (Kr=0.97).

JPEG compression. Fig. 6 presents RMSE and Kr dependencies on the degree of compression. Fig. 6a illustrates that RMSE changes in a range of 10^{-10}-10^{-7} for E_{FS} (**3**) and MSE (**4**). The other informative parameters in the range of K_{comp} from 0 to 50 have RMSE=10^{-7}-10^{-5}. With the further increase of compression degree the RMSE rises to 0.01-0.1 for H (**1**) and H-MSE (**9**). After the analysis of Kr we can conclude that criteria can be divided into two groups: 1) with high Kr>0.94 – H (**1**), E_{FS} (**3**), MSE (**4**) and H-MSE (**9**); 2) with 0.8<Kr<0.87 – F_d (**2**), PSNR (**5**), VSNR (**7**) and Sq (**8**). UIQ has a very low Kr~0.77, therefore it cannot be attributed to any group.

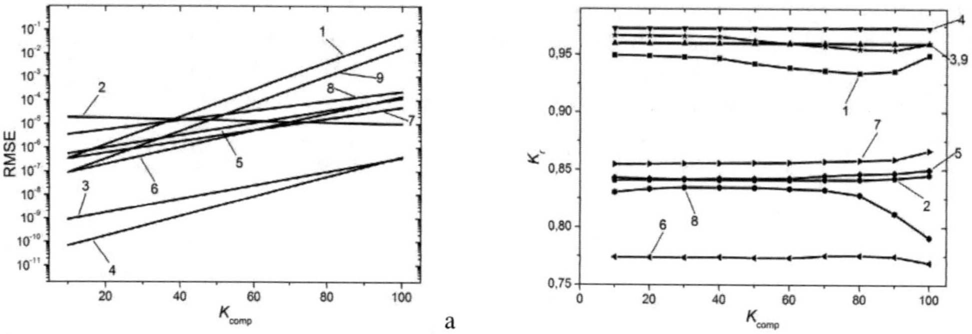

Fig. 6. The graphs for different compression degree: a) RMSE; b) correlation coefficient Kr. 1) H; 2) F_d; 3) E_{FS}; 4) MSE; 5) PSNR; 6) UIQ; 7) VSNR; 8) Sq; 9) H-MSE

Joint analysis of graphs on Fig. 6 results to a conclusion that the most suitable and preferable criteria for processing of compressed images are MSE and E_{FS}, because the Kr is higher than for the other criteria (Kr~0.975 for MSE and Kr~0.96 for E_{FS}), moreover RMSE is the least. Criteria H and H-MSE show the high correlation (Kr~0.95–0.94), however the error is the highest. The second group of criteria has almost the same values: VSNR (Kr~0.85-0.87), PSNR (Kr~0.84-0.85) and F_d (Kr~0.84), also they are characterized by moderate error. The Sq parameter is the most sensitive to distortion influences, because it has average values of correlation coefficient Kr~0.79-0.84 and RMSE in comparison to the other criteria. The plot character of UIQ parameter could be assumed as the most nonlinear (Kr~0.77-0.78) from all of them, besides it has average RMSE.

4. Results discussion

As a result of the investigation of all informative parameters the most insensitive to different distortions are H, MSE, E_{FS} and H-MSE. They are recommended for assessing the relief on the aluminum foil during cyclic tests. Fig. 7 presents these parameters with different distortions.

The Shannon entropy H is practically insensitive to all types of distortions; therefore this criterion is reliable and adequately indicates the changes in deformation relief. The disadvantage is low sensitivity to deformation after N=10 000 cycles. The MSE and E_{FS} don't have such drawback (they can characterize the relief after 10 000 cycles; the upper limit for E_{FS} is 20 000 cycles, for MSE 45 000 cycles). But they are slightly sensitive to image blurring. So the H-MSE, which combines the advantages of two parameters, is offered. It is reliable like the Shannon entropy H and changes in wide range of cycles (up to 45 000 cycles). Based on this research the H-MSE can be recommended for practical purposes for estimation of a specimen lifetime during cycling load.

The obtained results (the tests of foil sensors for CFRP fatigue evaluation) should be compared with the results received in our previous works [4] and other authors [1,3,8] (the metals were tested). The investigated objects are steel [4] and aluminum alloys [1,3,8] and their cyclic test was performed within the elastic region (~0.2% of elongation). So the rate of relief formation was slower than the rate obtained in the present work (~1.2% of elongation for testing of CFRP). The criteria in mentioned articles are fractal dimension [1,4,8] and the ratio of white and black pixels [3], which made possible to describe the changes on the foil surface in a wide range (up to 80 000 cycles [3]).

The Fractal dimension and the ratio of white and black pixels are not suitable for CFRP fatigue evaluation, because the rate of relief formation is higher than in a case of metallic specimen. Therefore, these parameters have upper limit of ~10 000 cycles after which the fatigue state cannot be assessed. It was a reason for searching of new criteria to be presented in this work. All parameters, used in the present work can be applied for metal testing with the same illumination setup as well.

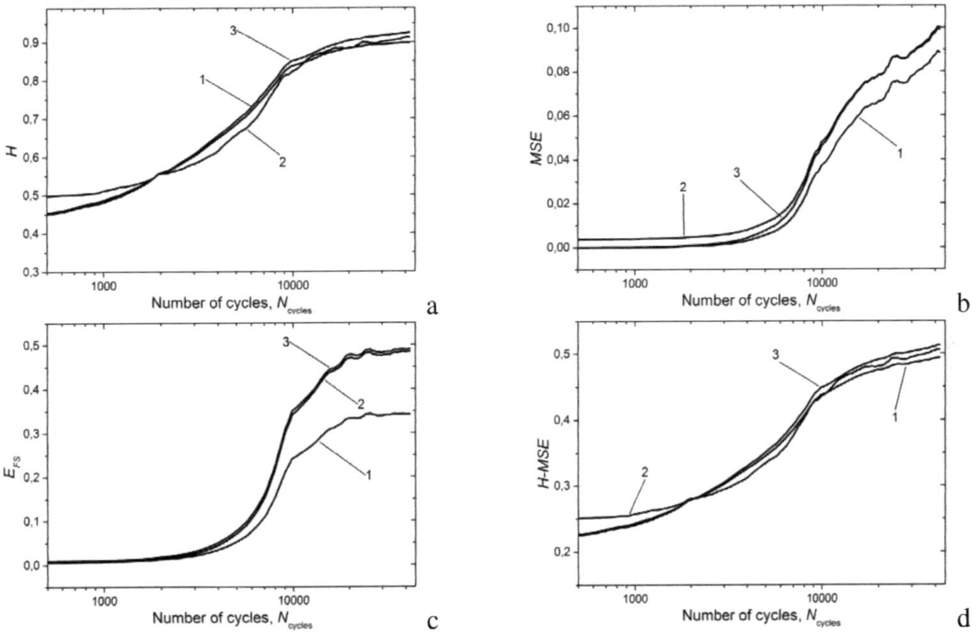

Fig. 7. Graphs of a) H, b) MSE, c) E_{FS}, d) H-MSE with applied Gaussian blurring $\sigma b=5$ (line 1), Gaussian noise $\sigma_n=15$ (line 2), and JPEG compression $K_{comp}=60$ (line 3)

5. Conclusion

The investigation of information parameters for strain relief assessment of aluminum foil glued to CFRP specimen was performed. The most sensitive parameters to relief formation were determined and tested in a case of different distortions influence: blurring, nose and compression. The most reliable informative parameters are H, MSE, E_{FS} и H-MSE.

It was shown that these parameters are changing insignificantly with rising of distortion degree. The authors suggest using a combined criterion – H-MSE calculated as the mean value of H and MSE. This complex parameter has wide range of applicability and it is "stable" in analyzed region of distortions influences (blurring, noise, compression).

References

[1] Paget C 2008 Bonded thin metallic foil sensor applications for structural health monitoring of aeronautical structures *Phys. Mesomechanics* **5-6** pp 308-313

[2] Panin S V, Altukhov Y A, Lyubutin P S et al 2013 Effect of bilateral filtration on fractal estimation of optical images of loaded material surfaces *Optoelectronics, Instrumentation and data processing* **49** pp 234-242

[3] Zmyzgova T R 2012 *Expo. Oil. Gas.* **25** pp 14-19

[4] Byakov A V 2011 *PhD thesis: 05.11.13* 181 p

[5] Panin S, Burkov M, Lyubutin P, Altukhov Yu 2014 Application of aluminum foil for "strain

sensing" at fatigue damage evaluation of carbon fiber composite *Sci. China Phys., Mech. and Astron.* **57** pp 59-64
[6] Shapiro L G and Stockman G C 2001 *Computer Vision* (Upper Saddle River, NJ: Prentice Hall)
[7] Pratt William K 2007 *Digital image processing* (Hoboken, NJ: A John Wiley & Sons)
[8] Cikalova U, Kroening M, Schreiber J et al 2011 *Phys. Mesomechanics* **5-6** pp 308-315

Bone substitute material on the basis of natural components

S Melnikova [1,a], E Zelichenko [2,b], B Zenin [1,c], V Guzeev [2,d], O Gurova [2,e]
[1] Tomsk Politechnic University, Tomsk, Russia
[2] Seversk Technological Institute – a branch of State Autonomous Educational Institution of Higher Professional Education «National Research Nuclear University «MEPhl» Seversk, Russia

E-mail: [a]miss.melsva@yandex.ru, [b]zelichenko65@mail.ru, [c]bosezen@tpu.ru, [d]uranilf@yandex.ru, [e]goa120@yandex.ru

Abstract. The creation of regenerative materials remains a problem for rehabilitation medicine, but the obtaining of initial substances that can cause bone tissue regeneration, possessing biological activity and creation on their basis of composite materials with specified physical and mechanical characteristics is an important scientific problem. This paper presents the investigation of physical-chemical and biological properties of bioresorbable composite material that can restore their own bone tissue of the body.

Keywords: hydroxyapatite, biopolymer, regeneration, bone resorption, composite material, natural components.

1. Introduction

Modern medical experience of traumatology and orthopedics is mainly based on a mechanical replacement implant bone defect with the creation of complex biomechanical structure. At the same time, intensive research is carried out to develop methods to restore damaged bone by physiological regeneration processes.

In past, materials, intended for implantation, are designed as "bioinert." Nowadays materials are increasingly used, which integrate with biological molecules or cells and regenerate tissue. Application of regenerative biodegradable composite material eliminates the necessity for re-operation to remove of the implant from the patient due to its resorption and eliminates the need for additional fixation of bone fragments (debris), as well as allows to regenerate lost or destroyed bone.

Despite the considerable amount of bone replacement materials are used in regenerative medicine in the world, searching and the development of new ones, is still an urgent problem. It is possible to solve it if compounds are natural and they can be a part of the metabolic processes of the human organism.

Bone tissue is a mineral-organic composite material with a porous structure, which is subjected by constant remodeling. Mineral matrix of bone is hydroxyapatite but larger part of organic matrix is collagen.

In order to create a material based on calcium salts and biological polymers with porous bone structure and which has the special strength, it's necessary to combine technology and the components so that the implant is made of this material does not destroy bone and bone does not destroy the implant. Moreover, in the case of loose implant contact and native bone regeneration process slows or stops altogether. To meet these conditions it is necessary that the material was a viscous fluid or plastic at the beginning of work and then at the end of work it became hard and strength.

It is established that increasing the surface area and porosity of biocomposite materials affects positively on the kinetics of bone formation and improves the bioactivity (complex of material properties, allowing creating a direct contact with natural bone [1, 2]). To understand the relationship between structure and bioactivity, as well as to design better implants is very important to control of total porosity, pore size and internal structure of the porous biomaterials [1, 2].

2. Material and methods

The structure of a bone replacement material includes only natural components. The basis of the composite material is ultrafine hydroxyapatite obtained by the original technology [3] from the bones of farm animals. Its main feature is the presence of trace elements equal to bone mineral, and stable chemical composition.

Biopolymer matrix collagen is used as the organic binder. Shredded tendons of cattle are used to make collagen. They were subjected to acid- saline treatment and purification of non-collagenous proteins. Then, an ultrafine powder of hydroxyapatite is mixed with the liquid polymer in specific ratios to form the plastic cement paste.

The technology permits to make a bone replacement composite material to fill bone defects by different ways. The finished product is self-curing cement mass. Seizing instantaneously occurs under room temperature. Pre-curing of the material occurs within 15-20 minutes, and full seizing occurs after 24 hours.

Studies of micro- and macrostructure are carried by a scanning microscope JSM-7500FA. Investigation of the phase composition and structure parameters was made by diffractometer XRD - 6000 CuKα - radiation. Analysis of the phase composition, the size of coherent scattering of internal elastic stresses is made by databases PCPDFWIN, as well as full-profile analysis program POWDER CELL 2.4. The composition was determined by X-ray fluorescence spectroscopy (XRF) with spectrometer VRA- 30. The specific surface area was determined with multipoint BET method by hydroxyapatite particles tool named Sorbtometer M ver 1.0.0.0. The mechanical properties of bone cement were tested by electromechanical testing machine Instron 3369.

Biological tests included preliminary toxicological and preclinical studies. Investigation of the cytotoxicity of cell culture was carried out by direct contact (ISO 10993-5, 1999) by using rabbit fibroblast cells.

Bone regeneration was investigated by rabbits weighing 1.8-4 kg. In the experiment, all requirements of the international standard of quality assurance GLP (Good Laboratory Practice) were taken into account. Under anesthesia by intramuscular injection into the femoral muscle and after skin treatment with 70% alcohol, skin incision was made parallel to the long axis of the femur to 10 cm. Then the thigh muscle bands were cut and muscle was diluted exposing the outer surface of the femur. A hole with diameter of 7 mm and depth of 0.5 cm was performed in the shaft of the femur. The hole of the animals from the first group was filled with composite material. Muscle tissue is sutured over the site of filling, and then the wound was sutured. Bone defect wasn't filled in the control group.

The level of bone regeneration was tested by X-ray computed tomography named Somatom AR.HP produced by "Siemens".

3. Results and discussion

Micrographs presented in Figure 1 clearly show that the bone replacement composite material is highly porous with different pore sizes and has a rough surface.

Figure 1. The micrographs of composite material, made with different zooming: a) x 2000; b) x 5000; c) x 10000

Research of bone replacement material composition indicates the presence only the main elements and trace elements. There are no uncharacteristic bone impurities. It is confirmed by spectral diagram in Figure 2.

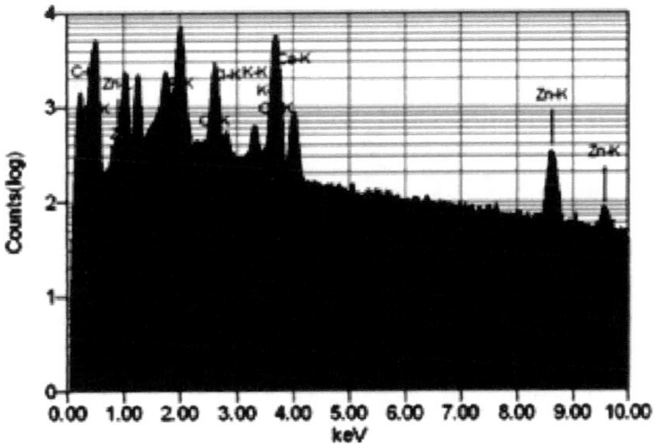

Figure 2. The spectral diagram of composite material composition

According to X-ray analysis results, which are given in Table 1, the composite material is represented by 30% amorphous phase; a crystalline phase is composed of hydroxyapatite monoclinic (about 23%) and hexagonal (about 47%) modifications. It corresponds to a ratio of hydroxyapatite-polymer in the structure of composite material. So, polymer and hydroxyapatite chemically unrelated, and biopolymer only envelops particles of hydroxyapatite.

Table 1. Results of X-ray diffraction

Sample	Detected phase	Contents phases of the %	Lattice parameters, A
Hydroxyapatite	$Ca_5(PO_4)_3(OH)$	89,4	A=9,522 C=6,875
	$Ca_{10}(PO_4)_6(OH)_2$	10,5	A=9,519 C=6,850 B=18,748 β =120,66

	$Ca(HPO_4)(H_2O)$ - - Brush	0,1	a=6,389 β=118,95 B=15,227
	Amorphous	1	–
	$Ca_5(PO_4)_3(OH)$	23	*a*=9,5218 *c*=6,9631
Composite material	$Ca_{10}(PO_4)_6(OH)_2$	47	*a*=9,3582 *c*=6,850 *в*=18,748 *β*=120,66
	Amorphous	30	–

Mechanical tests of composite material show high strength properties close to the dense bone strength. The results of mechanical tests are given in Table 2.

Table 2. Results of mechanical tests of composite material

Test	Strain rate, (mm/min)	Area, (mm^2)	Voltage, (MPa)	Load, (кН)	Young's modulus, (GPa)
1 Compression	0,2±0.001	1713±1	11,9±0.03	1,95±0.005	0,513±0.003
2 Bend	0,2±0.001	1951±1	40±0.2	106±0.3	8,605±0.04

To assess the effect of implant on a body, pre-tests were carried out on rabbits.

Figure 3 shows the results of histological studies.

Figure 3a (after 14 days) shows the traces of a polymer as homogeneous red- brownish masses, which have the formation of bone trabeculae on the periphery. There is the proliferation of osteoblasts among them. There are foci of fibrosis presented by fibroblasts layers of collagen in a side of the medullary canal. Bone marrow is celled; the cells are modified by decalcification. From the side of the periosteum there is phenomena osteogenesis presented by mature beams with layers of soft collagen fibers.

Figure 3b shows the histological section through the month. In the cortex of the femur there is a slight defect which is closed by newly bone beams formed perpendicularly to longitudinal axis of bone. There are newly formed vessels and fibroplastic reaction from the periosteum.

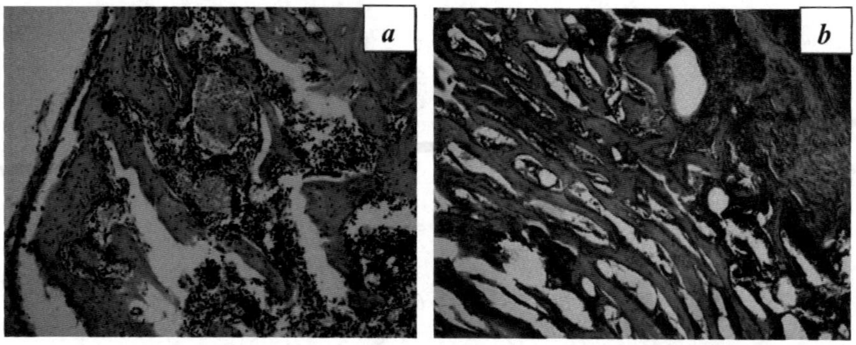

Figure 3. Staining by hematoxylin-eosin (uv. x 100):
a) mature bone beams; the period of 14 days; b) newly formed bone beams with large blood vessel between them, after1 month

As a result of histological examination of composite material, it was found that the newly formed tissue surrounding and sprouting in a matrix there are vessels of different diameters, and nerve fibers. Across the thickness of the implant newly formed bone trabeculae changed material in a result of biodegradation. The vessels are filled with blood. It indicates that the vessels are connected to the bloodstream. The results of X-ray studies are given in Table 3.

Table 3. Change the index of bone tissue regeneration

Research Group	Index bone regeneration		
	one week after the surgery	two weeks after the surgery	four weeks after the surgery
Control	28.2 ± 0.3	28.8 ± 0.2	30.7 ± 0.3
Composite material	29.2 ± 0.1	44.3 ± 0.2	46.8 ± 0.3

Conclusions

Thus, obtained composite material has high specific surface and porosity, which is one of the basic conditions for the formation of new bone tissue. The constituents of the composite material (hydroxyapatite and collagen) included in the connective tissues, such as bone, cartilage, skin, tendons, ligaments and blood vessels.

In vivo studies have shown that the investigated composite material proved to be a biocompatible, biodegradable, has osteogenic properties. Therefore, it may be used for bone tissue engineering.

References

[1] Barinov S M, Komlev V S 2005 *Bioceramics based on calcium phosphate*. - Moscow: Nauka,. – p. 204. ISBN 5-02-033724-2.

[2] Tadic D, Epple M, Thorough A 2004 *Physicochemical Characterisation of 14 Calcium Phosphate-Based Bone Substitution Materials in Comparison to Natural Bone* // Biomaterials. — V. 25. — pp. 987 - 994.

[3] Zelichenko E A, Guzeev V V, Rogulina A S, Gurov O A, Kowalska Ya B 2012 *Study properties of hydroxyapatite derived from bone farm animals* // Chemistry for Sustainable Development. – № 5. – pp. 1-6.

Authors:

1. Svetlana A. Melnikova, graduate student of the Department of Materials Science in Mechanical Engineering of National Research Tomsk Polytechnic University.
2. Elena A. Zelichenko, senior teacher of chair of state Autonomous Educational Institution of Higher Professional Education «National Research Nuclear University «MEPhl»
3. Boris S. Zenin, associate professor of the Materials Science in Mechanical Engineering Department of National Research Tomsk Polytechnic University
4. Vitaliy V. Guzeev, Doctor of technical sciences, professor of "Chemistry and Technology of Materials of Modern Power" chair of state Autonomous Educational Institution of Higher Professional Education «National Research Nuclear University «MEPhl».
5. Oksana A. Gurova, postgraduate student chair of state Autonomous Educational Institution of Higher Professional Education «National Research Nuclear University «MEPhl».

Influence of chemical composition of initial powders on structure and properties of «Ti-Ta-Zr» coatings fabricated on cp-titanium substrates by electron beam cladding

V V Samoylenko[1], **D V Lazurenko**[1], **O G Lenivtseva**[1], **I A Polyakov**[1]
[1] Material Science in Mechanical Engineering Department, Novosibirsk State Technical University, 20 Prospekt K.Marksa, Novosibirsk, 630073, Russia

E-mail: samoylenko.vitaliy@mail.ru

Abstract. In this study «Ti-Ta-Zr» corrosion resistant coatings fabricated by electron beam treatment of Ti substrates were investigated. The series of experiments on varying Ta and Zr concentrations in initial powder mixtures aimed on identifying influence of the alloying components ratio on structure and properties of coatings were carried out. It was found that during the electron beam cladding the high-quality coatings with a dendritic structure exhibited uneven distribution of alloying components in a cross section were formed. Moreover scanning electron microscopy (SEM) revealed the formation of needlelike structure. However difference in Ta and Zr concentrations in initial powder mixtures did not cause significant structural transformations in the coatings. Meanwhile the direct dependence between microhardness of the surface layers and Zr content in it was revealed. The maximal microhardness level was detected in the coating containing 60 % Zr. The impact strength test indicated decrease of impact toughness of titanium samples after cladding by a factor of 1.5…2.

1. Introduction

Titanium alloys are considered to be the prospective structural materials which are widely-used in chemical and nuclear industries, aircraft and rocketry as well as in medicine due to their high specific strength in combination with corrosion resistance in the majority of corrosive media [1].

However titanium is exposed by intensive corrosion in boiling concentrated acids. In [2-4] it was shown that adding tantalum to titanium alloys allows increasing significantly their corrosion resistance which can reach in some cases the pure tantalum level. However applying Ti-Ta alloys is restricted by a high cost of Ta and complexity of capacity workpieces formation consisted in repetitive arc remelting. Herein highly doped Ti-Ta alloys are of a high density level, wherefore the weight of details and constructional element made of this alloys significantly increase. In order to save expensive tantalum and improve corrosion resistance of titanium it is reasonable to produce coatings containing tantalum on titanium workpieces.

The results of previous investigations showed that the most prospective technology of coatings formation is electron beam cladding which allows obtaining surface alloyed wear resistant [5-7] and corrosive resistant [8, 9] layers with a thickness of 2 mm on steel and titanium workpieces.

In this study «Ti-Ta-Zr» coatings fabricated on surfaces of cp-titanium substrates were investigated. The cladding process was realized in the Institute of Nuclear Physics SB RAS by the industrial accelerator ELV-6. Such accelerators are equipped with the system of differential pumping which allows injecting the electron beam in the air [10].

2. Materials and methods

Before the electron beam treatment a powder mixture consisted of Ta, Zr and a flux agent was poured on a working surface with dimensions of 100x50 mm. A thickness of titanium plates was 12 mm. The powder mixture weight was equal to 22.5 g. Adding Zr in an initial powder mixture allows replacing partially expensive tantalum. Moreover at the initial processing stage zirconium possessed a lower melting temperature comparing to tantalum acted as a wetting agent. The accelerator capacity reaches 33.5 kW. In order to increase efficiency the electromagnetical scanning coil was inserted at the outlet. Electromagnetical scanner was set up in such manner as to provide the peak-to-peak value equaled to 50 mm at the distance of 90 mm from the outlet.

During the experiment the series of samples was fabricated. With increase of tantalum in coatings zirconium amount decreased (table 1). A flux volume in a powder mixture in relation to a volume of alloying components was constant.

Table 1. Weight fraction of powders in initial mixture.

Sample №	Weight fraction of powders in initial mixture, wt. %.				Chemical composition of coating after cladding, wt. %.		
	Ta	Zr	CaF_2	LiF	Ta	Zr	Ti
1	0	60	30	10	0	52.87	47.13
2	10.5	52	28.13	9,37	5.25	27.63	67.12
3	21	44	26.25	8.75	10.76	22.89	66.35
4	31.5	36	24.38	8.12	18.30	21.93	59.77
5	42	28	22,5	7,5	22.43	15.49	62.08
6	52.5	20	20.63	6.87	27.43	10.99	61.58

Metallographic investigations were carried out using an optical microscope Carl Ziess Axio Observer A1m and a scanning electron Carl Zeiss EVO 50 XVP. Chemical composition was estimated by energy-dispersive X-ray spectroscopy (EDX) using an analyzer INCA X-ACT (Oxford Instruments). Microhardness of coatings was determined by a tester Wolpert Group 402MVD. The load on a diamond indenter was 0.98 N. The distance between impressions was 200 μm. Impact strength test was carried out by a vertical drop machine MetroCom. The samples dimensions were 50x10x10 mm. V-notch was cut perpendicular to a cladded layer. Fracture surfaces were analyzed by scanning electron microscope.

3. Results and discussion

Electron beam cladding of Ta and Zr powders in the air atmosphere led to the formation of a graded structure. In the cross section of samples three main zones were identified: a coating, a heat affected zone and a base material.

The coatings obtained by the electron beam cladding of powder mixtures of a different Ta-Zr ratio were structurally similar. During the crystallization of cladded layer a cast dendritic structure was formed. The direction of dendritic crystals growth is coincided with the grains orientation (Fig. 1, a, b). At a high magnification the alloy structure consists of fine needlelike crystals (Fig. 1, c, d). Formation of such a structure is typical for quenched titanium alloys. During the electron beam cladding a treated surface was heated up to the melting point while a lower part of a titanium plate was kept at the ambient temperature and as a result heat was abstracted from a workpiece surface into a depth of titanium substrate. This process led to suppression of equilibrium processes and realizing the diffusionless transformation.

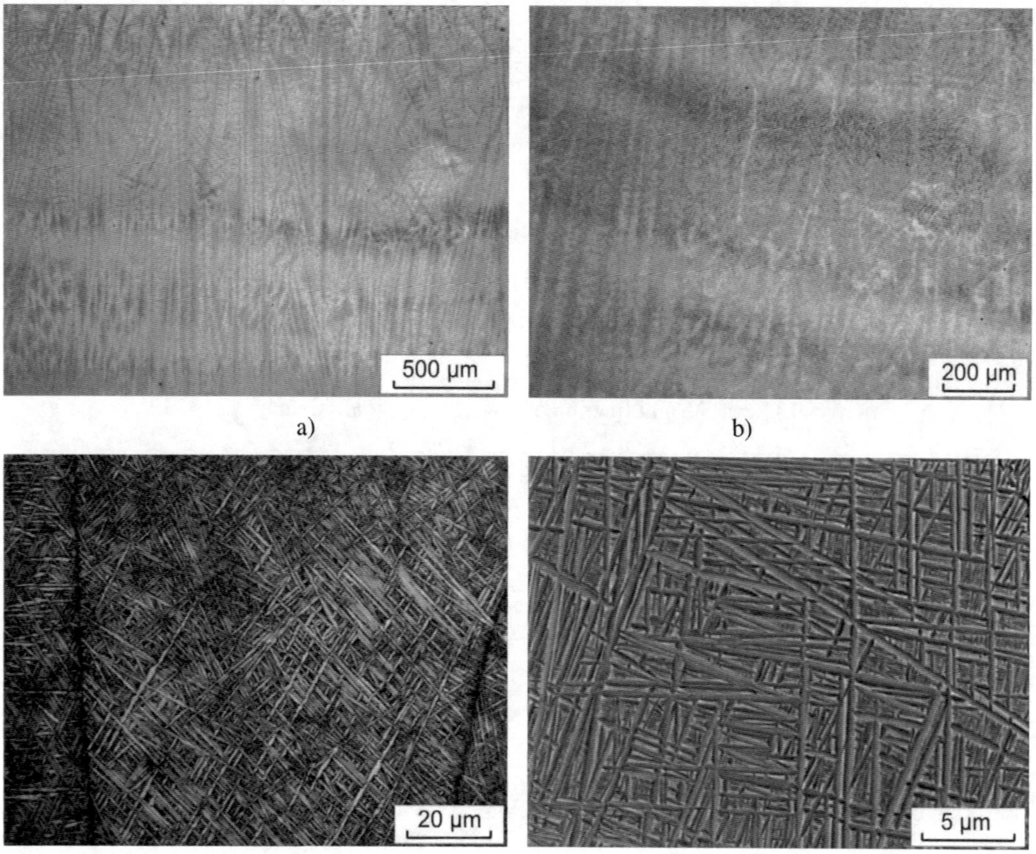

Figure 1. Typical microstructure of the coating after cladding: a, b – dendritic structure; c, d - needlelike structure.

EDX analysis results gave the evidence of increasing the tantalum volume fraction in coatings corresponding to its increasing in initial powder mixtures. The maximal tantalum concentration equaled to 27 wt. % was observed in the sample 6 (table 1). Herein the zirconium content in this sample was minimal and not exceeded 11 %. The maximal zirconium content (52.9 wt. %) was observed in the case of cladding the powder mixture containing only zirconium as an alloying component.

The presence of a contrast on SEM microphotographs made in the back scattered electrons regime indicated indirectly the difference in chemical composition between dendritic axes and interdendritic space. The elemental analysis of the sample containing 27 wt. % Ta and 11 % wt. Zr (Fig. 2) provided evidence of the enhanced tantalum content while in interdendritic space the increased content of zirconium was observed (Table 2).

Table 2. EDX-analysis of the coating

№	Contente, % (wt)		
	Ta	Zr	Ti
Spectrum 1	29.62	8.68	61.70
Spectrum 2	18.22	11.48	70.30

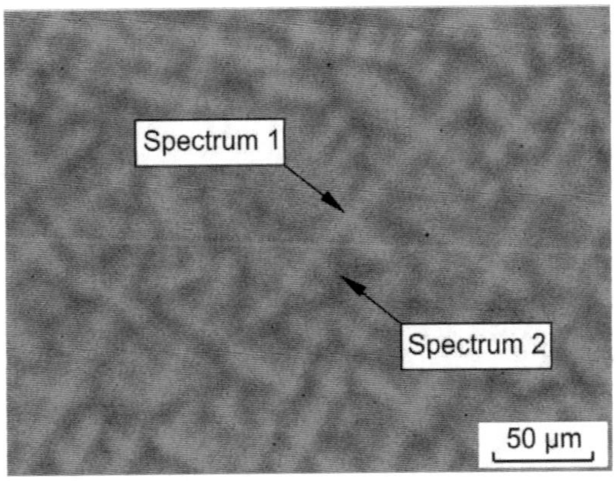

Figure 2. EDX-analysis of the coating

Microhardness of the cladded layers did not varied in the cross section. Microhardness level in the sample containing only zirconium as an alloying component was equal to 6000 MPa and decreased gradually corresponding to increasing tantalum in the initial powder mixtures (Fig. 3). At the distance of ~ 50 µm from a surface microhardness value of some coatings reached 6500 MPa. Hardness increase is explained by the aeration of subsurface layers. In going from coating to titanium substrate microhardness decreased to 1700 MPa.

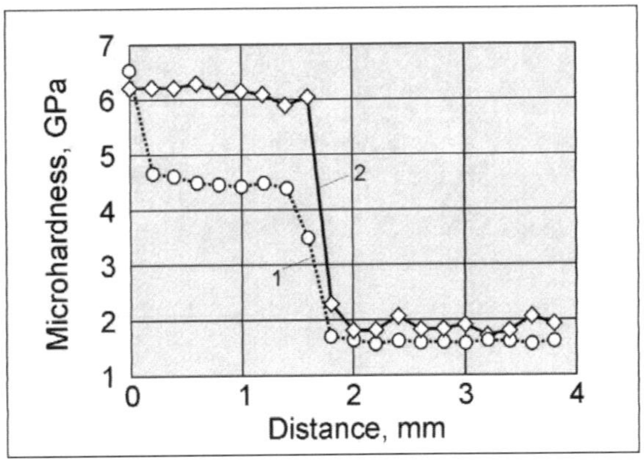

Figure 3. Microhardness distribution in the cross section of the sample containing: 1 – 22.43 wt. % Ta, 15.49 wt. % Zr; 2 – 52.87 wt. % Zr.

Impact toughness of bimetallic composites was approximately 1.5…2 – fold lower comparing to cp-titanium. The minimal fracture energy equaled to 81.2 J/cm^2 was obtained for the sample containing 5.25 wt. % Ta and 27.63 wt. % Zr. The sample containing 18.3 wt. % Ta and 21.93 wt. % Zr showed the highest impact toughness level (113.8 J/cm^2). The impact value of cp-titanium was equal to 160 J/cm^2. Decrease of the fracture energy is associated with the presence of a wide heat affected zone and a hard coating in samples. Fractographic analysis revealed the fracture of coatings according to the quasi-brittle mechanism. Elongated dimples were observed at the fracture surface (Fig. 4 a). Moreover the elements of intercrystalline fracture were revealed (Fig. 4 b). It should be noticed that there were no any cracks and fractures at the titanium-coating interface due to the good adhesion between the coating and the substrate (Fig. 4 c).

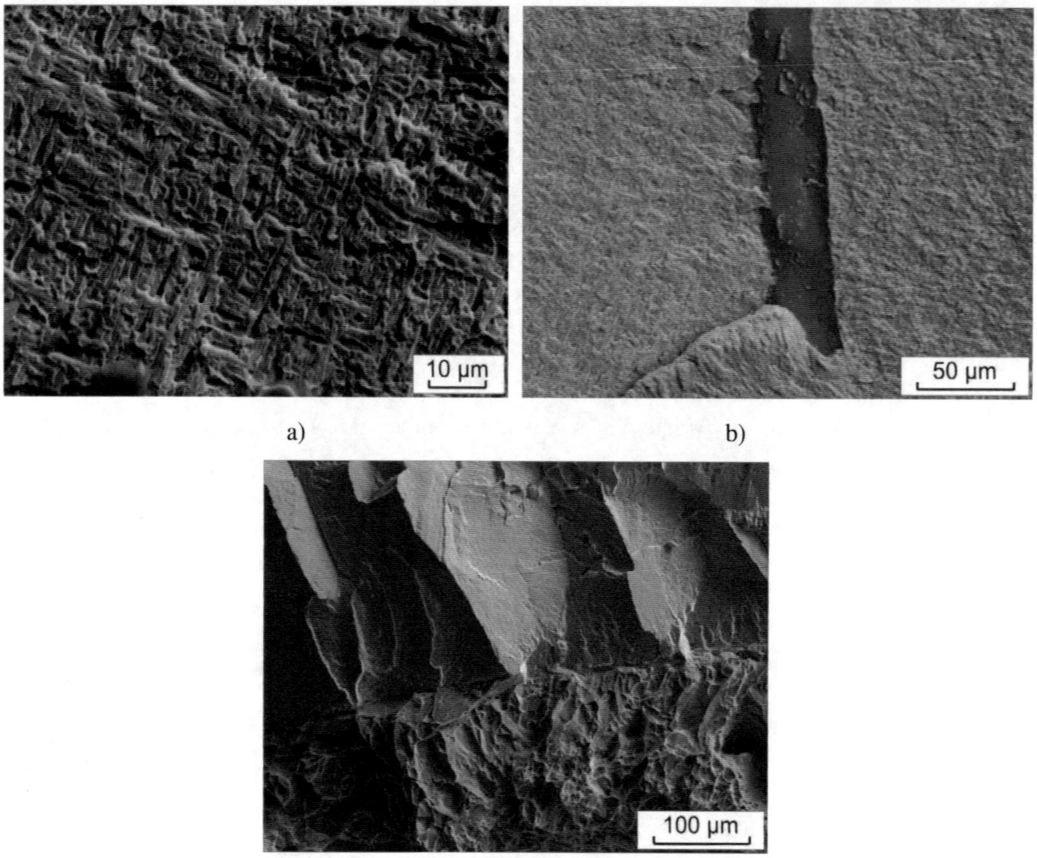

a) b)

c)

Figure 4. Fracture surfaces of coatings (a, b); fusion zone (c).

4. Conclusions

The electron beam cladding of the powder mixtures containing Ta and Zr on Ti substrates allowed fabricating the high-quality defect-free coatings with high adhesion characteristics. Optical microscopy and SEM investigations revealed that the concentration of Ta and Zr in the initial powder mixtures had no significant influence on the coatings structure. All the coatings were characterized by the formation of dendritic structure and microsegregation. The fine structure of coating was represented by needlelike crystals. Microhardness measurements showed that the coating strength increased from 4500 MPa to 6000 MPa in accordance with increasing the zirconium content from 20 % to 60 % (in initial powder mixture). Meanwhile there was no revealed any dependence between impact toughness of samples and tantalum and zirconium content. Fracture energy of Ti after the electron beam treatment decreased by a factor of 1.5…2.

References

[1] Lütjering G and Williams J C 2007 *Titanium (Engineering Materials and Processes)* second ed Springer p 442
[2] De Souza K A and Robin A 2007 *Mater. Chem. Phys.* **103** pp 351–360
[3] Golkovski M G *et al* 2013 *Mater. Sci. Eng., A* **578** pp 310-317
[4] Ahn H *et al* 2008 *Surf. Coat. Technol.* **202** pp 5784-5789
[5] Lenivtseva O G *et al* 2013 *Obrabotka metallov* **61** pp 49-57

IOP Conf. Series: Materials Science and Engineering **66** (2014) 012026 doi:10.1088/1757-899X/66/1/012026

[6] Mul D O *et al* 2013 *Obrabotka metallov* **60** pp 115-120
[7] Bataev I A *et al* 2013 *Appl. Surf. Sci.* **284** pp 472-781
[8] Bataev I A *et al* 2012 *Obrabotka metallov* **54** pp 90-95
[9] Golkovski M G *et al* 2013 *Obrabotka metallov* **61** pp 43-48
[10] Fadeev S N *et al* 2004 *Probl. Atom. Sci. Technol. Ser. Nucl. Phys. Investigat.* **1** pp 178–180

DESIGN OF THE FILTER TO STATICALLY STABILIZE POSITIONS OF THE ANTHROPOMORPHOUS ROBOT USING ZMP METHOD

E Shelomentcev and T Alexandrova

Tomsk Polytechnic University, Tomsk, Russia

E-mail: see4me@mail.ru

Abstract. Motion capture control methods including the Kinect control method are needed to stabilize the robot in a vertical position in order to prevent the robot from falling. The process of design of the position filter to perform this task is described in the paper. The application of such a filter enables to control the anthropomorphous robot using motion capture technology even if the connection with the robot has lags.

1. Introduction

Anthropomorphous robotics is an actively developing field of the robotics. The anthropomorphous robot is more effective through the maintenance process because it can replace the human to operate in different environments including hazardous environments for the human. The analysis of the teleoperating control systems of the anthropomorphous robot allows us to highlight two control methods which use the motion capture technology to get the control signals. The first method is the control method using mechanic sensors (exoskeleton) and the second one is the control method, which uses optic sensors (Kinect). Both methods have different advantages and disadvantages. The first method gives us more stable and simple data of the sensors than the second one, however, it is more expensive, and there is a need for an operator to wear the exoskeleton that constrains the limb movements of the operator. The second method, an anthropomorphous robot control method using Kinect sensor [1,2,3], is simpler to operate through maintenance process, but it needs to ensure stable vertical position of the robot in the course of operation of the control algorithm. The stability of static positions for the robot is important, but actually, some actions of the robot operator, for example, a fast movement of hands or body tilts can lead to destabilization of the robot and its damages. In this paper we consider the technology to provide the stable vertical position of the robot in the case of applying the method of the Kinect control.

2. Evaluation of robot stability

There exist different ways to ensure the stable vertical position of the robot, which include the stabilization of the static position with or without accounting inertial forces effect. The most common method for solving the problem of static stabilization is the ZMP (Zero Moment Point) method allowing to determine whether the position of the robot is stable or not accounting the effect of all forces [4,5].

IOP Conf. Series: Materials Science and Engineering **66** (2014) 012027 doi:10.1088/1757-899X/66/1/012027

The method comprises calculating the ZMP position, where both inertial and gravity forces do not produce any moment in a horizontal direction (Figure 1). To evaluate the position stability of the robot it is necessary to determine that the ZMP is located in a supporting polygon (the area of a foot in the single support phase or the area limited by two feet in the double support phase). The moment of the gravity forces has a significant role when we calculate the position of the ZMP. To calculate this moment it is necessary to determine the coordinates of the CoM (Center of Mass) by the formula:

$$\bar{r}_c = \frac{\sum_i m_i \bar{r}_i}{\sum_i m_i}, \quad (1)$$

where m_i is a mass of ith point that is located in the space with $\bar{r}_i$ coordinates, and $\bar{r}_c$ is a vector with coordinates of the CoM.

If to neglect the influence of the inertial forces, the ZMP is a projection of the CoM on a supporting plane (Figure 1). In this case the vector of the gravity force crosses the X and Y axes, what means that the moment produced by this force is zero. Thus, to assess the position stability of the robot we need to ensure that the projection of the CoM is located in the supporting polygon.

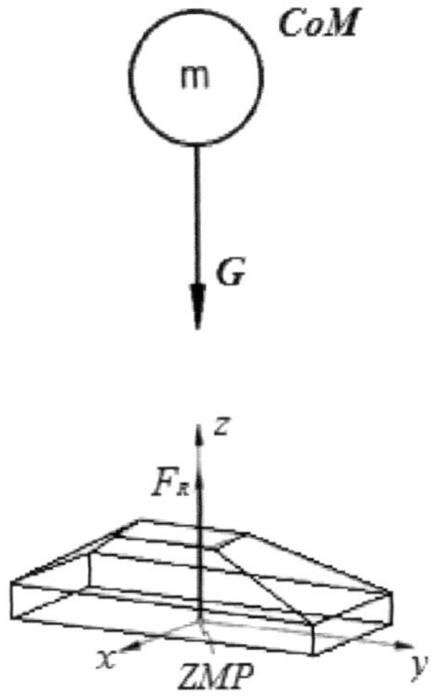

Figure 1. Position of the ZMP

In Biomechanics [6,7] the level of the position stability in the double support phase is a stability angle (the angle formed by vertical line dropped from the CoM of the body and straight line drawn from the CoM to the boundary of the supporting area (Figure 2). The greater is the stability angle the greater is the position stability.

3. Design of the position filter

The simplest way to stabilize robot controlled by the Kinect sensor is a design of the position filter that allows the control system to discard unstable positions of the robot. If the control algorithm starts from the predetermined stable position, the application of the filter lets the robot save stability when it changes its position.

We take the ZMP method as a basis of the filter. To do calculations we need to know the coordinates of the CoM that we can find using formula (1). Assuming that the overall dimensions, the mass of the joints, and their CoM coordinates relative to the attachment point are known, we only need to calculate the coordinates in the coordinate system of the robot with the origin in the centre of the line segment connecting the shoulder actuators of the robot. It is necessary to create the function that calculates the coordinates of the all joints after processing the control signals that are formed with the help of the data of the Kinect sensor. Let us write the matrix of Denavit-Hartenberg to model the robot joints:

$$A_i = \begin{pmatrix} \cos\theta_i & -\cos\alpha_i \cdot \sin\theta_i & \sin\alpha_i \cdot \sin\theta_i & a_i \cdot \sin\theta_i \\ \sin\theta_i & \cos\alpha_i \cdot \cos\theta_i & -\sin\alpha_i \cdot \cos\theta_i & a_i \cdot \cos\theta_i \\ 0 & \sin\alpha_i & \cos\alpha_i & S_i \\ 0 & 0 & 0 & 1 \end{pmatrix}$$

where θ_i is an angle about previous Z, from old X to new X; α_i is an angle about common normal, from old Z axis to new Z axis; a_i is a length of the common normal; S_i is an offset along previous Z to the common normal.

Using Denavit-Hartenberg convention, we can get the formulas to calculate the coordinates of the joints if the angles of the movement are known. These formulas are placed in the 4th column of Denavit-Hartenberg matrix $A = A_1 \cdot ... \cdot A_n$, where n is a number of the joints. When we calculate the coordinates of the joints we need to transform them to the robot's coordinate system. After this we can calculate the position coordinates of the CoM of the robot.

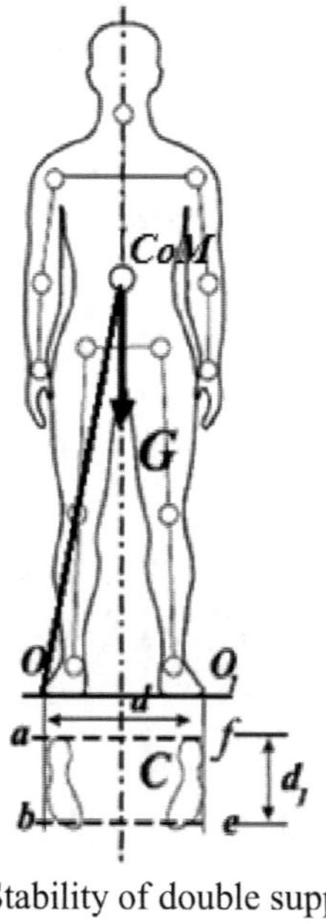

Figure 2. Stability of double support phase

Assuming that the robot is in the double support phase, the robot's position is stable, if the projection coordinates satisfies the condition:

$$\bar{a} \cdot \bar{b} \le 0,$$

where $\bar{a}$ is a vector with initial point at the left/right boundary of the supporting area foot of the robot and terminal point in the robot's CoM; $\bar{b}$ is a vector with initial point at the left/right boundary of the supporting area and terminal point in the point of the projection of the robot's CoM.
Dot product of these vectors in the coordinate form is below:

$$\bar{a} \cdot \bar{b} = (x_1 - x)(x_2 - x) + (y_1 - y)(y_2 - y) + (z_1 - z)(z_2 - z)$$

where x_1, y_1, z_1 are the coordinates of the robot's CoM; x_2, y_2, z_2 are the coordinates of the projection of the robot's CoM; x, y, z are the coordinates of the left/right foot.

In case the condition of the stability is satisfied, we send the robot the command to move in this stable position, in another case, we warn the operator that the position is unstable, and the movement is dangerous, and this can lead to the fall of the robot.

4. Conclusions

The designed filter of the statically stable position allows the control system to ensure the stable vertical position of the robot through the operation process of the Kinect control algorithm. The application of this filter allows the operator to control anthropomorphous robot even if the connection with the robot has lags, for example, in case of wireless connection, since the robot is statically stable in every moment of time.

This filter uses only the static stable criterion; that is why it can be used only if the speed of the robot's motions is limited. The application of the dynamic criterion is more complex because it needs the robot to have accelerometers in each limb. It has been planned develop the position filter using the dynamic stability criterion to solve the problem of the robot stability through the teleoperating control process in further research.

5. References

[1] Shelomentcev E and Duseev V 2012 Comparative analysis of the different methods to control android robots *"Modern Technics and Technologies" XVIII Int. Conf. Proc.* (Tomsk, TPU) [In Russian].

[2] Shelomentcev E 2012 Control of the android robot using Kinect sensor *"Actual problems of the modern sciences-2012" Int. Conf. Proc.* (Przemysl, Nauka i studia) [In Russian].

[3] Shelomentcev E 2013 Practical aspects of the Kinect control realisation of the android robot Bioloid *"Modern Technics and Technologies" XIX Int. Conf. Proc.* (Tomsk, TPU) [In Russian].

[4] Vukobratovic M and Stepanenko Y 1972 On the Stability of Anthropomorphic Systems *Mathematical Biosciences* Vol.15 pp 1-37.

[5] Lozhkin P 2013 ZMP method for the task of the dynamic stability of the robot AR-600 *Journal of IKBFU* (Kaliningrad, IKBFU) [In Russian].

[6] Lamash B 1999 *Biomechanics lectures* (Vladivostok, DVFU) [In Russian]

[7] Maslennikov A 2008 *Biomechanics* (Veliky Novgorod, NOVSU) p 189 [In Russian].

Overview of frequency bandwidth determination techniques of useful signal in case of leaks detection by correlation method

V A Faerman[1a], V S Avramchuk[1b], E E Luneva[1c]

[1] Tomsk Polytechnic University, Tomsk, Russia
[a] e-mail: fcsevladfaerman@gmail.com
[b] e-mail: avs@tpu.ru
[c] e-mail: lee@tpu.ru

Abstract. In this paper an overview of useful signal detection methods on the background of intense noise and limits determination methods of useful signal is presented. The following features are considered: peculiarities of usage of correlation analysis, cross-amplitude spectrum, coherence function, cross-phase spectrum, time-frequency correlation function in case of frequency limits determination as well as leaks detection in pipelines. The possibility of using time-frequency correlation function for solving above named issues is described. Time-frequency correlation function provides information about the signals correlation for each of the investigated frequency bands. Data about location of peaks on the surface plot of a time-frequency correlation function allows making an assumption about the spectral composition of useful signal and its frequency boundaries.

1. Introduction

Recovery of useful signals on a background of random noise is one of the most actual problems of digital signal processing (DSP). Despite the fact that so far, many methods and techniques were developed in order to solve this problem, they all have significant limitations in use and none of them is universal. The problem of useful signals detection on the background of random noise was widely considerated in the middle of last century due to the rapid development of radio and radar systems [1]. Various method of mathematical statistics were the foundation of techniques aimed to solve this problem, and field of radio engineering related to the development of these techniques was called the theory of optimal signal reception [1]. Curently, the issue of signal detection with a priori unknown format is typical problem for radiolocation branch as well as for non-destructive testing [2, 3, 4]. Specifically, one of the problems of signal detection and determination of its unknown parameters is the problem of detecting leaks solved by the correlation method [2].

2. Setting of signal detection problem and estimating signal's parameters problem

In a generalized sense the problem of optimal signal reception can be formulated as follows: on the supposition of in a priori known some characteristics of the useful signal, noise, as well as their functional interaction, it is necessary to obtain optimal receiving or decision-making device. This device should in the best way reproduce a signal or make decision with the least errors [1]. Depending on the amount of a priori available data, several tasks of varying difficulty receiving signals were distinguished [1]:

- simple detection of the useful signal that is completely known;
- complex signal detection with unknown parameters without measuring these parameters;
- signal detection with measuring its parameters.

Traditionally, much attention was directed to the last group of tasks due its complexity, as well as practical significance for radiolocation branch.

A common radiolocation task of signal detection with measuring its unknown parameters is a task of correlation signal $s(t, \lambda)$ reception, having one unknown parameter λ. It is assumed that a receiver device takes up a signal $\xi(t)$, ant this signal is a mixture of a detectable signal $s(t, \lambda)$ and additive random noise $n(t)$. Bellow the case of discrete signal processing is considered. So signal $\xi(t)$ and $s(t, \lambda)$ are represented by time series containing the samples that are equally spaced in time by a value Δ: ${}_i = \xi(t_i)$, $s_i(\lambda_j) = \xi(t_i)$ $(t_i = i \cdot \Delta, i = 1,2, \dots, N-1)$.

Total number of available sample units is determined by the observation time T and the discretization interval Δ according to the following formula:

$$N = \frac{T}{\Delta}.$$

Methodics of solution described problem comes down to obtaining and analyzing the correlation function [1].

$$R_{s\xi}(j) = \sum_{i=0}^{N-1} {}_i s_i(j) = \sum_{i=0}^{N-1} \xi(t_i) s(t_i, \lambda_j), \qquad (1)$$

where $s(t_i, \lambda_j)$ – reference implementation of useful signal when ($\lambda = \lambda_j$); $r_{s\xi}(j)$ – correlation function of mixture of signals $\xi(t_i)$ and reference implementation of useful signal $s(t_i, \lambda_j)$. Further analysis of correlation funtion is reduced to obtaining its maximum, that, in simple case, corresponds to the value of requiered parameter λ. Usually under requiered parameters serves: time delay of arrival reflected signal τ_0, which is used in order to determine the distance to the object location; Doppler frequency f_0, which is used in order to determine the velocity of the object.

Despite the fact that Eq. 1 can be simplified in case of the signal $s(t, \lambda_1, \dots, \lambda_n)$ detection, that depends on many unknown parameters $\lambda_1, \lambda_2, \dots, \lambda_n$:

$$R_{s\xi}(j, k, \dots, m) = \sum_{i,=0}^{N-1} \xi(t_i) s\left(t_i, \lambda_{1_j}, \lambda_{2_k}, \dots, \lambda_{n_m}\right),$$

the complicity of procced operations is rising significantly and this leads to the need to increase number of calculation devices (correlators) as well as to make the receiver device more complicated [1].

3. Correlation method of leak location detecting in a pipeline

Description of the method. Schematically location leaks detection by correlation method is represented in Fig. 1. Since the sensors and signal source are lain out in single line, the location of signal source is uniquely determined by the values of distance from both sensors. The propagation velocity of the signal is determined dependant on solving task, and so usually it can be measured [5]. Thus, the problem reduces to determining the delay signal on sensor by correlation method.

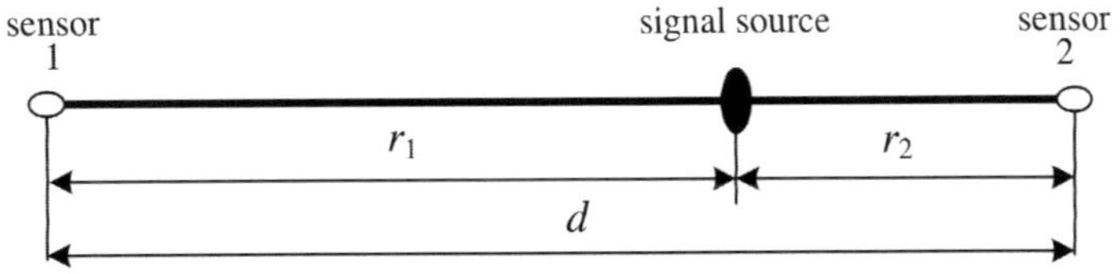

Figure 1 – Scheme of location leaks detection by correlation method
r_1, r_2 – distance from the signal source to the sensor; d – distance between sensors

The main feature of the correlation method lies in simultaneous use of a pair of sensors. Suppose the sensors, during the interval T, fix signals that are arrived from equally separated moments of time Δ. Then in any discrete time t_i $(i = 0,1, \dots N-1$, where N – whole number of sample units) a signal that registered by the first sensor can be represented as following:

$${}_1(i) = s_0(i - j_1) + n_1(i), \qquad (2)$$

where $n_1(i)$ – noise component of signal from the first sensor; $j_1 = \tau_1/\Delta$ – signal lag from the first sensor τ_1 that is measured in sample units. In a similar way, for second sensor signal:

$${}_2(i) = s_0(i - j_2) + n_2(i). \qquad (3)$$

Correlation function should be calculated as following [3]:

$$r_{12}(j) = F^{-1}(F[\xi_1(i)]F^*[\xi_2(i)]), \quad (4)$$

where F - direct discrete Fourier transform (DFT); F^{-1} – inverse DFT; F^* - complex conjugate representation of the direct DFT results. Suppose that noise components of signal from sensors n_1 and n_2 are not correlate that allow representing Eq. 4 as a following sum:

$$r_{12}(j) = r_{s_0s_0}(j) + r_{n_1s_0}(j) + r_{n_2s_0}(j) + r_{n_1n_2}(j).$$

It is known that the following maximum value $max(r_{s_0s_0}(j))$ corresponds to $r_{s_0s_0}(j_2 - j_1)$ [2, 5] in other words to the difference in arrival time of signals to sensors.

Limitation of the method. In order to detect a maximum of the correlation function, and this maximum is informative, the following condition has to be fulfilled:

$$r_{s_0s_0}(j_2 - j_1) \gg \sqrt{D\left[r_{n_1s_0}(j) + r_{n_2s_0}(j) + r_{n_1n_2}(j)\right]},$$

where D – variance of random variable. Thus, an unknown parameter is defined by the ratio of peak determinate component value of correlation function ($r_{s_0s_0}(j)$) and root mean square (RMS) value sum of its random components ($r_{n_1s_0}(j) + r_{n_2s_0}(j) + r_{n_1n_2}(j)$).

Techniques that reduce RMS value of random components. In practice there are different techniques of RMS value reducing of signal random components. The obvious technique is consist in increasing sample size (*N*), that allow weakling random components owing to averaging properties of the correlation sum. However, increasing the sample size in some cases is impossible because of fast Furrier transform limitation on number of sample units, which is used as the fastest way of correlation function calculation [2].

Another simple and effective technique to reduce the RMS value of signal random components at the output of a correlator is a digital filtering (usually band or band-rejection filtering) at the input of the correlator. But, this technique implementation has some difficulties, since signal spectrum usually is unknown [6]. Therefore diverse techniques of identifying bandwidth of spectral localization of signal are used for following validated filter settings.

3. Frequency bandwidth determination techniques of useful signal.

The simplest frequency bandwidth determination techniques consist in obtaining cross-amplitude spectrum [4]. Cross-amplitude spectrum $S'_{12}(k)$ can be obtained by multiplying frequency samples and amplitude signal spectrums of sensors $|X_1(k)|, |X_2(k)|$:

$$S'_{12}(k) = |X_1(k)| \cdot |X_2(k)| \quad (5)$$

where k – number of spectrum sample ($k = 0,1, \ldots, N/2 + 1$). In practice instead of Eq. 5 is widely used the following equation:

$$S'_{12}(k) = |F[\xi_1(i)]F^*[\xi_2(i)]|. \quad (6)$$

The advantage of Eq. 6 is the use of intermediate data that obtained in the correlation function calculation according to Eq. 4. Noticeable peaks as well as , as well as high values of cross- amplitude spectrum are signs of the presence of the useful signal in the considered frequency band. Normally, cross-amplitude spectrum is more informative than the amplitude spectra of signals from each of the sensors. In the Fig. 2 amplitude spectra of signals from sensors and cross-amplitude spectrum are represented. Test signal is the sum of normally distributed random quasi-white noise and narrowband signal with frequency range 10…12 kHz.

The disadvantage of this technique is its sensitivity to the spectrum of the noise component at the input of the correlator device. If the noise energy is not uniformly distributed, but concentrated in some parts of the spectrum, it cannot be excluded the wrong choice of frequency bands for further analysis.

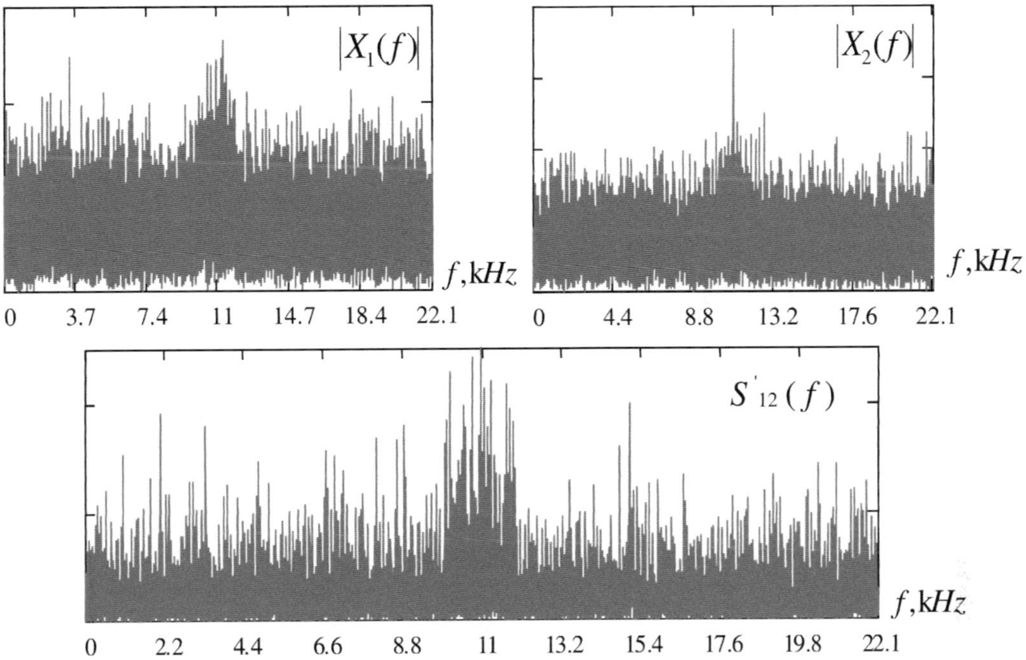

Figure 2 – Amplitude spectra of signals from sensors $|\boldsymbol{X_1(f)}|$, $|\boldsymbol{X_2(f)}|$ and their cross-amplitude spectrum $\boldsymbol{S'_{12}(f)}$

4. Coherence function analysis

Another common way to obtain a frequency bandwidth of useful signal is based on calculating coherence function. The coherence function is usually defined as the ratio of signals cross spectrum and root of the product of their own spectrums:

$$\gamma(k) = \frac{|E(S_{12}(k))|}{\sqrt{E({X_1}^2(k))E({X_2}^2(k))}}, \tag{7}$$

where E – averaging operator; ${X_1}^2(k), {X_2}^2(k)$ – eigen spectrum of signals $\xi_1(i)$ and $_2(i)$ properly; $S_{12}(k)$ – cross spectrum of signals ($S_{12}(k) = F[\xi_1(i)]F^*[\xi_2(i)]$). Also in practical spectral analysis is often the square of the coherence function calculated. When coherence function is calculated according to Eq. 7 or square coherence function, the value of signal spectrum should be averaged by some sample. If the sample is represented by a single set of spectrum, Eq. 7 takes the value 1.

Therefore, in practice of coherent analysis, the square of the coherence function is calculated by the following equation:

$$\gamma^2(k) = \frac{\left|\sum_{q=0}^{Q-1} S_{12}(k)\right|^2}{\sum_{q=0}^{Q-1} {X_1}^2(k) \sum_{q=0}^{Q-1} {X_2}^2(k)}, \tag{8}$$

where Q – number of spectrum sets of signals, applicable in coherence function calculating. Each set of spectra requires N samples of the measured signal, thus total sample size required coherence function calculation is Q.

In the fig. 3 coherence function plot for different value of Q is represented. Test signal is the sum of normally distributed random quasi-white noise and narrowband signal with frequency range 14…17.25 kHz. The coherence function in Fig. 3a contains no useful information. On the contrary, a region of spectrum localization of useful signal, as shown in Fig. 3b, is well defined.

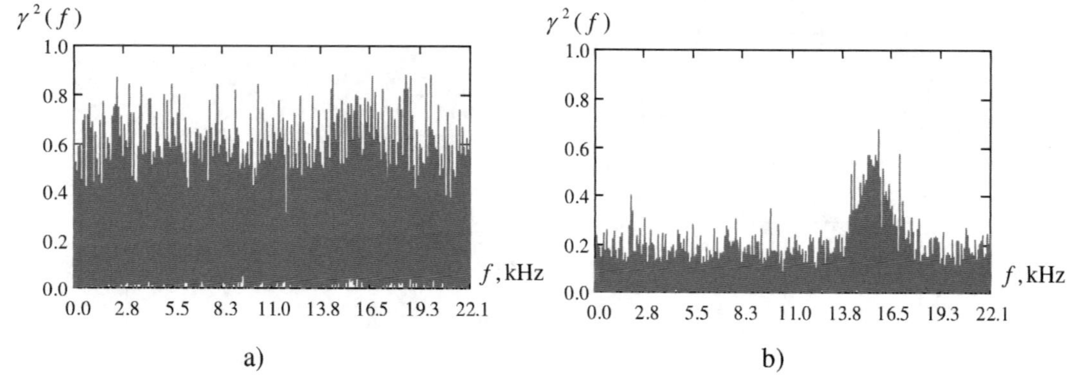

Figure 3 – Coherence function for deferent number set of short-range spectrum of signals $\boldsymbol{Q}$:

a) $\boldsymbol{Q} = \mathbf{5}$; b) $\boldsymbol{Q} = \mathbf{20}$.

Thus, coherence function is an analogue of the correlation function, in case of representing the correlation of processes in frequency domain [7]. In addition, the region of high values of a coherence function indicates the presence of related and not random signals. This fact can be used in problems of signal detection [6].

There are following disadvantages of analysis approach based on coherence function: requirement in the accumulation of additional data about signals; possible occurrence of false high values of the function in case of some problems determined by the existence of several vibration modes [8].

5. Time-frequency correlation analysis

Another way of correlation analysis that allows extracting and visualizing information about spectral composition of a signal is the use of time-frequency correlation functions [6]. Unlike with common functions, described above, time-frequency correlation functions have the additional argument – frequency. Thus, time-frequency correlation function represents a correlation of signals for each studied frequency band. The spectral composition of a useful signal is estimated by the location of peaks on the time-frequency correlation function plot [9]. For example, time-frequency correlation function plot was obtained for test signal that contains white noise and narrowband useful signal. As it shown in Fig. 4, the useful signal was located in frequency band 4…5,2 kHz.

The peaks of time-frequency correlation function become more noticeable at frequency bandwidth of useful signal. This can be explained by the fact that the algorithm for obtaining a time-frequency correlation function, represented [6], is equivalent to the simultaneous use of a number of correlators with band-pass basic frequency filter at the input.

IOP Conf. Series: Materials Science and Engineering **66** (2014) 012028 doi:10.1088/1757-899X/66/1/012028

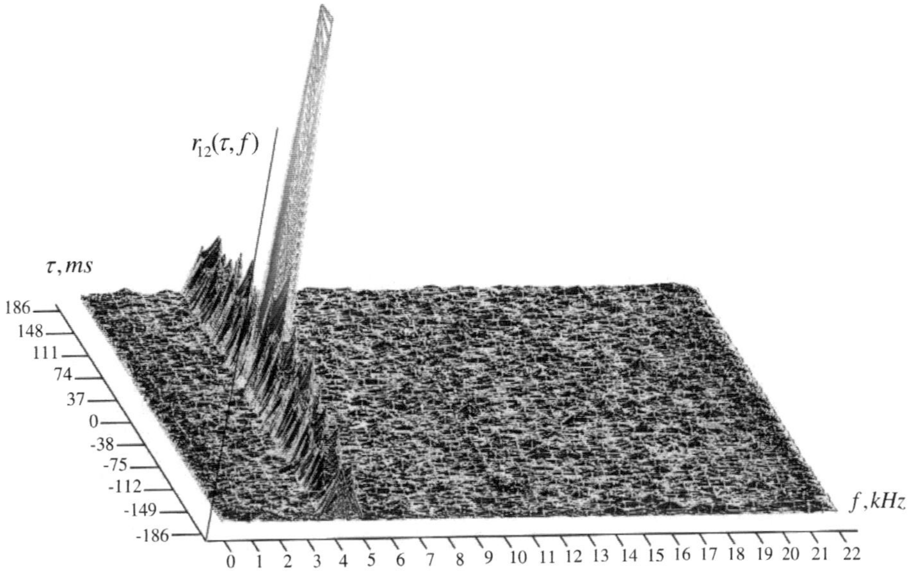

Figure 4 – Time-frequency correlation function plot obtained for test signal

6. Summary

Due to the fact that the major correlation method limitation in case of signal detection with a priori unknown format is impossibility to manage with high intensity noise at the input of the correlator. So the most important step of solving this problem is filtering of an input signal.

In order to obtain validate filter selection, spectral bandwidth of useful signal should be localized. In this case the same approaches to perform this step, give different efficiencies depending on a field of application (i.e. leak detection ofoil pipelines, water pipelines, etc.). Therefore, the best result can be achieved by combined use of approaches that were described in this work.

References

[1] Tikhonov V I 1983 *Optimal Signal Reception*, Radio i Svyaz, Moscow.

[2] Ifeachor E C, Jervis B W 2011 *Digital Signal Processing: practical approach*, John Willey & Sons Ltd., West Sussex.

[3] Beck S B, Curren M D, Stanway R 2005 Pipeline Network Features and Leak Detection by Cross-Corellation Analysis of Reflected Waves, *J. Hydralic Engineering*, 131, 715-723.

[4] Bendat J S, Piersol A G 2013 *Engineering Applications of Correlation and Spectral Analysis*, Wiley-Interscience.

[5]Huandi O, Chue W, Wang A, Guan W 2000 Detecting leaks in plastic pipes, *J. AWWA*, 92, 82-94.

[6] Avramchuk V S, Goncharov V I, Chan Viet T 2009 Russian Patent 2405163.

[7] Matz G, Hlawatsch F 2000 *Time-frequency coherence analysis of nonstationary random processes*, proc. of 10th IEEE-SP Workshop on Statistical Signal and Array Processing, Pocono Manor, 554-558.

[8] Ovchinnikov A L, Lapshin B M 2008 Experience of using a leak detector TAK-2005 for service of the municipal pipelines, *Bulletin of Tomsk Polytechnic University*, **312**, 2, 196-202.

[9] Avramchuk V S, Goncharov V I 2013 Time-frequency Correlation Method for Improving the Accuracy in Detecting Leaks in Pipelines, *Advanced Materials Research*, Trans Tech Publications, Switzerland, 650, 443–446.

Computer simulation of processes in the dead-end furnace

A S Zavorin, Tomsk Polytechnic University, Tomsk, Russia
S A Khaustov, Tomsk Polytechnic University, Tomsk, Russia
N A Zaharushkin, Tomsk Polytechnic University, Tomsk, Russia

Email: khaustovSA@tpu.ru

Abstract. We study turbulent combustion of natural gas in the reverse flame of fire-tube boiler simulated with the ANSYS Fluent 12.1.4 engineering simulation software. Aerodynamic structure and volumetric pressure fields of the flame were calculated. The results are presented in graphical form. The effect of the twist parameter for a drag coefficient of dead-end furnace was estimated. Finite element method was used for simulating the following processes: the combustion of methane in air oxygen, radiant and convective heat transfer, turbulence. Complete geometric model of the dead-end furnace based on boiler drawings was considered.

Introduction

Several advantages of Decentralized heating jointly with other economic and organizational causes led to the overall use of fire-tube boilers. The performance coefficient of the fire-tube boilers reach 95%, and there are additional advantages – best maintainability, full automation and low price in comparison to water-tube analogues.

The fact that the fire-tube boiler furnace normally has a water-jacket draws attention to questions connected with optimal temperature distribution along the furnace volume. Because of high volumetric thermal stress in fire-tube furnace the flame profile becomes the main characteristic of burner device. Estimated size of the flame should exclude every local contact of fire with the metal for all modes of boiler operation. Irregular and fluctuating sweep of gases may provoke excess of acceptable local thermal loads and cause reliability degradation of boiler operation.

It is well known fact that appropriate organization of furnace aerodynamics can influence burning stability, lead to required speed and temperature values in the furnace volume and reduce the level of toxic emission into environment. Therefore, improving aerodynamic characteristics of the fire tube is a main task in elaborating the fire-tube boiler construction. It is particularly relevant to the dead-end furnace types where aerodynamics of the reverse flame includes interaction between the straight-through flow and reversed flow.

The lack of a regulatory framework for the dead-end fire-tube furnace calculations and design indirectly indicates the complexity of the real thermal processes occurring in its volume. Full-scale multiple-factor experimental research of three-dimensional dynamics of turbulent burning flare is quite labor-consuming. Therefore, nowadays computational modeling with the use of wide approved application packages is used to solve complex problems of combustion aerodynamics optimization [1].

The present article is aimed at computer simulation of the turbulent combustion of methane-air mixture in a fire-tube boiler. The flame structure has been determined computationally, contours of average speed, temperature and concentrations have been acquired.

The subject of studying: complex of characteristics with space-time fields of the aerodynamic structure, range and aperture angle of the flame, pressure distribution, turbulent intensity and quite a number of other operation factors. Quantitative dependence of construction characteristics and operation conditions on aerodynamic parameters should be also estimated.

Principles and object of study

For the computational research, a full geometrical model of the dead-end fire-tube furnace is implemented. A computational grid designed by means of the ANSYS Workbench based on the boiler drawings contains about $8 \cdot 10^5$ control volumes. A diagram with the boiler's construction scheme is presented on Figure 1.

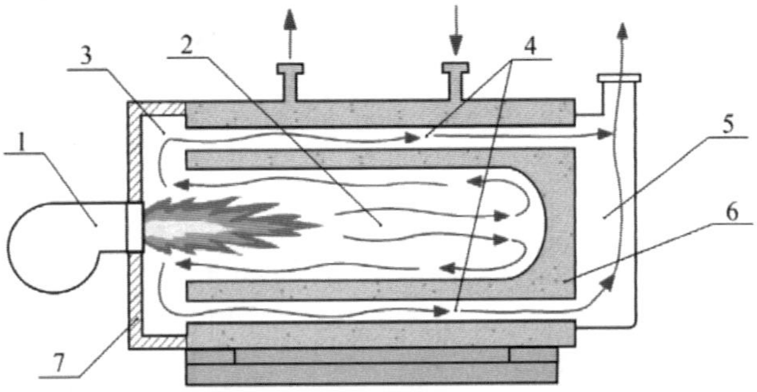

Figure 1. Fire-tube boiler's construction scheme
1 – burner; 2 – dead-end fire-tube ; 3 – turning chamber; 4 – convection fire-tube bank; 5 – collecting box; 6 – water-jacket; 7 front cover

The boiler has a dead-end fire-tube 2 with a bottom that causes the flame to turn 180 degrees in the direction of the boiler's front door 7. Then, moving along the periphery of the furnace, the combustion products flow into the turning chamber 3 where they turn 180 degrees again and flow into the convection fire-tube bank 4 with the water-jacket 6. Next moving inside these tubes, the combustion products flow into the collecting box 5, then to the chimney.

The fuel-air mixture enters the furnace through the burner 1 under the pressure of 200 PA which allows overcoming of the aerodynamic resistance of the gas path inside the boiler. The furnace length along the shaft from the burner to the bottom equals 1800 mm, a diameter of the fire tube is 884 mm.

A complete geometric model of the furnace, including the burner slot and the reversing chamber is used for computer simulation.

Once-through fuel supply is considered for better calculation performance since the influence of the twist parameter on the heat exchange in the reversing flame is insignificant [2].

Turbulent combustion of natural gas in the reverse flame of fire-tube boiler was studied by means of the ANSYS Fluent 12.1.4 engineering simulation software. The following processes were modelled through the finite-element method: methane burning in atmospheric oxygen, turbulence, convectional and radiant heat exchange. The gas-phase was simulated according to the Euler's approximation. Aerodynamic calculations were performed using laws of flow continuity and energy conservation.

IOP Conf. Series: Materials Science and Engineering **66** (2014) 012029 doi:10.1088/1757-899X/66/1/012029

The radiant heat exchange calculation was conducted by the method of spherical harmonics in 1 approximation (P_1-model). Differential equations used by the given radiative heat transfer model are presented in [3].

RNG k-ε model was used to simulate turbulence. This model is justified for fully advanced turbulence flow, i.e. for high Reynolds numbers, where the direct viscosity influence on the turbulence structure is small to negligible.

For the numerical simulation of chemical processes in turbulent stream, the Spalding model (Eddy-Break-Up) and the kinetic model were implemented jointly. It was supposed that fuel oxidation occurred irreversible in two stages: $2CH_4 + 3O_2 = 2CO + 4H_2O$; $2CO + O_2 = 2CO_2$. The accuracy of the chosen mathematical model was confirmed by the field tests [2, 4].

The initial data for calculation included: fuel – methane (100%); oxidant – atmospheric oxygen; theoretically required air volume for complete combustion – 9.52 m^3/m^3, excess air coefficient 1.03; mass flow of fuel-air mixture at 0.2 kg/sec; its temperature 20 °C (which equals the air temperature in the boiler room); average water temperature 92.5 °C.

Calculation results

The computation results display the processes within the furnace to be axisymmetric. Pathlines in the furnace medium (Fig. 2) indicate its double change in the direction. When moving towards the back part of the furnace, fresh air-fuel mixture heats up, ignites and burns out, afterwards combustion products move from the furnace centre to the periphery and at last smoke fumes along the furnace sides return to the front part and leaves the fire-chamber.

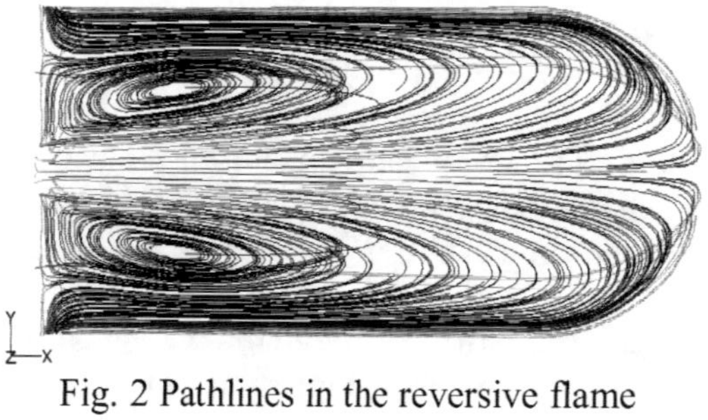

Fig. 2 Pathlines in the reversive flame

The hot combustion products make the first pass in the boiler, i.e. on the near-wall zone of the furnace, flowing in a circumferential direction of the fire tube from the heel to the front. The flue gas stream thickness on the near-wall zone lies between 130 and 135 mm. The chamber bottom is normally considered to be the twist point for the flame and the starting point for the initial path of smoke gases. However, pathlines (Fig. 2) and vector field of average speed (Fig. 3) indicate that turn of the combustion products holds the whole flare length.

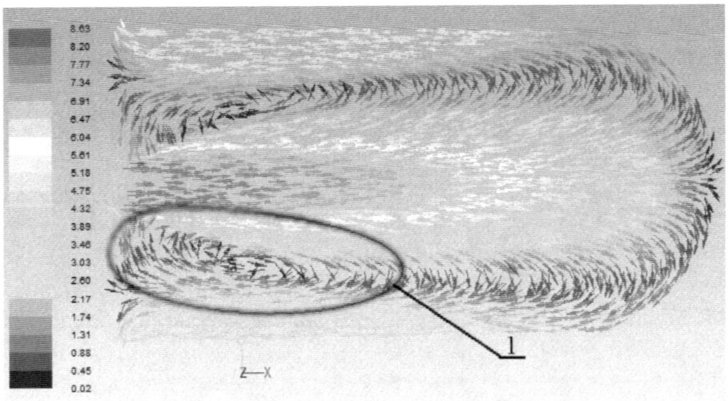

Fig. 3. Vectors of furnace medium velocity with color indexation according to the velocity magnitude (m/sec):
1 – area with recirculation

Low inlet pressure at the furnace entrance provokes formation of scorching furnace gas vortexes, circulating to the burner slot (Fig. 2). Some furnace gases do not leave the combustion chamber, rather revert to the burner and mix with the fresh air-fuel mixture. Thereby, an area for recirculation of some combustion products toward the burner root occurs in furnaces of the reverse flame type with an external formation of air-fuel jet. On the axial plane of the fire tube, this zone is provided by elliptical vortexes along both flare sides (Fig. 3), whereas spatially it is a single vortex in a toroidal form whose rotation axis coincides with the burner's central axis (Fig. 4).

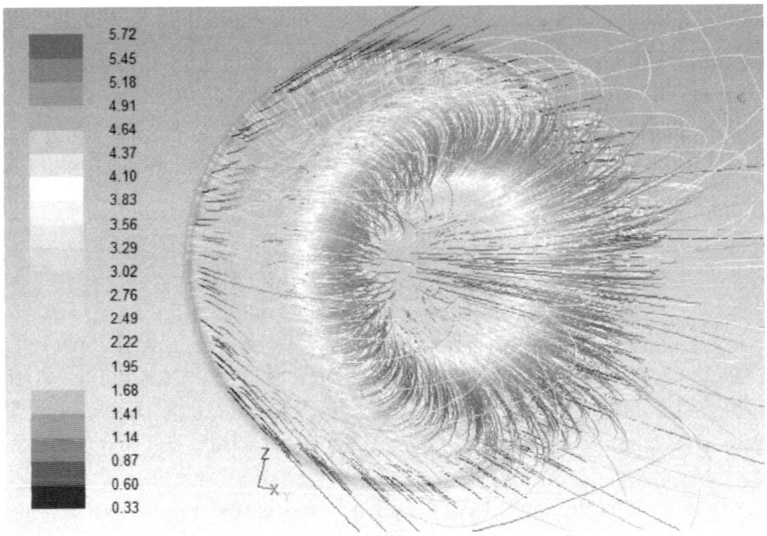

Fig. 4. Pathlines in the recirculation area with color indexation according to the kinetic energy of turbulence (m^2/sec^2)

The gas mixture in the recirculation area consists of hot combustion products: carbon dioxide, nitrogen and its oxides, water vapor. Their temperature is close to that in the burner core far exceeding 1000 °C. For this reason, the new fuel-air mixture, blending with recirculation gases, almost immediately heats up to 270 °C. The rotation frequency of the whirl recirculating towards the burner slot is approximately 2.5 revolutions per second. The mass flow of whirl gases is 0.057 kg/sec making 30% of the mass entering through the fuel-air mixture burner.

Isolines of static gauge pressure (Fig. 5) quantitatively demonstrate its decline in longitudinal section of the furnace. The total pressure drop along the length of the flare is 34 Pa (from 200 Pa to 166 Pa) meanwhile more that 70% from this value is lost in the first section beyond the burner caused by form loss. At the periphery of the flare the pressure declines from 166 Pa to 156 Pa (about 23% from total aerodynamic furnace resistance).

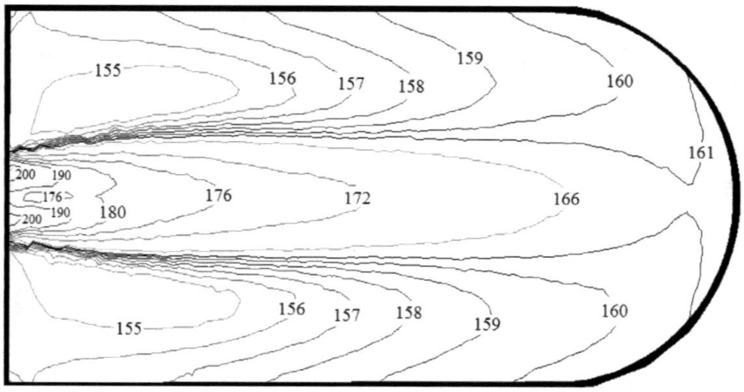

Fig. 5 Isolines of static gauge pressure (Pa) in the furnace

As can be concluded from above that resistance in the fire tube with the reverse flame can generally be estimated by form loss at the output of burner and by resistance created by vortex from the recirculation area. Turn and friction resistance values along the length of the fire tube are not significant.

The aerodynamic resistance of the fire-tube (Δp) with reversive flame can be determined by calculations according to the normative method [5] through the average speed of air-fuel jet ϖ :

$$\Delta p = \zeta \frac{\varpi^2 \rho}{2}, \text{ Pa}$$

Use of vortical burner devices intensifies burning and enables the process at a low air excess, but increases aerodynamic furnace resistance in comparison with the direct-flow organization of fuel supply. Quantitatively, the turbulence intensity is estimated by the n twist parameter, which is calculated from velocity field and furnace environment pressure values. Calculations and computer simulations have displayed that in case of direct-flow and weakly swirled air-fuel mixture supply ($n \leq 0,9$) the aerodynamic resistance coefficient ζ turned into a function little depending on the swirling degree, and thus, with acceptable accuracy, it can equal 1.35 in a wide range of design characteristics. With an increase in swirling $n > 0.9$, increase of aerodynamic resistance coefficient occurs in a fire tube with various intensity depending on different design characteristics. This is caused by gas recirculation intensification in the reverse flame.

Changes in air excess in the interval of values approaching one do not influence the coefficient of aerodynamic resistance.

Summary

Using the ANSYS software obtained a detailed picture of the turbulent combustion in dead-end firetube, which is suitable for qualitative analysis of singularities in the reversible flame.

In accordance with the computer simulation results, it is possible to conclude that the processes occurring in the furnace of the boiler under study are axially symmetric. Gases pathlines of the furnace medium change its direction twice. First, the new fuel-air mixture moving in the direction of the furnace heel, warms up, inflames and burns off; then combustion products flow from the center of the chamber to the periphery, and finally, flue gases along the fire-tube wall return to the front segment of the boiler and leave the furnace chamber. Aerodynamic resistance in the fire tube with the reverse flame can generally be estimated by form loss at the output of burner and by resistance created by vortex from the recirculation area. Turn and friction resistance values along the length of the fire tube are not significant.

For further development required to investigate a dependence of construction and mode characteristics of the boiler on aerodynamics and thermal parameters applied to dead-end firetubes.

The research project was conducted within the framework of Government Order 2.1322.2014.

References

[1] Gil A V and Starchenko A V 2012 *Thermophysics and Aeromechanics* vol. **19**, Issue 3, pp. 503-519.

[2] Khaustov S A Zavorin A S and Fisenko R N 2013 *Bulletin of Tomsk Polytechnic University* **4** pp.43-47.

[3] Taylasheva T S 2009 *Bulletin of Tomsk Polytechnic University* **4** pp. 42-47.

[4] Konyashkin V F Modeling of physical processes in the annular furnace with the aid of the FLUENT program, in: *Solid Fuel Combustion*: Proc. VI All-Russian Conf., Part 1, Inst. Thermophys. SB RAS, Novosibirsk, 2006, P. 170-178.

[5] *Aerodinamicheskiy raschet kotlov (Normativny metod)*. Saint-Petersburg: Energiya, 1977. 256 p. ("*Aerodynamic Boiler Design (The Normative Method)*")

IOP Conf. Series: Materials Science and Engineering **66** (2014) 012030 doi:10.1088/1757-899X/66/1/012030

Study of accuracy and surface roughness of holes in comparative testing of small diameters gun drills

S V Kirsanov [1 a], **A S Babaev** [1 b]
[1] *Tomsk Polytechnic University, Main Building of TPU, 30 Lenin Avenue, Tomsk Tomsk, Russia*
e-mail: [a]tmri@tpu.ru, [b]temkams@mail.ru

Abstract. The results of industrial tests of gun drills with the diameter of 2.05 mm with carbide and with steel stems are given. It is shown that the gun drills with cemented carbide stem and especially AlTiN coating have a greater resistance and greater accuracy diametrical holes sizes in comparison to gun drill with steel stem.

Introduction

In the process of manufacturing fuel injection equipment for diesel engines there is a need for deep holes drilling of small (≤ 5 mm) diameters. The mentioned above holes have high demands of accuracy (IT 6-9), surface roughness (Ra 0,08 ... 1,25) and drill runoff (0,01 ... 0,03 mm/100 mm). Such small diameters of holes and the specified requirements for them are achieved by the use of tools with certainty based - gun drills. Gun drills with steel (GDSS) or cemented carbide (GDCCS) stems and deposited on the working parts of the wear-resistant coatings are mostly used for this purpose nowadays [4-7].

THEORETICAL BACKGROUND

GDSS consists of carbide cutting tip 1, the steel stem 2 (steel 30HGSA) and the shank 3 (Figure, 1a). GDCCS tip and stem are a single cemented carbide rod 2 (produced by extruding) soldered to the shank 3 (Figure, 1b), or carbide rod that is both the stem and the shank [1, 3].

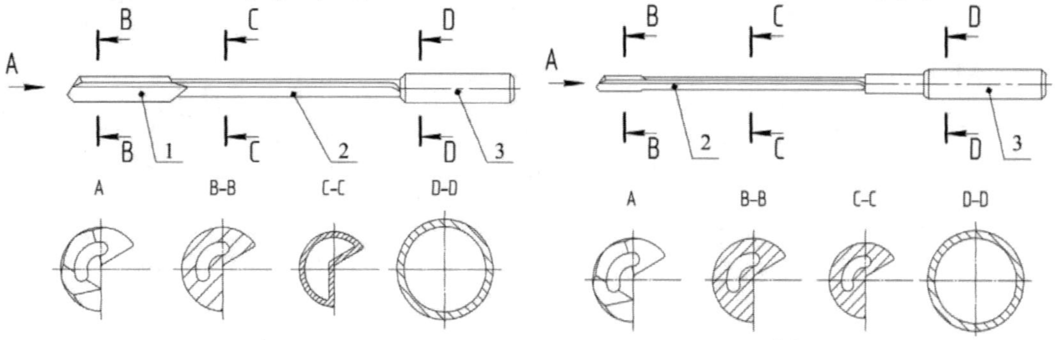

a) b)

Figure 1. Gun drills with steel a) and cemented carbide b) stems.

Thanks to the use of cemented carbide (E=6,20…6,45·10^5 MPa) [2] having the modulus of elasticity of more than 3 times higher than the modulus of elasticity of steel 30HGSA (Table 1) (E=1,98…2,04·10^5 MPa) GDCCS as compared with GDSS possess greater strength under torsion and bending as well as greater rigidity and resistance in the transverse direction. These advantages have made the company «Botek» (Germany) produce GDCCS the smallest diameter of 0.47 mm (GDSS from it is 2.0 mm), while the drill diameter> 1mm - significantly improve the processing performance by increasing drilling regimes [3]. Cylindrical cemented carbide rods with eccentrically located kidney-shaped direct channel, which serves to supply cooling (Figure, 1b) are used for manufacturing GDCCS [1].

Table 1. Alloy compositions in the stem of GDSS.

C, %	Si, %	Mn, %	Ni, %	S, %	P, %	Cr, %	Cu, %	Fe, %
0,28…0,34	0,9…1,2	0,8…1,1	<0,3	<0,025	<0,025	0,8…1,1	<0,3	~96

EXPERIMENTAL DETAILS

A special study was made to assess the technological capabilities of the drill resistance and accuracy of diametrical sizes. It was conducted in a production environment by drilling in automatic calcium steel brand AC41Cr4 of holes $2^{+0,1}$ mm with the length 40 mm on the quadruple machine brand ML 250-4-850 Tiefbohrtechnik (Germany). Gun drills with 2,05*h*5 mm in diameter with steel and cemented carbide stems produced by «Botek» and Transet (Russia, Saint-Petersburg) were used as cutting tools. The drawings and technical requirements of the mentioned instruments have been developed in Tomsk Polytechnic University. Working parts of 9 GDCCS were deposited with wear-resistant coatings *TiN*, *AlTiN*, *TiCN* (Table 2).

Table 2. Gun drills used in the tests.

The type of a gun drill	Coating	The number of units
Gun drill with cemented carbide stem (GDCCS)	Uncoated	3
	TiN	3
	AlTiN	3
	TiCN	3
Gun drill with steel stem (GDSS)	Uncoated	3

Drilling regime: spindle speed n=14000 min^{-1} (cutting speed V=90 m/min); feed S_M=64 mm/min; oil pressure coolant brand MP-3 with Technical Condition 0258-041-57518521-2011 p=9…13 MPa.

RESULTS AND DISCUSSION

37552 holes were drilled by 15 drills during the test. It indicates the high quality of the testing instruments (Figure 2).

So, the average limit (until catastrophic wear) tool life of GDCCS without coating was 3128 holes, with TiN coating – 1885 holes, with AlTiN coatings – 2567 holes and TiCN coatings – 3152 holes, and uncoated GDSS – 1770 holes. Moreover, all the GDSS tests were withdrawn due to breakdown while only 5 drills (42%) with cemented carbide stem were destructed. The remaining 7 drills (58%) were not used in the test due to the appearance of burrs at the exit holes and could drill more.

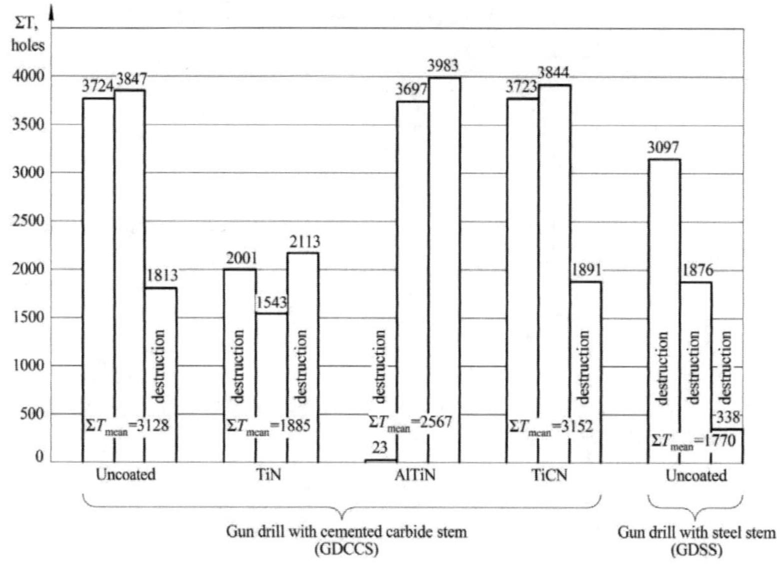

Figure 2. The influence of the gun drill type and the kind of wear-resistant coating on the total ΣT and the average ΣT_{mean} tool life.

After the resistance test each first and the last one in a thousand of the drilled workpieces was selected to cutoff into the rings with the width of 7 mm. On the both sides and in the middle of each ring limit (D_{max} and D_{min}) diameters of holes were measured by the high accuracy hole-gauge Mitutoyo with the precision of 1 μm.

Measurements showed that as the number of drill holes (drill wear) increases the average diameter $D_{av} = (D_{max} + D_{min}) / 2$ in each section of the holes also increases, and its fluctuations on length holes grow and as we approach the time of limit (catastrophic) wear the fluctuations reach the maximum (Figure 3).

In this case, the greatest diameter of the holes fluctuations occur during drilling with GDSS, and the lowest - in drilling with GDCCS with AlTiN coating.

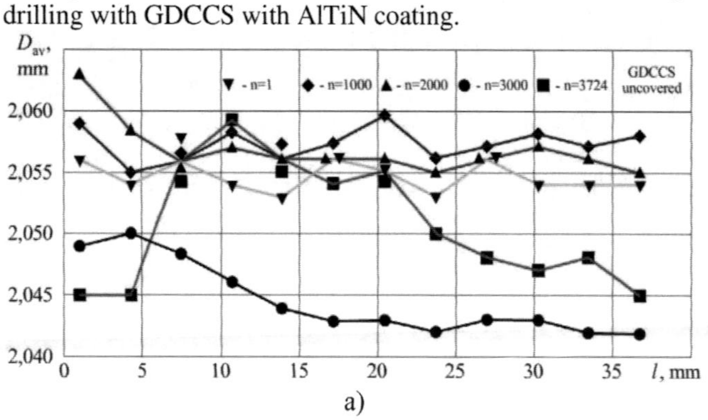

a)

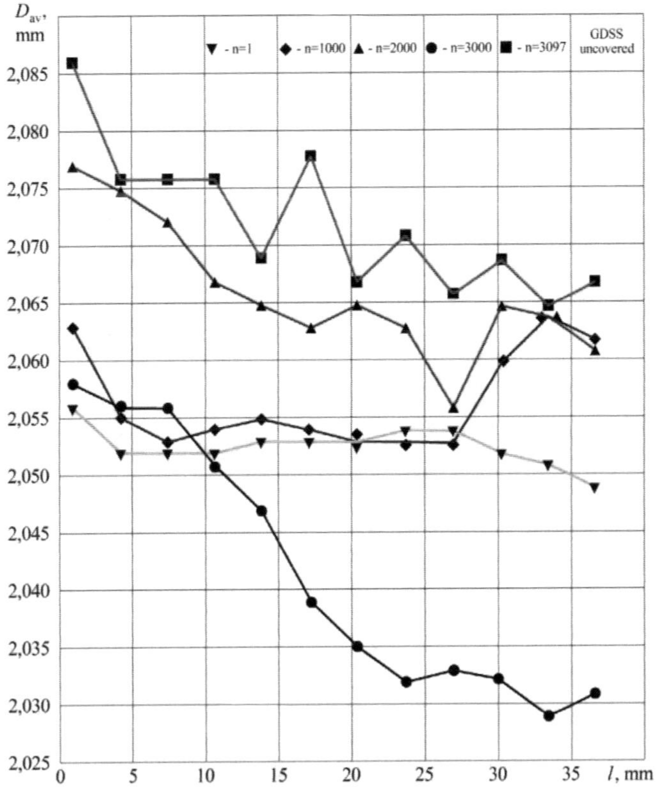
D_av, mm
- n=1
- n=1000
- n=2000
- n=3000
- n=3097
GDSS uncovered
2,085
2,080
2,075
2,070
2,065
2,060
2,055
2,050
2,045
2,040
2,035
2,030
2,025
0 5 10 15 20 25 30 35
l, mm

b)

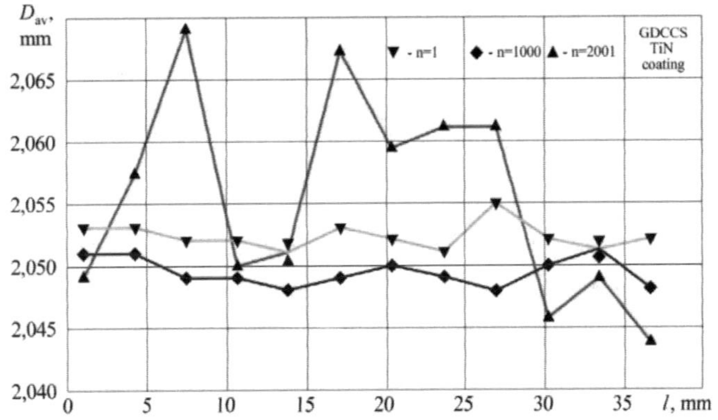
D_av, mm
- n=1
- n=1000
- n=2001
GDCCS TiN coating
2,065
2,060
2,055
2,050
2,045
2,040
0 5 10 15 20 25 30 35
l, mm

c)

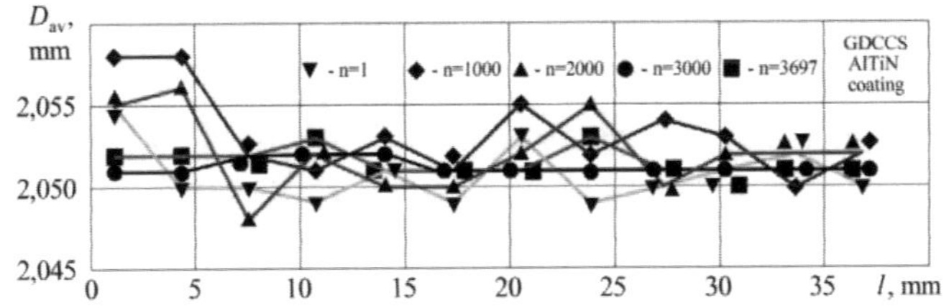
D_av, mm
- n=1
- n=1000
- n=2000
- n=3000
- n=3697
GDCCS AlTiN coating
2,055
2,050
2,045
0 5 10 15 20 25 30 35
l, mm

d)

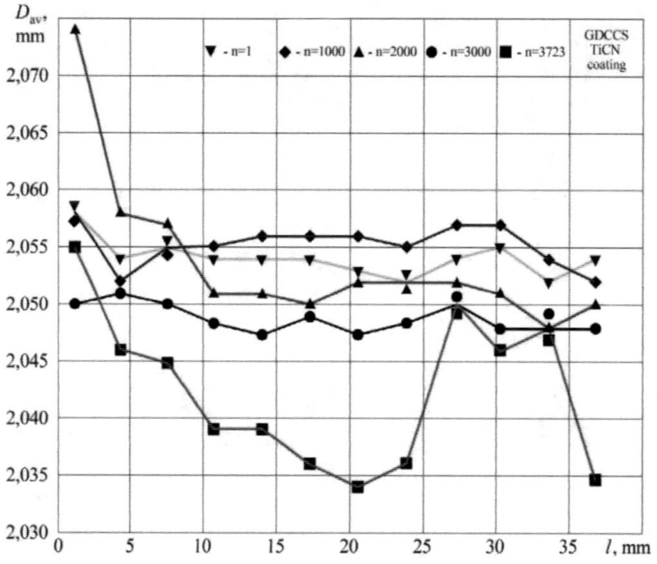

e)

Figure 3. The fluctuations of average diameter holes D_{av} along its length l depending on the amount of the number of drilled preforms, gun drill design and the type of the wear-resistant coating.

After all of this milling workpiece for measuring arithmetic mean value Ra on non-contact profilometer Micro Measure 3D Station produced by STIL (Figure 4). All measurements of roughness were performed in CMPM in Tomsk Polytechnik University. For the analysis was used special program Surface Map 1.0.3. The resolving power of the laser was set at 1 μm, the speed of movement of the laser beam is 0,03 mm / sec. Tracing length was 2 mm and was chosen from recommendations by ISO 4288-1996 and ISO 16610. In order to isolate the roughness used filtering Gaussian function with a cutoff value equal to 0,08 mm.

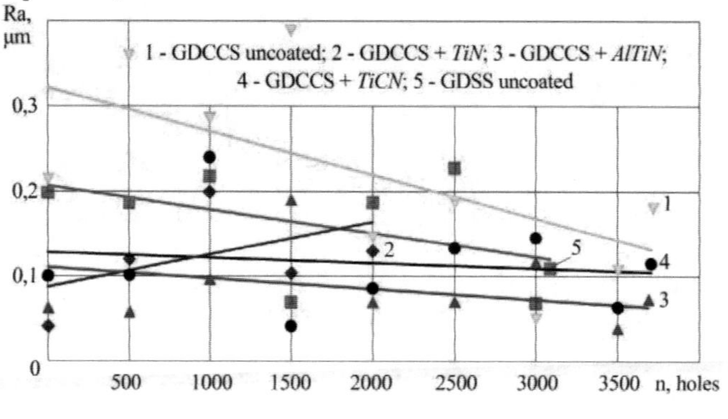

Figure 4. Fluctuations the surface roughness Ra of holes depending on the number n drilled holes, design of gun drills and the type of wear-resistant coating.

SUMMARY

1. Tests showed the high quality of gun drills made by «Botek».
2. Tool life of GDCCS due to the greater stiffness and strength of 1,5 ... is 1,8 times higher than the tool life of GDSS.
3. GDCCS with AlTiN coating provide the greatest accuracy and stability of diametrical hole sizes.
4. Wear-resistant coatings are reduces holes roughness Ra of 1,5… is 3,0 than uncoated.

The authors express deep gratitude to the leadership of the company Botek and Transet for the provided gun drills for testing.

REFERENCES

[1] Kirsanov S V, Babaev A S 2013 Small-diameter hard-alloy bits for gun production *Russian Engineering Research.* **33** (3) p. 170-171.

[2] The designer's guide to tungsten carbide 2007 Group of authors of company General Carbide (USA) August, 52 p.

[3] Eichler R 1996 Prozesssicherheit beim Einlippenbohren mit kleinsten Durchmessern. Diss. – Stuttgart.: Institut für Werkzeugmaschinen 156 p.

[4] Astakhov V P, Galitsky V V 2005 Tool life testing in gundrilling: an application of the group method of data handling (GMDH). *International Journal of Machine Tools and Manufacture* **45** 4–5 p. 509–517.

[5] Astakhov V P, Galitsky V V 2008 The combined influence of various design and process parameters of gundrilling on tool life: experimental analysis and optimization. *The International Journal of Advanced Manufacturing Technology* **36** 9 64 p.

[6] Y Wang, X Yan, B Li, G Tu 2012 The study on the chip formation and wear behavior for drilling forged steel S48CS1V with TiAlN-coated gun drill. *International Journal of Refractory Metals and Hard Materials* **30** 1 p. 200-207.

[7] T Cselle, A Barimani 1995 Today's applications and future developments of coatings for drills and rotating cutting tools *Surface and Coatings Technology* **76-77** p. 712-718.

MTT'2014 IOP Publishing
IOP Conf. Series: Materials Science and Engineering **66** (2014) 012031 doi:10.1088/1757-899X/66/1/012031

The Radiation Dose Determination of the Pulsed X-ray Source

I Miloichikova, S Stuchebrov, G Zhaksybayeva, A Wagner

Tomsk Polytechnic University, Tomsk, Russia

E-mail: ircha1988@mail.ru

Abstract. In this paper the radiation dose measurement technique of the pulsed X-ray source RAP-160-5 is described. The dose rate measurement results from the pulsed X-ray beams at the different distance between the pulsed X-ray source focus and the detector obtained with the help of the thermoluminescent detectors DTL-02, the universal dosimeter UNIDOS E equipped with the plane-parallel ionization chamber type 23342, the dosimeter-radiometer DKS-96 and the radiation dosimeter AT 1123 are demonstrated. The recommendations for the dosimetry measurements of the pulsed X-ray generator RAP-160-5 under different radiation conditions are proposed.

1. Introduction

At present time there is a wide range of various X-ray sources for the radiographic analysis in the nondestructive and medical examinations [1]. However, all these techniques have limitations related to the radiation dose for the biological objects. Nowadays the issue of the radiation doses estimation of the biological objects in the radiographic examination is relevant for different types of the X-ray diagnostics. One of the dose reduction techniques is use of the pulsed X-ray source synchronized with the detecting device.

Such X-ray visualization devices based on the different linear detectors were developed in the Applied Physics Department at Tomsk Polytechnic University. The emitting source in both devices is a pulsed X-ray generator RAP-160-5. These setups can be used for receiving shadow projection images and tomograms. The purpose of developing experimental setups is using obtained results in medicine [2].

One of the main aims of the development is to find out the radiation burden estimation methods of the pulsed X-ray sources.

The dose measurement from a pulsed source has a number of difficulties. Due to the fact that the X-ray tube RAP-160-5 emits only 2% of the time the current pulse is relatively high – about 50 times more integral values. Consequently, the dosimetric equipment should be capable of working with a sufficiently high dose rates. In addition, the pulse-repetition rate is high enough and the profile of the beam intensity is fast-changing. Thus, specific requirement for response rate of the recording equipments are arisen in these conditions [3].

The research objectives are:

- to measure the radiation doses of the pulsed X-ray source using the thermoluminescent detectors, the plane-parallel ionization chamber, the dosimeter-radiometer DKS-96 and the radiation dosimeter AT 1123;
- to compare obtained results;

- to elaborate recommendations for future dosimetry measurements of the pulsed X-ray generator RAP-160-5.

2. Materials and methods

2.1. Emitting source
The irradiation was performed using the pulsed X-ray generator RAP-160-5. The main parameters of the X-ray generator are: the anode voltage varies from 40 to 160 kV; the anode current varies from 0.4 to 5 mA; the peak X-pay tube power is 0.6 kW; the impulse frequency radiation varies from 60 to 700 Hz; the duration of one pulse is about 140 μs; the pulse form is nearly rectangular; the focal spot size is 1.2 ×1.2 mm. The X-ray generator RAP-160-5 can be synchronized with the other devices [4].

2.2. Dosimetry equipment
The main problem associated with the dosimetry of the pulsed X-ray radiation is the response rate of the dosimeters. This problem can be solved by using storage detectors. One of this can be the thermoluminescent detectors, which can be used for accurate measurements of dose distribution and estimate the integrate dose in the medical examination.

In the experiment the solid thermoluminescent detectors DTL-02 (thermoluminescent material – LiF: Mg, Ti) designed for personal dosimetry of staff were used. These dosimeters comply with requirements of the task. Operating limits of the dose rates and absorbed doses correspond to the operation conditions. These detectors DTL-02 can work in the energy range 15 keV – 10 MeV of gamma-radiation, the dose range is 20 μSv – 10 Sv [5].

For diagnostic radiological measurement of air kerma the most common type is a plane parallel chamber. Plane-parallel ionization chambers are acceptable for absolute dosimetry of soft X-ray beams. In the experiment the universal dosimeter for radiation therapy and diagnostic radiology UNIDOS E equipped with a PTW soft X-ray plane-parallel ionization chamber type 23342 (the energy range is 7.5 – 70 keV; the dose range is 3 mGy – 30 Gy; the measuring volume – 0.02 cm^3) was used [6, 7].

In the experiment the dosimeter-radiometer DKS-96 was used as a dosimetry equipment. This dosimeter is used in medical institutions for task-related and routine monitoring of radiation environment. The dosimeter-radiometer provides the measurement of exposure dose rate of continuous and pulsed X-ray and gamma-radiation. The dosimeter-radiometer DKS-96 can work in the energy range 15 keV – 10 MeV of X-ray and gamma-radiation, the dose range is 0,1 μSv – 10 Sv [8].

In the experiment the radiation dosimeter AT 1123 was used as a dosimetry equipment. This dosimeter is tissue-equivalent detector. The dosimeter can be used for measuring continuous, short-term and pulsed X-ray and gamma radiation. The radiation dosimeter AT 1123 can work in the energy range 15 keV – 10 MeV of X-ray and gamma-radiation, the dose range is 10 nSv – 10 Sv [9].

2.3. Experimental setup
The irradiation was produced by the X-ray generator RAP-160-5 according to the scheme shown in figure 1.

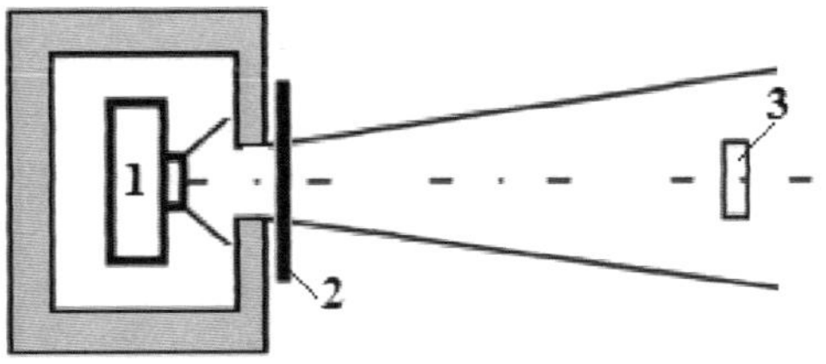

Figure 1. The scheme of irradiation of the dosimeters at the X-ray generator RAP-160-5.

Between the X-ray source (1) and the detecting device (3) an aluminum filter (2) was installed (3.0 mm thick) to reduce the contribution of soft X-rays of the spectrum, which did not correspond to the dosimeters operating modes (figure 1). The distance between the X-ray source focus and the detector varied from 40 cm to 60 cm.

The radiation was produced by the following parameters of the X-ray tube:

- the anode voltage was 70 kV;
- the anode current varied from 0.6 to 3.5 mA.

2.4. The radiation doses measurements of the pulsed X-ray source

According to the scheme on figure 1 the thermoluminescent detectors DTL-02 was used for the dose measurement of the pulsed X-ray source. The dose rate values were calculated. The measurements were carried out by three different DTL-02 dosimeters at the equal parameters of the X-ray tube, and then the average values were calculated.

In the second part of the experiment the dose and the dose rate measurements were produced by the dosimeter UNIDOS E (the measurement mode was "LOW") equipped with an ionization chamber type 23342.

According to the experimental set up in the figure 1 the dosimeter-radiometer DKS-96 was used for the dose and the dose rate measurement of the pulsed X-ray source. The regimes with the X-ray tube anode current equal to 3.0 and 3.5 mA at the distance between the X-ray source focus and the detector equal to 50 cm were out of operating mode of the dosimeter-radiometer DKS-96. The results received with these regimes have shown poor accuracy.

In the last part of the experiment the dose rate measurement was produced by the radiation dosimeter AT 1123. The dosimeter was used in the pulsed measurement mode. The sensitivity range switching occurs automatically. The resulting dose rate data from the dosimeter AT 1123 are over 3-fold more than the data from the other types of detectors.

3. Results and discussion

On figure 2 and figure 3 the dose rate dependences on the anode current of the pulsed X-ray tube are presented. The results were obtained with the help of the solid thermoluminescent detectors DTL-02, the plane-parallel ionization chamber type 23342, the dosimeter-radiometer DKS-96 and the radiation dosimeter AT 1123. The measurements were carried out in the air. The dose and the dose rate measurement results for the equal X-ray tube operating mode are average over three different DTL-02 dosimeters for accuracy increase. Results are presented for the distance between the X-ray source focus and the detector equal to 50 cm (figure 2) and 60 cm (figure 3).

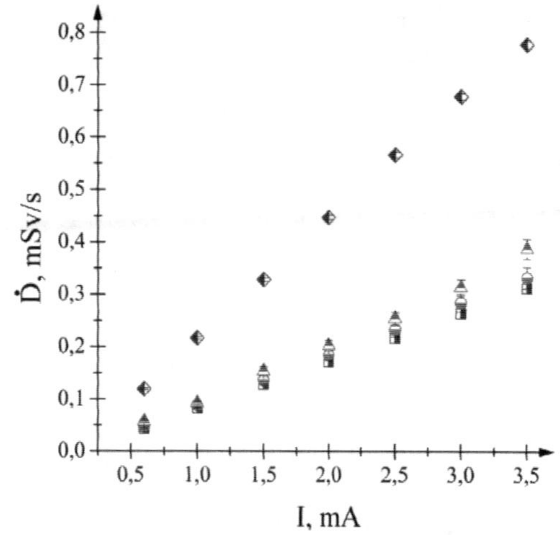

Figure 2. The dependence between the dose rate of the pulsed X-ray beam and the anode current at the distance between the X-ray source focus and the detector equal to 50 cm (the anode voltage equal to 70 kV): ▲ – the detector DTL-02 results; ◨ – the plane-parallel ionization chamber results; ◒ – the dosimeter-radiometer DKS-96 results; ◈ – the radiation dosimeter AT1123 results.

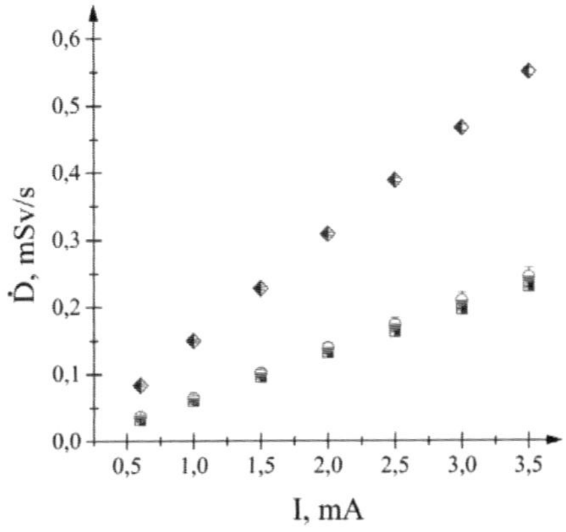

Figure 3. The dependence between the dose rate of the pulsed X-ray beam and the anode current at the distance between the X-ray source focus and the detector equal to 60 cm (the anode voltage equal to 70 kV): ◨ – the plane-parallel ionization chamber results; ◒ – the dosimeter-radiometer DKS-96 results; ◈ – the radiation dosimeter AT1123 results.

The results show that the dose rate measurements obtained with the help of the thermoluminescent detectors, the plane-parallel ionization chamber and the dosimeter-radiometer DKS-96 are in a good agreement. One can see that the plane-parallel ionization chamber can be used for dosimetry measurements of the pulsed X-ray generator RAP-160-5 under demonstrated radiation conditions.

However, for obtaining better results, the presented X-ray tube parameters of the pulsed X-ray generator RAP-160-5 using the dosimeter-radiometer DKS-96, the distance between the X-ray source focus and the detector should be more than 50 cm. The dose rate measurements obtained with the help of the radiation dosimeter AT 1123 show that other X-ray tube parameters should be used.

4. Summary

In this study, some methods of radiation dose measurement of the pulsed X-ray generator RAP-160-5 are presented. Experimental results show the advisability of using the solid thermoluminescent detectors DTL-02 and the plane-parallel ionization chamber for future dosimetric measurements of the pulsed X-ray generator RAP-160-5 under different radiation conditions. For example, the solid thermoluminescent detectors DTL-02 can be used for the measurement of the radiation burden of the X-ray source in the process of stabilization and the plane-parallel ionization chamber type 23342 can be used for the measurement of the dose rate spatial distribution produced by the pulsed X-ray source.

Acknowledgements
This work was partly supported by Competitiveness Enhancement Program of Tomsk Polytechnic University.

References

[1] Cherepennikov Yu and Gogolev A 2014 Device for X-ray spectral absorption analysis with use of acoustic monochromator *J. Phys.: Conf. Ser.* **517** 012037
[2] Stuchebrov S, Batranin A, Verigin D and Wagner A 2014 Setups for tomographic imaging with submillimeter spatial resolution *J. Phys.: Conf. Ser.* **517** 012046
[3] Stuchebrov S, Wagner A, Suhih E 2013 The radiation doses estimation of the biological objects in the radiographic examination *Izvestija vuzov. Fizika* **56** 11/3 pp 269-272 [in Russian]
[4] Operations manual 2008 Portable X-ray device for the industrial utilization 34 p. [in Russian]
[5] Technical specification ZhBIT2.805.006RJe 2012 Thermoluminescent dosimeters DTL-02 13 p. [in Russian]
[6] Instruction manual D545.131.00 2006 UNIDOS E PTW – universal dosimeter 61 p.
[7] Instruction manual D564.131.0/1 2006 Soft X-ray chamber Type 23342 15 p.
[8] Operations manual TE1.415313.003RJe 2005 Dosimeter-radiometer DKS-96 62 p. [in Russian]
[9] Operations manual 2009 X-ray and gamma-radiation dosimeters DKS-AT1121, DKS-AT1121A, DKS-AT1123, DKS-AT1123A 53 p. [in Russian]

IOP Conf. Series: Materials Science and Engineering **66** (2014) 012032 doi:10.1088/1757-899X/66/1/012032

Researching the possibility to use Fricke dosimeter for measurement of absorbed dose generated by pulse electron beam

Alena Khusnulina

Tomsk Polytechnic University, Tomsk, Russia

E-mail: alenaxyc@gmail.com

Abstract. The focus of this article is the working range of Fricke dosimeter and modified Fricke dosimeter. This study is motivated by two research questions: (1) Is it possible to use Fricke dosimeter or modified Fricke dosimeter as an instrument for dose measurements when absorbed dose is generated by pulse electron beam? (2) What limitations do these methods have? This article tells in detail about the solutions preparing and chemical analyses. It shows that modified Fricke dosimeter may be used for measurements of absorbed dose generated by pulse electron beam up to 11 kGy. And Fricke dosimeter leads to incorrect results when local dose in solution upper than 500 Gy.

1. Introduction

A successful result of every radiochemical investigation depends, first of all, on right measurement of the absorbed dose as dosimentry represents a quantitative aspect of radiation chemistry. So, every radiochemical experiment starts with dose measurements [1]. An absorbed dose in the irradiated object that corresponds to the optimal yield for this radiation-chemical process, determines its energy capacity and consequently economic efficiency of process realization in industrial construction.

More often the absorbed dose of electron radiation is evaluated by measured beam characteristics in certain operating regime of an accelerator and known characteristics of irradiated material. But accuracy of the computational method is not enough when the beam utilization factor coefficient is less than 1. To increase accuracy of the absorbed dose measurements, standardized dosimetric systems are used.

Absorbed dose measurements by experimental methods have a set of advantages especially when the dosimetric system with its characteristics (state of aggregation, effective atomic number, depth of detecting layer and other) is close to the characteristics of irradiated object.

2. Theoretical background

As a rule, in radiation chemistry for dose measurements chemical methods are used and the ferrosulfate dosimetric system (Fricke dosimeter) that has especially extensive application. It is a water solution of (1-5)·10-3 M soured ferrous sulfate FeSO4·7H2O in 0.4 M H2SO4 saturated by air. Ferrous iron is oxidized to ferric iron when the solution is ionized by radiation [2].

For preparing 200 ml of the Fricke dosimeter solution 4.4 ml of high-concentration sulphuric acid H2SO4 were used and 56 mg of iron vitriol FeSO4·7H2O. It provides for 56 mg/l ions of Fe2+ and

0,4 M H2SO4 in the solution. For preparing 200 ml of the modified Fricke dosimeter solution 0.11 ml of high-concentration sulphuric acid H2SO4 were used, 56.6 mg of iron vitriol FeSO4·7H2O and blue copperas 5 g. It provides for 56 mg/l ions of Fe^{2+}, 0.01 M H_2SO_4 and 0.1 M Cu^{2+} in the solution.

For determining the concentration of ferrous iron the standard method is used. Using this method 1 ml of the interested solution was sampled then it was equalized to level 4.5 by adding 5 ml of acetate buffer (pH 4.5). After 1 ml of 0.1 % solution of o-phenanthroline is added and reaches 50 ml by pouring distilled water into the painted solution. 30 minutes later the test portion is exposed to colorimetry with the following parameters of photoelectric colorimeter: wavelength is 490 nm and cuvet depth – 3 cm [3].

Determining the ferric iron in the solution is carried out with the help of the standard method too. To the 1 ml of interested solution 1 ml of ammonium muriate (2 M solution), 1 ml of sulphosalicylic acid (20 % solution) and 0.1 ml of hydrochloric acid (3:2) are poured. 50 ml of the painted solution is obtained by pouring distilled water. Then the sample is exposed to colorimetry: wavelengths 540 nm and cuvet depth 3 cm [3].

Recomputation of optical density was done with the equation (1) [4-6].

$$C = \frac{D - D_0}{\Sigma} \cdot \frac{V_{retort}}{V_{sample}} \cdot M\ Fe\ , \tag{1}$$

where C – ferrum concentration, mg/l; D, D_0 – optical density of sample and blank corresponding; Σ – extinction coefficient; V_{retort} – volume of painted solution, ml; V_{sample} – sample volume, ml; $M[Fe]$ – ferrum molar weight, mg/mole.

The upper limit of dose measurements is determined by exhaustion of oxygen in the solution. Thus, for the saturated by air solution with $G(Fe^{3+})$ = 16 ion/100 eV the upper limit is 500 Gy. In this paper Fricke dosimeter was used for the dosimetry of pulse high-current electron beam, generated with the accelerator "Astra-M", with the following parameters: accelerating potential – 450 kV, pulse energy – 1.2 J, half-amplitude duration – 150 ns [2].

The absorbed dose was calculated with the equation (2).

$$D = \frac{cN \cdot 100}{G\rho}, \tag{2}$$

where D – absorbed dose, Gy; c –concentration, M; N – Avogadro constant; $G(Fe^{3+})$ – radiation-chemical yield, ion/100 eV; ρ – density, kg/l.

The absorbed dose may also be computed with the equation (3), given the mass of a sample and the transferred energy [4-6].

$$D = \frac{Q}{M}, \tag{3}$$

where M – mass of sample, kg; Q – energy that was transferred to the sample, J.

The sample mass could be calculated if given its density and volume (in this paper we refer to water solution). And the energy that was transferred by an electron beam could be measured with the help of a calorimeter.

3. Experiments

3.1. Fricke dosimeter

The experiment was carried out as follows. Firstly, the cuvet with the diameter 90 mm was filled with Fricke dosimeter solution with the layer depth range of 0.5 mm, 1 mm, 1.5 mm and 2 mm. Then the cuvet was disposed in front of accelerators output window and irradiated. For every depth of Fricke dosimeter solution the series of experiments were carried out with number of pulses varying from 1 to 5. In the experiment the energy in the pulse was measured by the calorimeter and was equal to 1.2 J.

After that chemical analysis was carried out to define the concentration of ferrous and ferric iron in the solution.

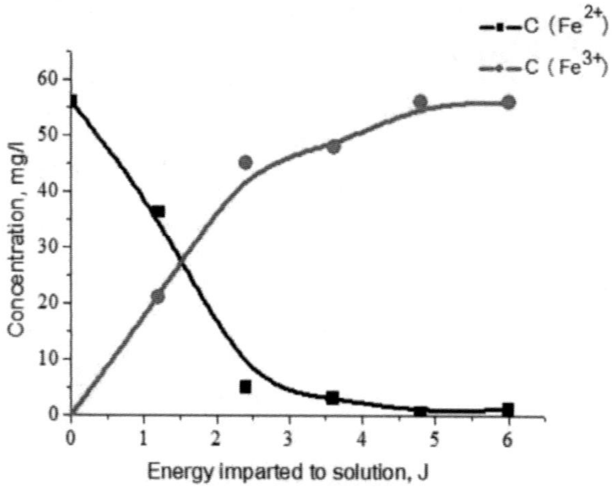

Figure 1. Concentrations Fe^{2+} and Fe^{3+} (solution depth is 0.5 mm)

In the figure 1 a graph of concentrations ferrous and ferric iron is shown as a function of energy transferred to solution when solution depth was 0.5 mm. Under the influence of pulse electron beam the concentration of ferrous iron decrease depends on the energy transferred to the solution. As the concentration of ferric iron increases the one of ferrous iron decreases by the same value.

Radiochemical yield $G(Fe^{3+})$ was computed based on experiment results and equal to (9 ± 2) ion/100 eV. When radiochemical yield is about this value, the absorbed dose may be measured up to 1000 Gy.

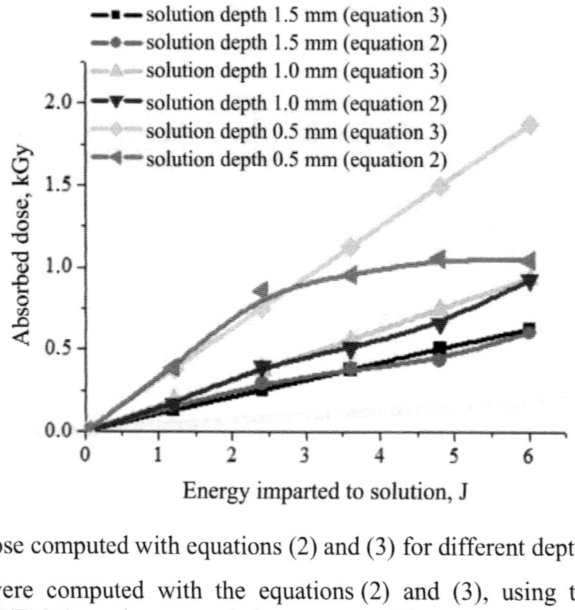

Figure 2. Absorbed dose computed with equations (2) and (3) for different depth of irradiated solution

Absorbed doses were computed with the equations (2) and (3), using the received values of radiochemical yield G(Fe3+), and measured the amount of beam energy in pulse. In the figure 2 absorbed doses calculated with equations (2) and (3) are shown. This plot confirms the possibility to use Fricke dosimeter for dose measurements of pulse electron beam but to a certain extent. The amounts of the absorbed dose for 0.5 mm solution depth become really different after the second pulse. It is connected with the conversion of ferrous iron to ferric, as was mentioned above.

3.2. Modified Fricke dosimeter

Then the experiment was carried out with modified Fricke dosimeter. It allows to compare results with literature data and to define the value of the absorbed dose in practice, which will show work limits of this dosimeter.

The experiment with modified Fricke dosimeter was carried out the same way as with Fricke dosimeter. And then by chemical methods the concentration of ferrous and ferric iron in the solution was also measured. But defining of ferric iron in this experiment, using the standard method, shows that its factor is absent in test samples: in the original sample as well as in radiation-exposed samples. It shows that for modified Fricke dosimeter this standard method is not applicable. Probably it is due to copper ions influence in the solution, because it is singular difference between modified Fricke dosimeter and Fricke dosimeter. So, in these experiments only the concentration of ferrous iron was defined.

Figure 4 shows that after the second pulse nearly all ferrous iron in Fricke dosimeter was oxidized to ferric iron. But in modified Fricke dosimeter complete consumption of ferric iron was not achieved.

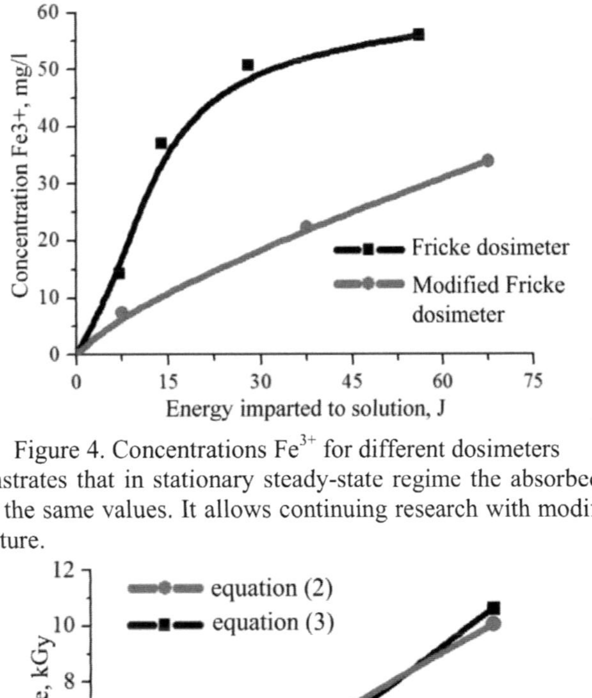

Figure 4. Concentrations Fe^{3+} for different dosimeters

The figure 5 demonstrates that in stationary steady-state regime the absorbed dose, calculated by different formulas, has the same values. It allows continuing research with modified Fricke dosimeter in flowing regime in future.

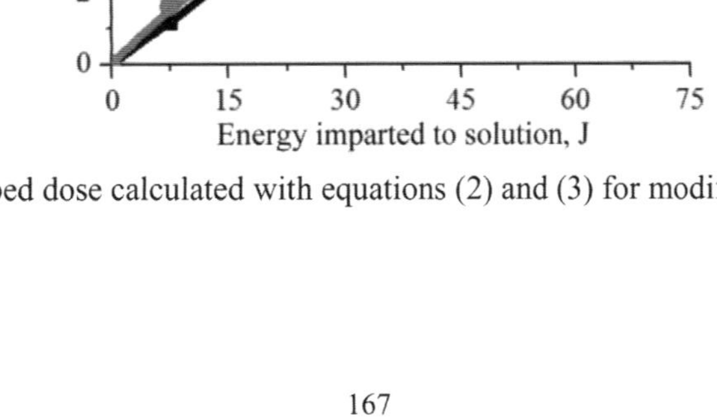

Figure 5. Absorbed dose calculated with equations (2) and (3) for modified Fricke dosimeter

4. Results

By the experiment values radiation-chemical yield was calculated and for modified Fricke dosimeter it was equal 0.62 ion/100 eV. In the literature [1, 4-5] when the absorbed dose rate is 10^{10} Gy/s the radiation-chemical yield for modified Fricke dosimeter equals 0.58 ion/100 eV. So the estimated value is relatively equal to the value from literature [1, 4-5].

For these conditions, as the experiment shows, Fricke dosimeter can not be used. For measurements of these absorbed dose values modified Fricke dosimetershould be used. The absorbed dose was calculated with equations (2) and (3), using calculated early radiation-chemical yield for modified Fricke dosimeter and measured with the calorimeter value of pulse energy. Figure. 5 shows absorbed doses and confirms that modified Fricke dosimeter can be used for dose measurements up to 11 kGy.

Produced results show that ferrosulfate dosimetric system, laced with copper ions, may be used for measurements of the absorbed dose, generated by the pulse electron beam. As well as using the ferrosulfate dosimetic system without copper ions leads to incorrect results when the local dose in solution is higher than 500 Gy. To prevent this, radiation-chemical yield should be controlled all the time. If it varies from 8 to 16 ion/100 eV Fricke dosimeter can be used.

References

[1] Pickaev A 1975 *Dosimetry in radiation chemistry* (Moscow: Science)
[2] Egorov I S 2011 Remnev G E Frequency-response pulse electron accelerator for radiation technology *Articles of 8th International conference "Nuclear and radiation physics* pp 549-555
[3] GOST 4011-72 *Drinking water. Measurement methods for mass concentration of total iron* (Moscow: GOST)
[4] Swallow A J 1973 *Radiation chemistry* (John Wiley & Sons Canada, Limited)
[5] Henly E, Jackson E 1974 *Radiation chemistry* (Moscow: Atomizdat)
[6] Chapel L 1975 Electron accelerators applying for radiation chemistry (Moscow: Atomizdat)

Catalytic decomposition of trilon B on the nickel-ferricyanide catalyst

Apalkov G.A.

Zheleznogorsk, 662970, Russia, The Mining and Chemistry Combine (GKhK)

E-mail: glebaldo@mail.ru

Abstract. The method of nickel-ferricyanide catalyst preparation is elaborated. The results of catalytic decomposition of trilon B on the prepared catalyst are presented. The developed method allows performing the efficient decomposition of trilon B (>99,9%) either in acid or in alkaline medium in temperature interval of 37-80°C. Technologically-fit dynamic conditions of the process, providing the intensity of a heterogeneous interaction, are selected and explored, and the nickel-ferricyanide catalyst resistance is achieved. The instrument flowchart of the developed method of catalytic decomposition is introduced for industrial application during the reprocessing of low- and medium-activity effluent, containing complexons.

1. Introduction

Organic complexons are widely used compounds in radiochemical technologies both during the decontamination processes and many other production activities.

In this case, huge amounts of radioactive waste (effluent), containing complexons such as ethylene diamine tetraacetate (EDTA), trilon B, oxalic and citric acids, surface-active substances, etc. are generated each year by radiochemical and nuclear power plants production. The problem is that such compounds form resistant complex compounds with radionuclides (^{60}Co, ^{137}Cs, etc.) (fig. 1), impede and reduce the efficiency of radioactive effluent reprocessing by traditional physicochemical methods (sorption, ion-exchange, precipitation).

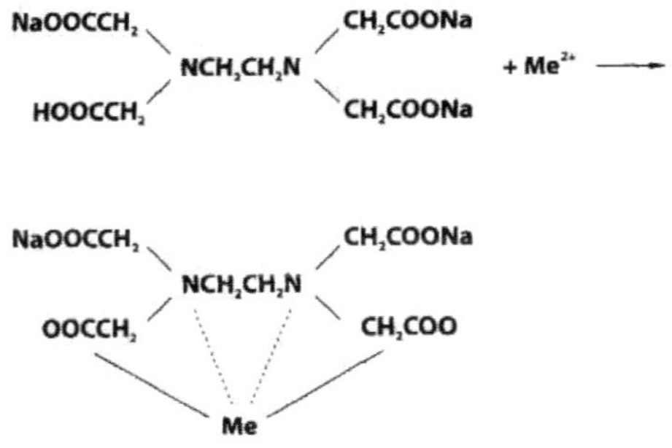

Fig. 1. – Trilon B as the decontaminating agent

Therefore, localization, concentration and reprocessing of the radioactive waste (effluent) can be considerably simplified after the decomposition and removal of the mentioned organic complexons from it.

Preliminary purification of the radioactive waste from complexons and its impurities also allows to reprocess and store the effluent from radiochemical and nuclear power plants more efficiently and to reduce the initial cost of production.

The oxidation (using O_3, H_2O_2) and photooxidation decomposition (using ultraviolet treatment and H_2O_2) are among the most efficient methods of complexons destruction [1,2]. However, the catalytic decomposition of complexons is more preferable for radiochemical technology than mentioned methods of decomposition, because it allows minimizing the power inputs, resource consumption, and amounts of secondary waste and additional operations during the processes of complexons decomposition. The methods of oxalic acid ($H_2C_2O_4$), trilon B ($C_{10}H_{14}N_2Na_2O_8$) and EDTA ($C_{10}H_{16}N_2O_8$) decomposition using the silica gel and anionite platinum-covered catalysts, activated carbons, fibrous activated carbon-base materials were explored by the authors of [3,4,5,6].

Nevertheless, it should be noted that platinoids are rather expensive metals, besides practically it is impossible to avoid metal loss during the decomposition process, and when using activated carbons the catalyst degradation was registered [4].

2. Experimental part

In present work the ferricyanide catalyst covering the "VP-1AP" anion-exchange resin surface was made (fig. 2). It was used as the heterogeneous solid catalyst for the catalytic decomposition of complexon trilon B. Ferricyanide was chosen because of its ability of catalytic decomposition of hydrogen peroxide with the formation of a hydroxyl radical (powerful oxidizing agent) and $[Fe(CN)_6]^{3-}$ anion resistance, ability of ferricyanide fixation on the anionite surface with the formation of sparingly soluble transition metals salts, and because of ferricyanide low-cost. The amount of the transition metals ferricyanide on the anionite "VP-1AP" surface was optimized by the determination of the sufficient contact time for the complete complexons decomposition using technologically-fit dynamic mode.

The catalyst preparation process included the following operations: the preparation of the ferricyanide solution ($K_3[Fe(CN)_6]$), the preliminary preparation of the "VP-1AP" surface by the alkaline treatment (NaOH solution, 40 gram per liter), mixing and heating (80°C) of the ferricyanide solution with the anionite (resin), second alkaline treatment (NaOH solution, 40 gram per liter). After prepared catalyst washing, the surface of the "VP-1AP", covered with the ferricyanide, was treated by the $Ni(NO_3)_2$ solution in order to fix the formed ferricyanide composition (a sparingly soluble nickel ferricyanide) on the surface of "VP-1AP".

Using a method of the nitrogen thermal desorption the determination of the complete specific surface of the prepared catalyst (5-7% $R\text{-}Ni[Fe(CN)_6]$, R – "VP-1AP" resin) was experimented and was 18,47±0,396 m^2 per gram. Earlier, in the article [6] it was determined that the time of the complexons catalytic decomposition (EDTA) is influenced by the available for the adsorption process surface. For the prepared nickel-ferricyanide catalyst the effective area was evaluated as the 75-85% of the complete specific surface, because of the "VP-1AP" resin application as the catalyst support.

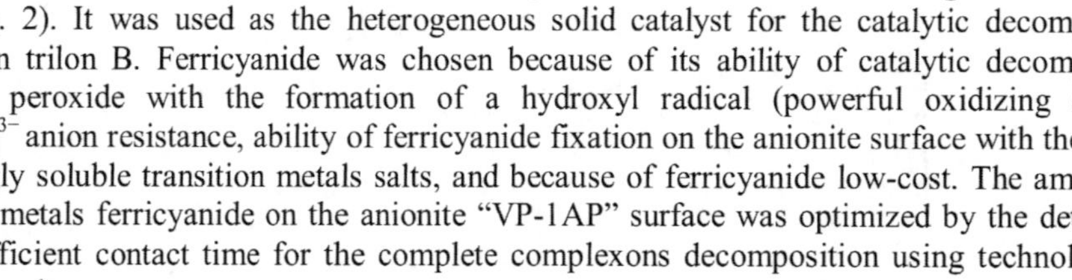

(a) (b)

Fig. 2. – "VP-1AP" anion-exchange resin (a) and nickel-ferricyanide catalyst (b)

During the experiments the following standard test technological solutions were used: solution №1 – trilon B 1,2 g/l, $NaNO_3$ 5,0 g/l, $Na_2C_2O_4$ 5,0 g/l, pH 2; solution №2 – trilon B 1,2 g/l, $NaNO_3$ 5,0 g/l, $Na_2C_2O_4$ 5,0 g/l, pH 9.

The catalytic decomposition of the complexon trilon B was implemented in a dynamic mode in the temperature-controlled column. The solutions were passed through the static bed of the nickel-ferricyanide catalyst at the rate of 7-8 columnar amounts per hour. The solution and catalyst contact time was nearly 50-60 seconds.

The catalytic column characteristics were the following: height of the static bed of the nickel-ferricyanide catalyst – 16 centimeters, inner diameter of the column – 2,7 centimeters. The columnar amount was about 90 milliliters. The ratio between the diameter of the catalyst bed to its height – 1:5,9.

The base framework of the used laboratory facility is shown in figure 3.

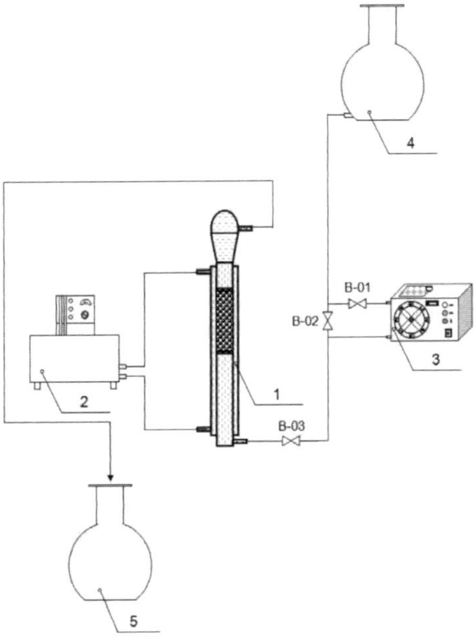

1 – Temperature-controlled column;

2 – Thermostat;

3 – Peristaltic pump;

4 – Stock solution capacitor;

5 – Receiver tank;

B-01 ÷ B-03 – needle valves.

Fig. 3. – Laboratory facility for the heterogeneous catalytic decomposition of trilon B method research

3. Results and discussion

Hydrogen peroxide, with optimal concentration of 6 gram per liter, was used as the oxidizing agent. The hydroxyl radical (•OH) with the oxidation potential of 2,8 V was formed during the interaction between hydrogen peroxide and the surface of the nickel-ferricyanide catalyst.

Table 1

Experimental results of the heterogeneous catalytic decomposition of trilon B on the nickel-ferricyanide catalyst in acid medium (solution №1)

T, °C	pH	[Trilon B], g/l	[Fe], mg/l	[Ni],mg/l	Decomposition degree of trilon B, %
-	2,0	1,1	-	-	-
37	1,7	0,056	< 0,60	46	94,90
45	1,8	0,017	< 0,60	11	98,46
52	1,8	< 0,001	< 0,58	1,7	> 99,91
60	1,9	< 0,001	< 0,56	< 0,1	> 99,91
70	2,0	< 0,001	< 0,33	< 0,1	> 99,91
80	2,0	< 0,001	< 0,30	< 0,1	> 99,91

Table 2

Experimental results of the heterogeneous catalytic decomposition of trilon B on the nickel-ferricyanide catalyst in alkaline medium (solution №2)

T, °C	pH	[Trilon B], g/l	[Fe], mg/l	[Ni],mg/l	Decomposition degree of trilon B, %
-	9,0	1,1	-	-	-
37	8,7	0,093	< 0,30	< 0,1	91,54
45	8,8	0,038	< 0,30	< 0,1	96,55
52	8,9	< 0,001	< 0,30	< 0,1	> 99,91
60	8,9	< 0,001	< 0,33	< 0,1	> 99,91
70	9,0	< 0,001	< 0,30	< 0,1	> 99,91
80	8,8	< 0,001	< 0,30	< 0,1	> 99,91

Despite the fact that the process of trilon B decomposition flows during the whole temperature interval, though a rise of temperature more than 85-90°C can cause a destruction of the organic matrix of the catalyst support ("VP-1AP" resin). That's why, the optimum condition of the heterogeneous catalytic decomposition of trilon B using the nickel-ferricyanide catalyst, concerning temperature factor, was determined as 50-60°C. Catalyst resistance was estimated relatively to the concentration of iron and nickel in the solution after the process of trilon B decomposition. The detected amounts of iron and nickel in the solution were below the detection limit of the used ICP-spectrometer. This fact confirmed the resistant stabilization of the nickel-ferricyanide catalyst on the "VP-1AP" resin surface.

Thereby, the prepared during the research nickel-ferricyanide catalyst (5-7% R-$Ni[Fe(CN)_6]$, R – "VP-1AP" resin) has demonstrated a high efficiency during the process of the thermal oxidation decomposition of the trilon B complexon in the interval of pH 2-9 and temperature interval 37-80°C. But, it must be admitted that during the process of catalytic decomposition, along with the complete destruction (mineralization) of trilon B, the formation of the most likely intermediate decomposition products, such as nitrilotriacetic (NTA) and hepato-iminodiacetic acid (HIDA), is possible, which can potentially effect the following sorption and precipitation processes of radionuclides (^{60}Co) extraction by complex compounds formation. At present, the decomposition products chromatograms and infrared spectra, obtained during the oxidizing catalytic process, are being explored. Future tasks are connected with the identification of intermediate decomposition products after the catalytic decomposition of trilon B on the nickel-ferricyanide catalyst, its complexing properties investigation, assessment of its influence on the following radionuclides extraction, concentration determination of the referred above intermediate decomposition products provided that they will not effect the process of radionuclides (^{60}Co, etc.) extraction.

The installed productivity of the catalytic column indicates about the possibility of the technological application of the explored catalyst during the process of the complexons oxidation decomposition. For example, the developed method can be used during the recycling of the process liquors of radiochemical plants, when the intensification of the oxidation process is needed, in particular, during reprocessing of the radioactive effluent, containing ^{60}Co in the form of complex compounds. As a consequence of the developed research, the invention application was processed and registered [7].

Also, the instrument flowchart of the developed catalytic process during the recycling of low- and medium-activity waste, containing complexons (trilon B), was introduced (fig. 4). The aggregative design engineering of the catalytic column for the developed process was made. Critical parameters of the column: height – 1900 millimeters, diameter – 100 millimeters; the volume of the piled catalyst bed – 3,9 liters; productivity – up to 1,5 m^3/day.

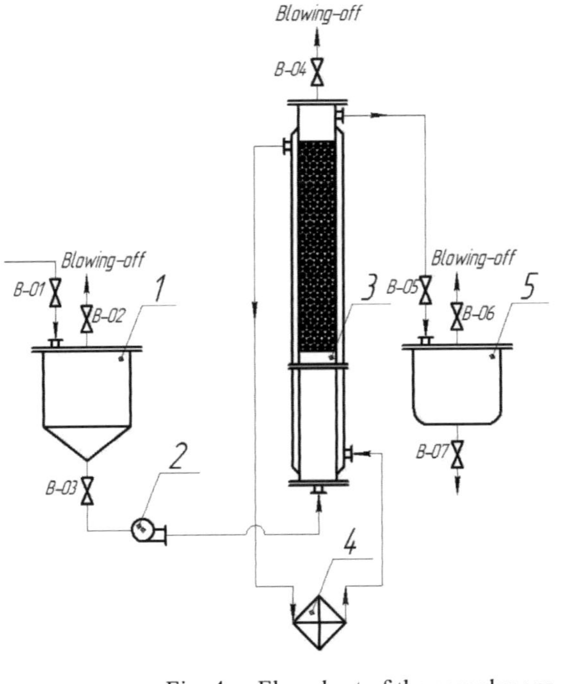

1 – Storage capacitor with radioactive effluent;
2 – Membrane pump;
3 – Temperature-controlled column;
4 – Thermostat;
5 – Receiver tank;
B-01 ÷ B-07 – adjusting valves.

Fig. 4. – Flowchart of the complexons catalytic decomposition process

At present, the developed catalyst efficiency during the process of real radioactive effluent (containing ^{60}Co, high content of trilon B, oxalic ion, surface-active substances) reprocessing is being experimented. Future tasks will be connected with technological approbation of the developed method in real radioactive effluent (containing complexing compounds) reprocessing technological framework.

References

[1] Lagunova Y O, Seliverstov A F, Ershov B G, Anguladze S 2006 *Materials of the 5th international scientific and technical conference: "Radioactive waste handling"* pp 39–40

[2] Lagunova Y O, Seliverstov A F, Ershov B G, Morozov P A, Kamrukov A S, Shashkovskiy S G 2008 *Radiochemistry* vol **50** **№ 1** pp 62-66

IOP Conf. Series: Materials Science and Engineering **66** (2014) 012033 doi:10.1088/1757-899X/66/1/012033

[3] Krot N N, Shilov V P, Dzuybenko V I, Matuykha V A, Malkova N N 1994 *Radiochemistry* vol **36** **№ 1** pp 19-24

[4] Shilov V P, Dzuybenko V I, Astafurova L N, Krot N N 1990 *Journal of applied chemistry* № **2** pp 348-353

[5] Apalkov G A, Efremov I G, Zhabin A Y, Dyachenko A N, Kokarev G G 2013 *Power engineering: efficiency, reliability, security: materials of the XIX All-Russian scientific and technical conference* Tomsk polytechnic university: publishing house "Scan" pp 537-541

[6] Zheleznov V V, Kovshun A A, Avramenko V A, Voyt A V 2008 *International symposium of sorption and extraction: materials* / edited by doctor of science Medkova pp 245-250

[7] Apalkov G A, Efremov I G, Zhabin A Y, Kokarev G G 2014 *The method of heterogeneous catalytic decomposition of complexons and surface-active substances of process liquors on the nickel-ferricyanide catalyst* Invention application № 2014118810 from 08.05.2014

IOP Conf. Series: Materials Science and Engineering **66** (2014) 012034 doi:10.1088/1757-899X/66/1/012034

Complex utilization of snf processing wastes in air plasma of high-frequency torch discharge*

A G Karengin [1, a], A A Karengin [1, b], O D Podgornaya [1, c] and E E Shlotgauer [1, d]
Tomsk Polytechnic University, Tomsk, Russia

E-mail: [a]karengin@tpu.ru, [b]karenginaleksey@gmail.com, [c]shahmatovaol@tpu.ru, [d]shlotyara@mail.ru

Abstract. We present results of complex spent nuclear fuel wastes utilization process in air plasma of high-frequency torch discharge in form of dispersed water-organic compositions. We demonstrate the possibility to apply magnetic separation for effective extraction of obtained dispersed solid products including magnetic iron oxide from water suspension.

1. Introduction

The Russian Federation first in the world began to creating closed nuclear fuel cycle. It provides for delivery spent nuclear fuel (SNF) from NPP, its storage, recycling and extraction Pu-239 and U-238 to produce MOX fuel based on these elements and further delivery back to NPP. Closed nuclear cycle conception also provides for utilization all the nascent radioactive wastes, thus providing long time storage and further using of them.

Tadiochemical plants have SNF recycling technologies based on PUREX process. It provides high level extraction of uranium and plutonium (>99,9%) with high level purification of them from fission products [1-3].

SNF extraction technological schemes provide for application tributylphosphate (TBP) with different diluents (kerosene, refined hydrocarbons, carbon tetrachloride, hexachlorobutadiene (HCBD), etc.) as an extractant for uranium and plutonium extraction. For natural and low enriched uranium concentration of TBP is 30÷40%. In the case of highly processed U-235 fuel or fuel with high plutonium concentration TBP content is reduced to 2,5÷5% in order to avoid the formation of dangerous concentrations of fissile nuclides in the extract [2].

Under the influence of radiation exposure due to a high content in the solution of fission products, plutonium and transplutonium elements extractants over time lose their effectiveness and become inflammable recycling waste (IRW SNF).

After uranium and plutonium extraction from dissolved SNF there are also recycling wastes (RW SNF) are formed as low-concentration solutions with next modelling consist [4]: HNO_3 – 18,0%, H_2O – 81,43%, Fe – 0,07%, Mo – 0,1%, Nd – 0,11%, Y – 0,06%, Zr – 0,058%, Na – 0,04%, Ce – 0,039%, Cs – 0,036%, Co – 0,031%, Sr – 0,026%.

Next these wastes go to evaporation and after addition chemical reagents (silicates, phosphates, borates, etc.) go to verification with further landfilling [5]. Such technology is multistage and it requires significant energy, maintenance, chemicals and time.

In article [6,7] the processes of plasma utilization of SNF in the form of optimal inflammable water-organic compositions (WOC) on the basis of acetone and ethyl alcohol.

These WOCs have an adiabatic combustion temperature at least 1200°C and thus give the possibility of a significant reduction in energy consumption for the process of recycling.

*This work was financially supported in the framework of the state task of the Ministry of Education and Science of Russian Federation for 2014÷2016 (Topic number: 2031).

In this paper we consider the possibility of separate and joint IRW SNF and RW SNF plasma utilization as inflammable water-organic compositions.

2. Calculation of optimal consist of water-organic compositions

At the first stage calculations were made for burning indicators of different WOCs based on IRW SNF and RW SNF.

Figure 1 shows influence of content of TBP and water (a) and also TBP and RW SNF (b) on adiabatic combustion temperature (T) in water-organic compositions with different consists

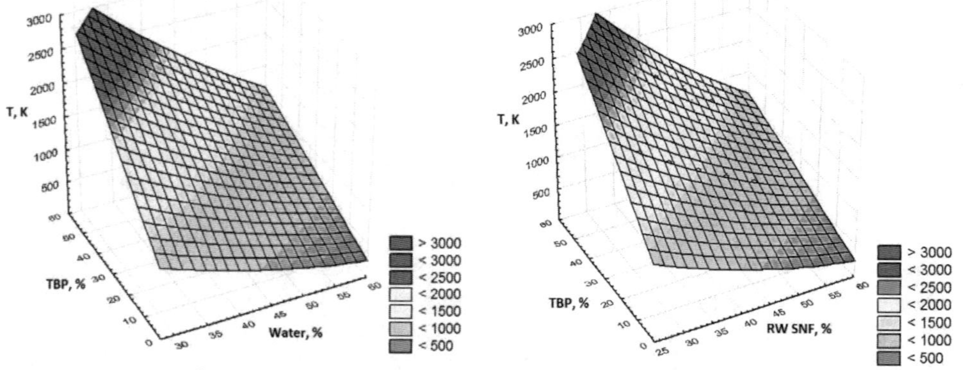

a) b)

Figure 1. Influence of content of TBP and water (a) and also TBP and RW SNF (b) on adiabatic combustion temperature in water-organic compositions

Analyzing the results should be that replacement of water in inflammable WOC with low-concentration RW SNF solutions does not influence on T of WOCs. It points on possibility to utilize of IRW SNF and RW SNF joint in plasma.

Considering all the findings the next optimal consists can be recommended for practical application:

- WOC-1 (50% water : 17,5% TBP : 32,5% HCBD);
- WOC-2 (50% RW SNF : 17,5% TBP : 32,5% HCBD).

Equilibrium consist calculations of gaseous and condensed plasma utilization products were done for recommended WOC-1 and WOC-2. These calculations were done by means of licensed program for PC called «TERRA».

3. Calculation of optimal plasma utilization regimes of water-organic compositions

Figure 2 shows characteristic equilibrium consists of the main gaseous (a) and condensed (b) products after joint plasma utilization of IRW SNF and RW SNF in for of WOC-2 with mass proportion of air plasma coolant 65%.

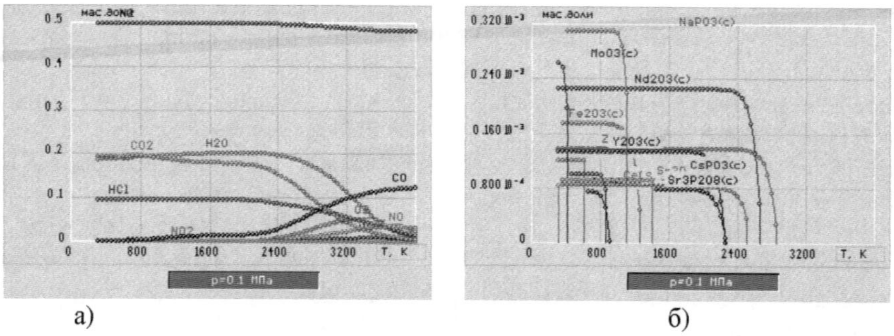

a) б)

Figure 2. Equilibrium consists of the main gaseous (a) and condensed (b) products after joint plasma utilization of IRW SNF and RW SNF in air plasma (65% air : 35% WOC-2)

Analyzing the results should be that with mass fraction of air coolant is 65% the main gaseous products (temperature under 1500 K) of RW SNF plasma utilization in form of WOC-2 are N_2, H_2O and CO_2.

At the temperatures up to 800 K the main condensed products are metal chlorides such as $FeCl_2$(c), $CoCl_2$(c) и $SrCl_2$(c) effectively attaching chlorine. In temperature interval 800-1500 K simple metal oxides form such as MoO_2(c), Nd_2O_3 (c), ZrO_2(c), Y_2O_3(c), CeO_2(c). Also in this interval complex phosphorus metal oxides form ($NaPO_3$(c), $CsPO_3$(c), $Sr_2P_2O_7$(c) и $Sr_3P_2O_8$(c)). They attach phosphorus effectively.

Small amount of soot C(c) and CO, NO, NO_2, HCl indicates that plasma utilization process of wastes in form of WOC-2 takes place in environmentally safe mode. It is worth noting formation iron oxide in condensed phase. Next increasing of air mass fraction to 90% reduces process productivity.

Thus joint plasma utilization of RW SNF and IRW SNF in form of water-organic composition with optimal consist eliminates evaporation stage, significantly reduces specific energy consumption on this process and also allows to use magnetic separation for effective extraction of particulate solid products of plasma utilization.

Table 1 shows calculated results of energy costs of the waste plasma utilization in form of optimal WOC-2 with with optimal air plasma coolant mass ratio (65%).

Table1. Energy costs of WOC-2 plasma utilization

T, K	500	1000	1500	2000	2500
Energy costs, MJ/kg	0,24	0,91	1,66	2,50	3,76

Considering all the findings the next optimal modes cam be recommended for practical application of plasma utilization process of SNF processing wastes in air plasma:

- WOC consist (50% RW SNF : 17,5% TBP : 32,5% HCBD);
- weight ratio of phases (65% air : 35% WOC);
- temperature interval (1200±100) K.

4. Experiments

Figure 3 shows scheme of plasma module based on high frequency torch (HFT) plasmatron. It is intended for high frequency air plasma generation and application of this plasma for carrying out different plasma-chemical processes.

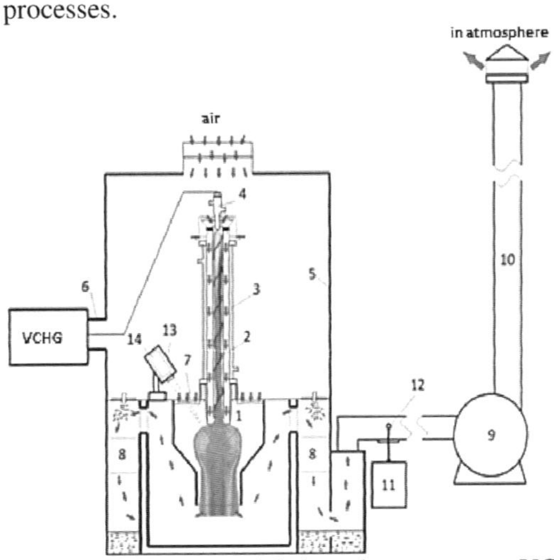

Figure 3. Scheme of plasma module based on generator VCHG8-60/13

1 – dispersant, 2 – HFT discharge, 3 – HFT plasmatron; 4 – copper electrode, 5 – case; 6 – coaxial output; 7 – reactor impeller; 8 – wet cleaning unit for exhausted gases; 9 – air exhauster (BP 12-26, №4), 10 – ductwork, 11 – gas analyzer «Quintox» KM 9106, 12 – sampler; 13 – pyrometer protective cover, 14 – pyrometer IPE 140/45

Plasma module includes high frequency generator VCHG 8-60/13-01 (oscillatory power up to 60 kW, operating frequency of 13.56 MHz), from which via coaxial output 6 high frequency energy is supplied to the water cooled copper electrode 4 of high frequency torch plasmatron 3 to generate air plasma jets with temperatures up to 4000 K. Air exhauster 9 provides constant air flow through the HFT-plasmatron 3 and impeller 7 in reactor.

During the plasma utilization process of modelling WOC dusty and steam mixture creates into the reactor. It includes metal oxides going through the centrifugal-bubbling device in wet cleaning unit for exhausted gases. At the outlet this mixture forms water suspensions from dispersed solid products included magnetic iron oxide. It's worth noting about interest of application of magnetic precipitation for dispersed solid products (with magnetic iron oxide) extraction.

Figure 4 shows experimental scheme for the comparative gravity and magnetic precipitation of dispersed solid plasma utilization products from water suspension.

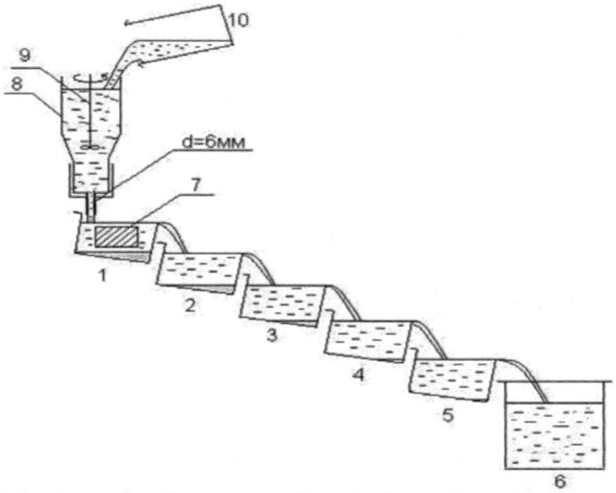

Figure 4. Experimental scheme for the comparative gravity and magnetic precipitation of dispersed solid plasma utilization products

For the precipitation of powders from water suspensions magnet M_1 (50x30x10 mm) was used. It is made from Fe-Nd-B alloy with magnetic induction 1,2 T.

Figure 5 shows the result of research gravity and joint gravity and magnetic precipitation dispersed solid plasma utilization products (with magnetic iron oxide) from water suspension.

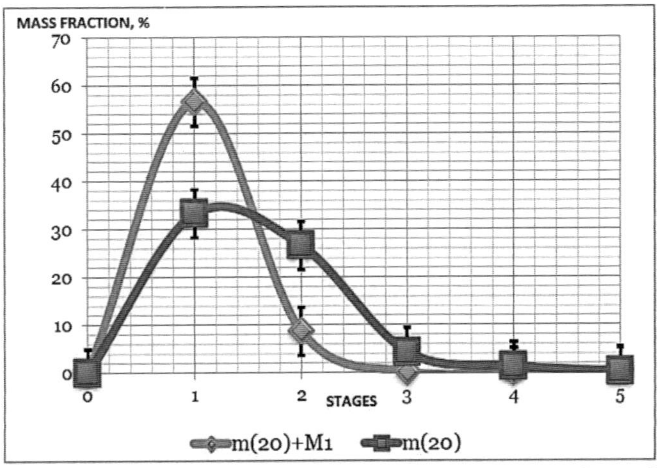

Figure 5. Comparative efficiency of gravity and joint gravity and magnetic precipitation dispersed solid plasma utilization products (with magnetic iron oxide) from water suspension

Analyzed the graph should be that when there are equal initial contents of ferriferous powders in water suspension (20 g/l) then application of magnetic precipitation considerably increases of extraction of powders from water suspensions (with magnetic iron oxide) in comparison with the only gravity precipitation. This indicates that there is possible to apply magnetic separators for effective extraction of solid SNF plasma utilization products from water suspension.

5. Conclusion

All the obtained results could be used in creating the commercial plants for complex plasma utilization of SNF processing wastes and other liquid radioactive wastes.

6. References:

[1] Information onhttp://nauka.relis.ru/06/0111/06111040.pdf

[2] Robert Alvarez 2005 Science and Global Security. Vol. 13. pp. 43–86.

[3] Kulagin V.A., Kulagina L.V., Kulagina T.A. 2013 Journal of Siberian Federal University. Engineering & Technologies 2 pp. 123–149.

[4] Panteleev Yu. A., Alexandruk A. M., Nikitina S. A., Makarova T. P., Petrov E. P., Bogorodickiy A. B., Grigorieva M. G. 2007 *Analytic methods of consist finding of liquid radioactive wastes.* (Leningrad, Proceedings of V. G. Khlopin institute of radium) **12** pp. 124-147

[5] Nikiforov A. S., Kulinichenko V. V., Zhiharev M. I. 1985 *Rendering harmless of liquid radioactive wastes* (Moscow, Energyatompablis) p. 184.

[6] Vlasov V. A., Karengin A. G., Karengin A. A., Shakhmatova O. D. 2012 Modeling of process of plasma utilization of spent nuclear fuel reprocessing wastes Rus. Phys. J. **55 11/2** pp. 377-382.

[7] Vlasov V. A., Karengin A. G., Karengin A. A., Shakhmatova O. D. 2013 *Research and optimization of process of plasma utilization of spent nuclear fuel reprocessing wastes in air-plasma of HFF-discharge* Rus. Phys. J. **56 11/3** pp. 201-205.

Tomsk Polytechnic University (Russia)
tel.: 701-777 (1) 2286; E-mail: karengin@tpu.ru.

Information about the authors:

IOP Conf. Series: Materials Science and Engineering **66** (2014) 012034 doi:10.1088/1757-899X/66/1/012034

Karengin Alexander Grigorievich: PhD, associate professor of Department of Applied Physics Engineering, Tomsk Polytechnic University;
Karengin Alexey Alexandrovich: postgraduate of Department of Applied Physics Engineering, Tomsk Polytechnic University;
Podgornaya Olga Dmitrievna: student of Department of Applied Physics Engineering, Tomsk Polytechnic University.
Shlotgauer Elena Eduardovna: student of Department of Applied Physics Engineering, Tomsk Polytechnic University.

IOP Conf. Series: Materials Science and Engineering **66** (2014) 012035 doi:10.1088/1757-899X/66/1/012035

Formation of failure matrix and failure-free control algorithm for multi-sectioned Switched-reluctance drive

G Odnokopylov[1, 2], **I Rozayev**[1, 3]

[1] Tomsk Polytechnic University, Tomsk, Russia
[2] E-mail: OGIz@yandex.ru.
[3] E-mail: RozayevIA@gmail.com.

Abstract. We review fault-tolerant switched reluctance drive with sectioning of the three-phase stator winding. In the operating process of an electric drive, there will be continuous monitoring of the operating state on the basis of a developed algorithm to analyse drive operability and formation tabulate a failure matrix. The paper introduces a failure-free control algorithm for multi-section switch - reluctance motor with formation the assignment values of amplitude phase currents taking into account the failure matrix. We show that in an emergency such single failure or multiple failure in switched-reluctance drive it is possible to provide reduction of torque fall and pro-gressively stock depletion with providing fault-tolerance of drive system. A method of residual life evaluation is proposed on the basis of calculating the coefficient of operability of the electric drive system that gives possibility to control the endurance of electric drive in real time from operational to completely unusable.

1. Introduction

In electric drives of industrial mechanisms of dangerous manufacturing is necessary to use failure-free control of electric drives with ensuring of fault tolerance. As known, the property of fault tolerance can be only in systems with redundancy, and this property ensured by partitioning the stator windings. The one way of ensuring failure-free control in switched-reluctance electric drive (SRD) in emergency situations like failure one or more phases of electric motors is both using multi-sectioned switched-reluctance motor and formation special algorithms of control. In this case the redundancy is formed by the complex using of both structural redundancy in multi-sectioned electric motor and functional backup, that is based on the possibility of work on remaining operational phases of electric motor after some failures. Required marginal current ensured by marginal load capacity that is giving possibility to work under nominal load when each sections work on the half of their nominal capacity [1].

2. Formation of failure matrix and failure-free control algorithm

The decision of issue ensuring failure-free control involve the real-time monitoring of technical drive status with creating appropriate failure bits in failure matrix in case of open-phase mode of operating with further full or partly operation recovery with using the special algorithm of failure-free control.

Required information and computing margin for realization of technical drive status monitoring and further operation recovery ensured by designed-in information reserve. The multi-sectioned electric motor created on the base of independent m-phase stator windings established on the one common shaft of electric motor. Structure scheme of two-sectioned switched-reluctance drive (m=2) shown on the figure 1.

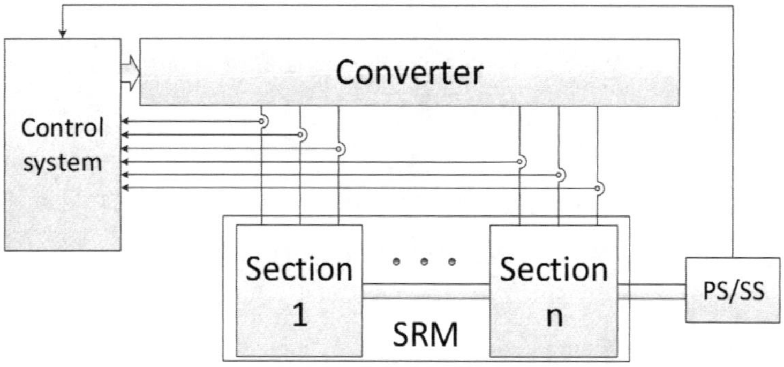

Figure 1. Structure scheme of two-sectioned SRD.

This two-sectioned switched-reluctance drive consist of control system, converter, two-sectioned SRM, position and speed sensor (PS/SS), current sensors each one on the each phase. Operation of this electric drive carry-out on account of the position and speed sensors signals which entered in the control system. In control system, on the base of appropriate current sensors signals of each phases are formed real-time assignment current values for converter. The converter performs switching transistors thereby ensuring feeding phases of each electric motor sections alternately [1, 2].

For ensuring fault-tolerance properties of electric drive, it is necessary to diagnose system in time based on the analysis of the phase currents. For functionality of failure-free algorithm are requires that each m-phase winding of two-sectioned SRM been alternately supplied unipolar current pulse. Separately for each section of m-phase two-sectioned switched-reluctance drive based on the signals from current sensors are formed failure bits: $\overline{a}_{1i}$, $\overline{a}_{2i}$, i=(X_1=A), (X_2=B), (X_3=C),…,(X_m), by analysis of the instantaneous amplitude values of phase currents in the motor sections, where X – phase index. For first section if $|I_{1in\ i}-I_{1i}|\geq\delta$, then system identifies the failure in the i-phase of first section and based on this analysis formed failure bit that is taking on the value either boolean zero or one: a_{1i}=1, $\overline{a}_{1i}$=0. This giving possibility to formation the failure matrix of electric drive. For the second section similarly if $|I_{2in\ i}-I_{2i}|\geq\delta$, then system identifies the failure in the i-phase of second section: a_{2i}= 1, $\overline{a}_{2i}$= 0. Where: $I_{1in\ i}$ - given values of instantaneous amplitudes of the phase currents formed for the i-phase of the first section; $I_{2in\ i}$ - given values of instantaneous amplitudes of the phase currents formed for the i-phase of the second section; δ – admissible error of current controller; I_{1i} – current feedback value from current sensor of i-phase of first section; I_{2i} - current feedback value from current sensor of i-phase of second section [1]. Based on failure bits is formed failure matrix of electric drive.

Failure matrix and the matrix of phase currents of two-sectioned switched-reluctance motor in normal operating mode has the form:

$$\alpha_{ni}=\begin{bmatrix} a & 0 & 0\\ 0 & b & 0\\ 0 & 0 & c\\ d & 0 & 0\\ 0 & e & 0\\ 0 & 0 & f \end{bmatrix}=\begin{bmatrix} 1 & 0 & 0\\ 0 & 1 & 0\\ 0 & 0 & 1\\ 1 & 0 & 0\\ 0 & 1 & 0\\ 0 & 0 & 1 \end{bmatrix};\quad I_{ni}=\begin{vmatrix} I_{1A}\cdot\alpha & 0 & 0\\ 0 & I_{1B}\cdot\beta & 0\\ 0 & 0 & I_{1C}\cdot c\\ I_{2A}\cdot d & 0 & 0\\ 0 & I_{2B}\cdot e & 0\\ 0 & 0 & I_{2C}\cdot f \end{vmatrix};$$

Where α_{ni} – matrix of failure bits; *a,b,c* – failure bits for phases of first sections of SRM, *d,e,f* - failure bits for phases of second sections of SRM. I_{ni} - i-phase current of n-section of SRD.

Based on the design features of the SRD and considering that taking into account the assumptions take torque on the motor shaft equal to the electromagnetic torque.

$$M_{ni}=\frac{\Delta M_{ni}}{\Delta\theta_r}=\frac{\partial L_{ni}(i_{ni},\theta_r)}{\partial\theta_r}\cdot\frac{i_{ni}^{\ 2}}{2} \tag{1}$$

Where: M_{ni} –torque formed by i-phase of n-section of SRD.

θ_r - rotation angle of the motor shaft.

L_{ni} - inductance i-phase of n-section of SRD. [3]

Then the matrix of torques formed by phases of the motor in two-sectioned switched-reluctance drive in normal operating mode has the form:

$$M_{ni} = \begin{bmatrix} M_{1A}(i_{1A}\cdot\alpha,\theta_r) & 0 & 0 \\ 0 & M_{1B}(i_{1B}\cdot\beta,\theta_r) & 0 \\ 0 & 0 & M_{1C}(i_{1C}\cdot c,\theta_r) \\ M_{2A}(i_{2A}\cdot d,\theta_r) & 0 & 0 \\ 0 & M_{2B}(i_{2B}\cdot e,\theta_r) & 0 \\ 0 & 0 & M_{2C}(i_{2C}\cdot f,\theta_r) \end{bmatrix};$$

The SRM torque is represents the sum of torques formed by phases of the motor, considering that the phases are independent and are not associated electric and magnetic interactions [3].

Therefore:

$$M_{SRM} = \sum_{j=1}^{m} M_{nj}\left(i_{nj},\theta_r\right) \quad (2)$$

For two-sectioned SRD in normal operating mode the torque of motor will be determined by the following expression:

$$M_{\partial\mathit{в}} = M_{nX_1} + M_{nX_2} + \ldots + M_{nX_m} =$$
$$M_{1A}(i_{1A}\cdot\alpha,\theta_r) + M_{1B}(i_{1B}\cdot\beta,\theta_r) + M_{1C}(i_{1C}\cdot c,\theta_r) + M_{2A}(i_{2A}\cdot d,\theta_r) + M_{2B}(i_{2B}\cdot e,\theta_r) + M_{2C}(i_{2C}\cdot f,\theta_r)$$

Consider the example when loss of phase A in first section of SRM. Timing chart of formation failure bit based on analysis of amplitude values of phase current in phase A first section of SRM shown on the figure 2. For this used the simulation model of SRD that discussed in detail in [4].

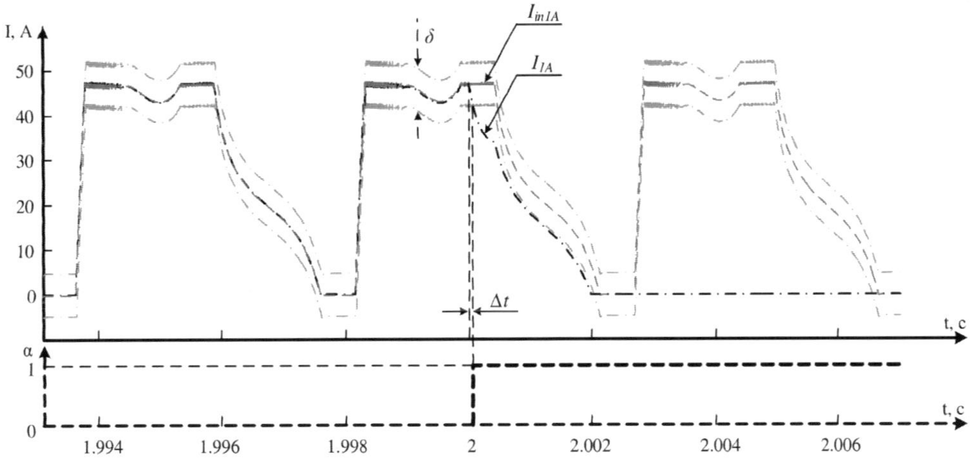

Figure 2. Timing chart of formation failure bit based on analysis of amplitude values of phase current in phase A first section of SRM. I – current, A; α – boolean signal characterizing the operation of phase (failure bit).

Where: δ – admissible error of current controller (10% zone of deviations), I_{inIA} - given value of instantaneous amplitude of the phase current formed for the A-phase of the first section, I_{1A} - current feedback value from current sensor of A-phase of first section, α - boolean signal characterizing the operation of phase (failure bit), Δt - the time interval from the start tripping motor phase to the end of the analysis the phase current and formation the failure bit. In this case, the failure matrix, matrix of phase currents matrix of torques formed by phases of the motor in two-sectioned switched-reluctance drive with loss of phase A in the first section has the form:

$$\alpha_{ni} = \begin{bmatrix} a & 0 & 0 \\ 0 & b & 0 \\ 0 & 0 & c \\ d & 0 & 0 \\ 0 & e & 0 \\ 0 & 0 & f \end{bmatrix} = \begin{bmatrix} 0 & 0 & 0 \\ 0 & 1 & 0 \\ 0 & 0 & 1 \\ 1 & 0 & 0 \\ 0 & 1 & 0 \\ 0 & 0 & 1 \end{bmatrix}; I_{ni} = \begin{bmatrix} 0 & 0 & 0 \\ 0 & I_{1B}\cdot\beta & 0 \\ 0 & 0 & I_{1C}\cdot c \\ I_{2A}\cdot d & 0 & 0 \\ 0 & I_{2B}\cdot e & 0 \\ 0 & 0 & I_{2C}\cdot f \end{bmatrix}; M_{ni} = \begin{bmatrix} 0 & 0 & 0 \\ 0 & M_{1B}(i_{1B}\cdot\beta,\theta_r) & 0 \\ 0 & 0 & M_{1C}(i_{1C}\cdot c,\theta_r) \\ M_{2A}(i_{2A}\cdot d,\theta_r) & 0 & 0 \\ 0 & M_{2B}(i_{2B}\cdot e,\theta_r) & 0 \\ 0 & 0 & M_{2C}(i_{2C}\cdot f,\theta_r) \end{bmatrix};$$

Motor torque with loss of phase A in the first section has the form:

$$M_{SRM} = M_{nX_1} + M_{nX_2} + \ldots + M_{nX_m} =$$
$$0 + M_{1B}(i_{1B}\cdot\beta,\theta_r) + M_{1C}(i_{1C}\cdot c,\theta_r) + M_{2A}(i_{2A}\cdot d,\theta_r) + M_{2B}(i_{2B}\cdot e,\theta_r) + M_{2C}(i_{2C}\cdot f,\theta_r)$$

With losses of other phases and sections the failure matrix will changes according to the current state of SRD. Two-sectioned switched-reluctance drive has 64 operating and emergency states. Therefore, the control system of electric drive already contains in a controller memory all operating and emergency states of SRD, that giving possibility during the diagnostic of electric drive system determine the current number of state from table and start the failure-free algorithm for this state of SRD system. Moreover, it helps to assess the current level of SRD redundancy and coefficient of operability of the electric drive system. Principle of operation of failure-free algorithm based on formation of value assignment of amplitude currents values of sections in accordance with the data obtained after monitoring of SRD system.

The two-sectioned SRD failure-free control algorithm has the form:

$$\begin{aligned} I_{1A} &= I_\omega \cdot a \cdot (1+d) \cdot \frac{2}{4^d} \cdot [1 + sign(\varphi(\theta_e, I) - \varphi(t))]; \\ I_{1B} &= I_\omega \cdot b \cdot (1+e) \cdot \frac{2}{4^e} \cdot [1 + sign(\varphi(\theta_e, I) - \varphi(t))]; \\ I_{1c} &= I_\omega \cdot c \cdot (1+f) \cdot \frac{2}{4^f} \cdot [1 + sign(\varphi(\theta_e, I) - \varphi(t))]; \\ I_{2A} &= I_\omega \cdot d \cdot (1+a) \cdot \frac{2}{4^a} \cdot [1 + sign(\varphi(\theta_e, I) - \varphi(t))]; \\ I_{2B} &= I_\omega \cdot e \cdot (1+b) \cdot \frac{2}{4^b} \cdot [1 + sign(\varphi(\theta_e, I) - \varphi(t))]; \\ I_{2C} &= I_\omega \cdot f \cdot (1+c) \cdot \frac{2}{4^c} \cdot [1 + sign(\varphi(\theta_e, I) - \varphi(t))], \end{aligned} \quad (3)$$

Where: I_{ni} – amplitude values of formed phase currents; $\theta_e = N_r\theta_r$ - electrical angle of rotor, rad; N_r - the number of rotor poles, θ_r - mechanical angle of rotation, $\phi(\theta_e,I)$ - assignment angle value, $\phi(t)$ - the instantaneous value of angle from the position sensor.

In the normal operation mode, consider the example of the algorithm operation (3) an example on the phase A:

$$I_{1A} = I_{\omega} \cdot 1 \cdot (1+1) \cdot \frac{2}{4^1} \cdot \left[1 + sign(\varphi(\theta_e, I) - \varphi(t))\right] = I_{\omega} \cdot \left[1 + sign(\varphi(\theta_e, I) - \varphi(t))\right];$$

$$I_{2A} = I_{\omega} \cdot 1 \cdot (1+1) \cdot \frac{2}{4^1} \cdot \left[1 + sign(\varphi(\theta_e, I) - \varphi(t))\right] = I_{\omega} \cdot \left[1 + sign(\varphi(\theta_e, I) - \varphi(t))\right];$$

In emergency mode in SRD based on the diagnostic of system are formed zero boolean failure bits for appropriate phase in sections of electric drive. As example consider situation when loses phase A in first section. Then a_{1A}=0, this phase A is blocked and further value assignment of amplitude current for this phase is not formed, simultaneously for same named phase of second section further value assignment of amplitude current is formed doubled. Thus based on (1 and 2) the resulting motor torque and motor speed maintained at the same level. It is possible on condition that multi-sectioned motor has margin in load capacity under nominal load.

When the phase A loses in firs section of SRM, value assignment of amplitude current for this phase according to the failure-free algorithm (3) take the form:

$$I_{1A} = I_{\omega} \cdot 0 \cdot (1+1) \cdot \frac{2}{4^1} \cdot \left[1 + sign(\varphi(\theta_e, I) - \varphi(t))\right] = 0;$$

$$I_{2A} = I_{\omega} \cdot 1 \cdot (1+0) \cdot \frac{2}{4^0} \cdot \left[1 + sign(\varphi(\theta_e, I) - \varphi(t))\right] = 2 \cdot I_{\omega} \cdot \left[1 + sign(\varphi(\theta_e, I) - \varphi(t))\right];$$

Based on the proposed failure-free algorithm is provided control with single and multiple failures in SRD without loses in torque and speed. And is possible to calculate the coefficient of operability of the electric drive system, that gives possibility to assess the residual service life of SRD up to full incapacitation. From equation (2) implies that the output motor torque will vary directly with the number remaining operation phases. For normal operating state: $M_{SRM} = M_{\mathrm{SRM.nominal}}$, for state with losing one phase in one section

$$M_{SRM} = (1 - \frac{1}{6}) \cdot M_{SRM.nominal}.$$

If we take redundancy (s) for number remain operation phases, and take index of operability (p) that is show in relative units the current level of (residual) torque m-phases motor relative to the initial nominal value, it is possible to calculate the coefficient of operability SRD system:

$$k = \frac{(s \cdot p)}{m}. \quad (4)$$

Below examples of some emergency states of SRD in table 1.

Table 1. Examples of some emergency states of switched-reluctance drive.

Failure	№	a	b	c	d	e	f	Redundancy, S	Index of operability, p		Coefficient of operability, k=(s*p)/m	
									With algorithm	Without algorithm	With algorithm	Without algorithm
Without failures (Normal operation state)	1	1	1	1	1	1	1	6	1	1	1	1
Loses of phase A in first section	2	0	1	1	1	1	1	5	1	0,833	0,833	0,694
Loses of phases A and B in first section	5	0	0	1	1	1	1	4	1	0,66	0,667	0,44
Full failure of first section	8	0	0	0	1	1	1	3	1	0,5	0,5	0,25
Full failure of first section and loses of phase A in second section	58	0	0	0	0	1	1	2	0,66	0,33	0,22	0,11
Full failure of first section and loses of phases A and B in second section	61	0	0	0	0	0	1	1	0,3	0,167	0,05	0,028
Full failure of SRD (electric drive is unworkable)	64	0	0	0	0	0	0	0	0	0	0	0

The coefficient of operability of switched-reluctance drive system gives possibility to assess the current state of electric drive and calculate residual service life of SRD up to full incapacitation after one or more failures.

3. Conclusions:

1. Developed the continuously monitoring algorithm of switched-reluctance drive on the base of analysis of the instantaneous values of the phase currents with formation the appropriate failure bits tabulated in failure matrix.
2. Developed the failure-free control algorithm for three-phase two-sectioned switched - reluctance drive that is giving the possibility to provide fault-tolerance property at the single failure or multiple failures in switched-reluctance drive with progressively stock depletion of electric drive.
3. A method of residual life evaluation is proposed on the base of calculating the coefficient of operability of the electric drive system that gives possibility to control the endurance of electric drive in real time from operational to completely unusable.

References

[1] Rozayev I and Odnokopylov G 2013 *Bulletin of Tomsk Polytechnic University* **6** p. 138
[2] Bukreev V, Odnokopylov G and Rozayev I 2013 RF Patent № 128409
[3] Krishnan R 2001 Switched reluctance motor drives: modeling, simulation, analysis, design, and applications. *BocaRaton: CRCPress* **416**
[4] Irimia N, Simion A, Livadaru L et al 2001 Study of a 3 phase (6/4) switched reluctance motor control. *Iaşi: Buletinul AGIR* **6** pp 129-134, Oct./Nov.

Research of Electrical Response Communication Parameters on the Pulse Mechanical Impact with the Stress-Strain State of Concrete Under Uniaxial Compression

D.Dann[1,2], A.Demikhova[1], T. Fursa[1], M.Kuimova[1]
[1]Tomsk Polytechnic University, Tomsk, Russia

E-mail: dddann@tpu.ru

Abstract. The article presents the results of research of the electrical response on elastic impact excitation of heavy concrete samples under uniaxial compression. In this paper we recorded and analyzed electrical responses during uniaxial compression of concrete samples with a constant velocity. Studies have shown that in the process of uniaxial compression of concrete samples, the transformation of amplitude-frequency characteristics of the electrical response is observed. The stage of elastic deformation of concrete samples is characterized by a centershift of gravity of the spectrum of the electrical response towards lower frequencies. Dramatic centershift of gravity of the spectrum of the electrical response to the high frequency region characterizes the beginning of fracturing.

1.Introduction

Concrete is commonly used in power constructions where it is exposed to high mechanical loads. In operation, the construction has to withstand operational loading without causing residual deformation and fracture. Therefore there is a need for concrete constructions monitoring during the operation in order to determine the stage of stress-strain state. For determining the onset of fracture processes in materials and products, methods based on recording and analysis of acoustic and electromagnetic emissions are being developed. Papers [1, 2] present the results of researches on the use of acoustic emission to determine the start time of cracks formation in concrete. Also researches of electromagnetic emission in the process of cracks germination in various materials such as rocks [3-8], cement mortar [9-10], ice [11] and etc. are being conducted. The main drawback of emission methods lies in the fact that they can only be used for continuous monitoring of constructions during their operation. A possible solution to the problem of determining the technical state of the construction is the use of the method based on the phenomenon of mechanoelectrical transformations in low impact action [12]. In this case there is no need for continuous monitoring; control can be carried out periodically.

The purpose of this work is to find informative evaluation criteria for the stress-strain state of concrete by the parameters of the electrical response to elastic impact excitation.

2.Experimental techniques

[2] To whom any correspondence should be addressed.

IOP Conf. Series: Materials Science and Engineering **66** (2014) 012036 doi:10.1088/1757-899X/66/1/012036

The essence of the phenomenon of alternating electromagnetic fields generating in concretes under the shock excitation pulse is that the impact of acoustic waves on the sources of mechanoelectrical transformations leads to the formation of an alternating electric field. Acoustic waves are generated in the sample of finite dimensions under its impact excitation. The electric field is generated due to the charges on the faces of piezoelectric quartz contained in fine and coarse aggregate (bank sand and gravel), under its deformation due to these charges displacement and charges of electrical double layers located at the boundaries of cement matrix and inert aggregate with respect to the electrical receiver. Electrical measuring receiver is located in the immediate vicinity of a sample and in the coverage of this field. Therefore, the parameters of the electrical response are determined by elastic characteristics of the research object and securely reflect interprocesses of acoustic waves with internal structural inhomogeneous defects.
The researches were performed using the laboratory complex allowing to produce pulsed mechanical excitation of materials and register the electrical signal.
Figure 1 shows a photograph of a portable laboratory hardware-software complex which consists of an external measure probe, portable power supply, input-output board and a laptop computer.

Figure 1. Laboratory complex: 1 – sample, 2 – external measure probe, 3 – portable power supply.

External measure probe is a metal cup; an impact device is located inside the cup allowing to make a single impact and a differential electrical sensor. The methodology of measuring the electrical response to impact excitation is described in further detail here [12].
Loading on the compression of samples was performed using a computerized press SP-500 with a constant rate of 0.2 kN/s. In the process of loading, the registration load and discreet displacement equal to 1s was made with the help of a special program.
The measurements were performed in the following way. Measure probe with the help of rubber bands was attached to a side surface of the sample. Then the concrete sample with the probe was mounted on the bottom plate of the press, loading was done until damage. External measure probe is equipped with long wires that allowed the probe to be laid inside the press, and the power supply and the laptop computer were placed at a considerable distance from the press. Impact excitation and recording of electrical responses were performed every 10 kN.

3. Experimental results and their considerations

The measurements of electrical response to impact excitation from the samples of heavy concrete with the size of 100x100x100 mm^3 were made. For this research the samples of heavy concrete were made. Ratio of cement, sand and coarse aggregate was 1:2:4. Water-cement ratio was 0.35. We used two types of samples. After the process of hardening one part of the samples was polished on the grinding machine for the removal of weak surface layer. The second part of the samples was not subjected to any treatment.

A typical load curve of concrete which was not subjected to grinding is shown in Figure 2.

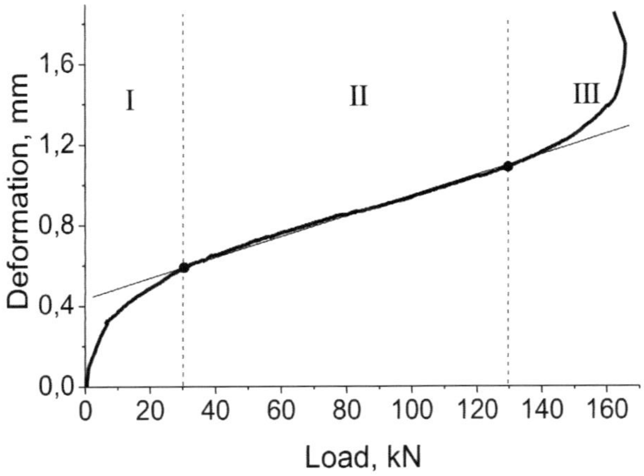

Figure 2. Dependence of concrete deformation on the magnitude of external mechanical load.

As shown in Figure 2 the strain dependence for concrete at a constant rate of loading is characterized by three stages.

The presence of the first nonlinear plot (I) occurs due to the surface compression of the sample which has a lower hardness and was not specially polished before testing. Then the plot (II) of linear anelastic deformation comes. Plot (III) is characterized by sharply nonlinear deformation associated with the process of crack initiation and penetration.

Figure 3 shows the amplitude-frequency characteristics of electrical responses to the impact excitation of the concrete sample during uniaxial compression.

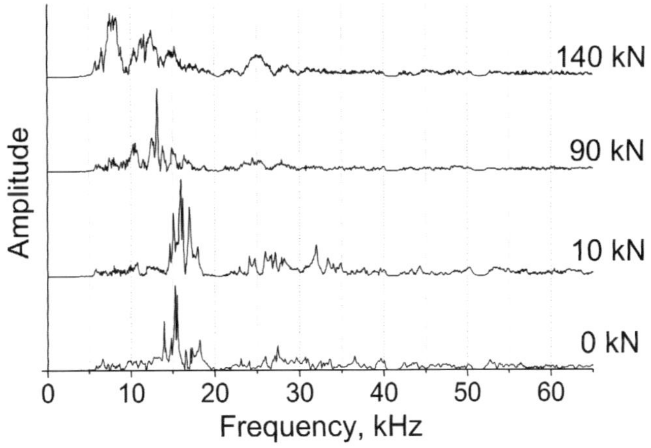

Figure 3. Change of the amplitude-frequency characteristics of the electrical response depending on the magnitude of the external load.

As shown in Figure 3 the bulk of high-amplitude spectral components for the unloaded sample is concentrated in the frequency range from 13 to 18 kHz. The dominant peak is at 15.8 KHz frequency. In the early stage, corresponding to the sealing area of the sample's surface layer, a slight shift of the dominant peak to the high frequency region and occurrence of significant peaks in the frequency range from 23 to 35 kHz appear. These changes of spectral distribution may be a consequence of surface disfunction in setting the shape of sample. Previously, it was shown on laboratory models of concrete that a change in artificial surface cracks concentration leads to the shift of electrical signal spectrum to the high frequency domain [13]. This shift may be due to the processes of reflection of acoustic wave excitation from surface cracks. With further loading (section II), there appears a successive shift of the dominant peak and the entire spectrum in the low frequency region. During intense fracturing at the late stage of loading (section III) the formation and development of deep cracks occur. Appearance and increase of low-frequency spectral components of the electrical signal may be due to rounding processes of deep cracks by acoustic waves. Spectrum shift of the electrical response in the low frequency region was observed earlier in the process of cracking caused by cyclic freezing and thawing of concrete [14]. As a criterion of changes of the spectral distribution of electrical responses occurring during loading, it is possible to suggest the use of the centershift of gravity of the spectrum.
Using special software program in LabVIEW programming environment, frequencies of center of gravity of the spectrum at different stages of deformation were identified. The analyzed signal spectrum is sent to the entry point. In signal spectrum using cursors, we choose the desired frequency band for the analysis. The program consistently calculates the integral of two parts of specified spectrum region and determines the frequency at which they are equal; this frequency is the center of gravity of the defined spectrum. Figure 4 shows the dependency graph connecting the center of gravity of the spectrum of electrical response to the magnitude of an external mechanical load for a normal sample of concrete.

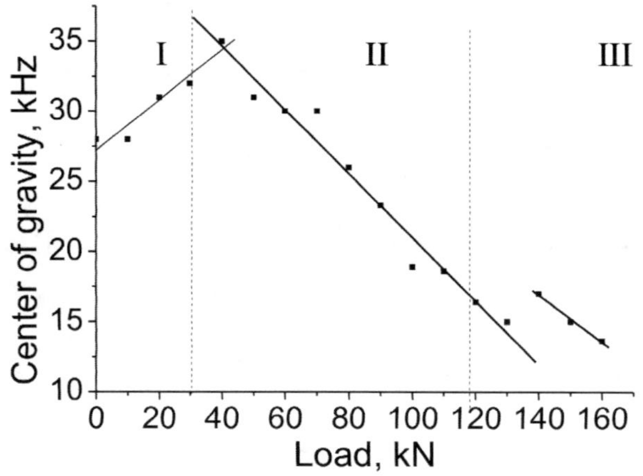

Figure 4. Changing the center of gravity of the spectra of the electric signals from concrete depending on the value of the external load.

As can be seen in figure 4 the stage of surface layer sealing is characterized by a linear increase of the center of gravity of the spectrum; the area of elastic deformation is characterized by linear decrease. The initial section of the transition in the cracks area is characterized by an abrupt increase in frequencies of the center of gravity.
Similar studies were conducted on polished samples of concrete.

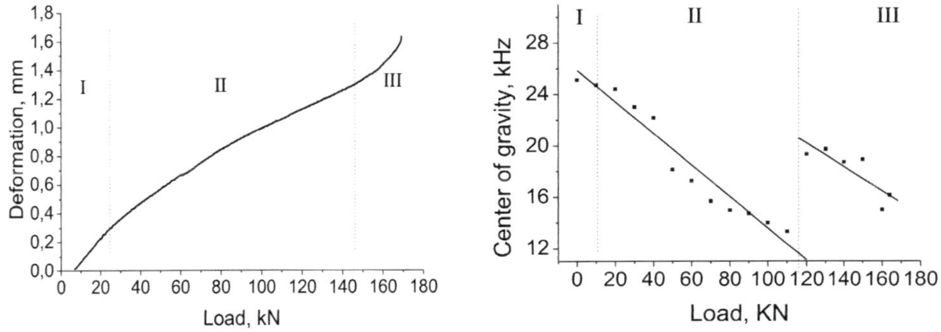

Figure 5. Typical distortion dependence (a) and the center of gravity of the spectrum of the electrical response (b) of the magnitude of an external mechanical load for the polished sample of concrete.

As shown in Figure 5a, the load curve for the polished sample has weakly expressed sealing area of the surface layer (I). As a result, there is no increase in frequencies of the center of gravity of the spectrum in this area (Figure 5b) compared with test results of the unpolished sample (Figure 4). Behaviour change of the center of gravity of the spectra in areas I and II for polished and unpolished samples has common patterns.

4. Conclusion

The purpose of the present research is to develop a non-destructive method for monitoring the stress-strain state of concrete products using the parameters of electrical response to elastic impact excitation.

The changing of the stress-strain state was carried out by uniaxial compression of concrete samples; the stages of the stress-strain state were defined on the basis of load registration and deformation. The process of uniaxial compression of concrete is characterized by three stages: the first – form installation, the second – quasi-elastic deformation and the third – fracturing.

It was established that the stage of form installation is characterized by electrical response of the spectrum shifting to higher frequencies. At the stage of quasi-elastic deformation there occurs a successive shift of the dominant peak and the entire spectrum to the low-frequency side. The initial stage of fracturing is characterized by the appearance of a significant proportion of high-frequency components in the spectrum of response.

Based on these studies it was suggested to use the center of gravity of the spectrum of the electrical response as a criterion for stages evaluating of the stress-strain state in concrete.

The work is done under the State task "Science".

References

[1] Carpinteri A, Xu J, Lacidogna G and Manuello A 2012 *Cem. Concr. Compos.* **34** pp 529-537
[2] Ohno K and Ohtsu M 2010 *Constr. Build. Mater.* **24** pp 2339-2346
[3] Koktavy P 2009 *Meas. Sci. Technol.* **20** 015704
[4] Bespalko A A, Yavorovich L V and Fedotov P I 2011 *Russian Journal of Nondestructive Testing* **47** pp 680-686
[5] Stavrakas I, Anastasiadis C, Triantis D, and Vallianatos F 2003 *Nat. Hazards Earth Syst. Sci.* **3** pp 243-247
[6] Koktavy P, Pavelka J and Sikula J 2004 *Meas. Sci. Technol.* **15** pp 973-977
[7] Aydin A, Prance R J, Prance H and Harland C J 2009 *Applied Physics Letters* **95** 124102
[8] Bespalko A A, Surzhikov A P, Yavorovich L V and Fedotov P I 2012 *Russian Journal of Nondestructive Testing* **48** pp 221-225

IOP Conf. Series: Materials Science and Engineering **66** (2014) 012036 doi:10.1088/1757-899X/66/1/012036

[9] Kyriazopoulos C, Anastasiadis D, Triantis D and Brown C J 2011 *Constr. Build. Mater.* **25** pp 1980-1990
[10] Triantis D, Stavrakas I, Kyriazopoulos A et al 2012 *Int. J. Fract.* **175** pp 53-61
[11] Fifolt D A, Petrenko V F and Schulson E M 1993 *Philosophical Magazine Part B* pp 289-299
[12] Fursa T V, Osipov K Y, Lyukshin B A and Utsyn G E 2014 *Meas. Sci. Technol.* **25** 055605
[13] Fursa T V, Surzhikov A P and Dann D D 2010 *Russian Journal of Nondestructive Testing* **46** pp 5-9
[14] Fursa T V, Osipov K Yu and Dann D D 2011 *Russian Journal of Nondestructive Testing* **47** pp 323-328

Development of control method for ferrite phase composition using thermomagnetometric analysis

A L Astafyev[1], E N Lysenko[2], A P Surzhikov[3] and N A Neudahina[4]

[1,2,3,4] Tomsk Polytechnic University,Tomsk, Russia.

E-mail:[1] astafeval@tpu.ru, [2] lysenkoen@tpu.ru, [3]surzhikov@tpu.ru

Abstract. In this paper we evaluated the phase composition of products in synthesis of $Li_{0.4}Fe_{2.4}Zn_{0.2}O_4$ and $Li_{0.3}Fe_{2.3}Zn_{0.4}O_4$ lithium-substituted ferrites using thermomagnetometric method. Results of thermomagnetometric analysis were compared with X-ray diffraction (XRD) data. It was found that thermomagnetometric analysis in comparison with XRD method better reflects the temperature of phase magnetic transitions and therefore provides a more precise estimate the phase composition of the samples. However, a complex analysis using of both X-ray and thermomagnetometry methods will be the most optimal in case of the formation of non-magnetic or weakly magnetic phases. By the thermomagnetometric analysis it was shown that the homogeneity of synthesized under the same temperature-time regimes ferrites decreases with increasing the concentration of zinc in the initial mixture. Using of intermediate grinding and mixing leads to homogeneity increasing of synthesized ferrite with higher content of the final phase.

Keywords: Ferrite; Thermomagnetometry method; X-ray diffraction; Lithium-substituted ferrospinels.

1 Introduction

During production multicomponent magnetic materials, including ferrites, the great attention is paid to maximization of powder homogeneity at the synthesis stage [1]. Of the existing methods for improving the homogeneity of powders, operations of intermediate grinding and mixing are the most simple and technological which included in the synthesis procedure [2]. The use of such technological operations requires careful phase control at all stages of the annealing.

X-ray diffraction analysis (XRD) is one of the traditional methods for the phase homogeneity control. However, with application to lithium-substituted ferrospinels, the XRD quantitative data and its reliability require additional verification, since it is difficult to separate different spinel phases that can be formed in multicomponent system [3, 4].

In [5, 6] the possibility of application thermomagnetometry method TG(M)/DTG(M) for analysis of phase homogeneity of lithium-substituted ferrites was investigated. This method is thermogravimetric analysis TG/DTG of the samples in magnetic field [7]. Results of investigations allow controlling phase analysis only qualitatively. These results allow us to study the phase composition analysis only on a qualitative level.

In this paper the phase composition of products in synthesis of $Li_{0.4}Fe_{2.4}Zn_{0.2}O_4$ and $Li_{0.3}Fe_{2.3}Zn_{0.4}O_4$ lithium-substituted ferrites widely applied in microwave technology was estimated using TG(M)/DTG(M) analysis. However, since the concentration of the introduced impurity of zinc into microwave lithium ferrites usually do not exceed the value of x = 0.4 (due to the reduction of the saturation magnetization of lithium-zinc ferrites) [8], it is not necessary to investigate the lithium-zinc ferrite synthesis with a higher content of zinc.

2 Experimental

The lithium-substituted ferrospinels were prepared by the solid phase synthesis in accordance to the reaction:

Li_2CO_3+Fe_2O_3 +ZnO → $Li_{0.5(1-x)}Fe_{2.5-0.5x}Zn_xO_4$ +CO_2, where x_{Zn} = 0.2, 0.4.

The initial reagents are commercial powders of Fe_2O_3, Li_2CO_3 and ZnO. The samples were compacted by single-ended cold pressing under a pressure of 200 MPa in the form of tablets with a diameter of 15 mm and thickness of 2 mm.

Then all samples were divided into two groups (samples A and samples B). A solid phase synthesis of the samples was carried out in the air atmosphere using a resistance furnace at the temperature of 800 °C during 360 minutes. Samples A were annealed with inclusion of intermediate grinding and mixing operations every 120 minutes during isothermal annealing. Samples B were annealed without any operations at the same time-temperature parameters.

The samples were investigated by XRD and TG(M)/DTG(M) analyses. XRD analysis was performed using an ARL X'TRA diffractometer with a semiconductor Peltier Si(Li) detector and CuK_α radiation. The Powder Cell 2.4 software was used for a full-profile analysis of the X-ray diffraction patterns. The phase composition was identified using PDF-4 powder database of the International Centre for Diffraction Data (ICDD).

Thermomagnetometric analysis was performed in air atmosphere using thermal analyzer STA 449C Jupiter (Netzsch). The magnetic assemblage of two permanent magnets creating a field of ~5 Oe was attached on the outer side of measurement cell for control of the magnetic state of samples. Determination of the phase composition was carried out using mathematical modeling and the software Peak Separation developed by Netzsch.

3. Result and discussion

X-Ray diffraction patterns for samples A and samples B were shown in Figure 1 and 2 for lithium-substituted ferrospinel with zinc concentration X_{Zn}=0.2 (Figure 1) and X_{Zn}=0.4 (Figure 2).

To analysis of the sample phase composition, a discrete set of lithium-zinc ferrite phases with X_{Zn}=0; 0.2; 0.4; 0.6; 0.8 was included in the full-profile analysis by means of Powder Cell 2.5 software. If necessary, a discrete set was expanded to include phase particles of the starting components. Phase concentration and lattice parameters were determined after fitting of calculated and experimental XRD patterns. Then, the values of X_{Zn} in spinel phases were refined using the dependence of lithium-zinc ferrite lattice parameters on the content of zinc.

The results of XRD analysis presented in Table 1 show that all observed reflections belong to the spinel phases. Phase identification detected the presence of transition lithium-zinc phases with different concentration of zinc for all samples, except for samples A of $Li_{0.4}Fe_{2.4}Zn_{0.2}O_4$, where there is a 100% yield of the final phase. Thus, the samples synthesized under different conditions, have demonstrated varying degrees of homogeneity of the phase composition. For samples A, which were synthesized using intermediate grinding and mixing operations, there is a large phase content with a concentration of zinc that is close to the final product. In contrast, samples B are characterized by a high content of transitional phases of lithium-zinc ferrite and the presence of pure lithium ferrite of $Li_{0.5}Fe_{2.5}O_4$.

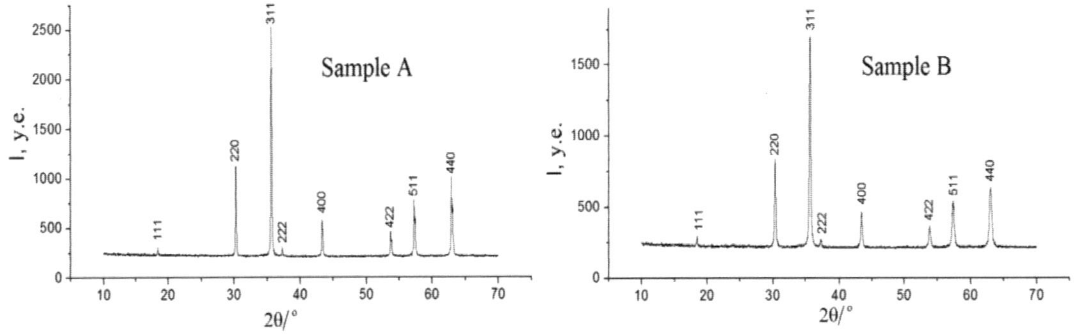

Figure 1. X-Ray diffraction patterns for $Li_{0.4}Fe_{2.4}Zn_{0.2}O_4$ after synthesis at different regimes.

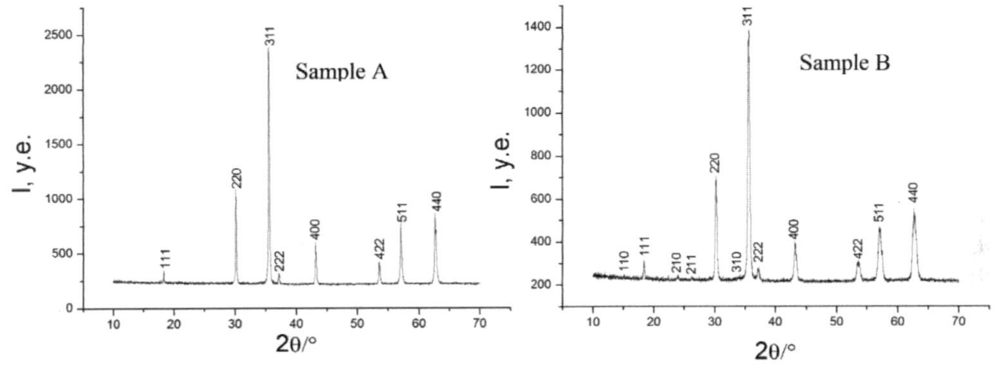

Figure 2. X-Ray diffraction patterns for $Li_{0.3}Fe_{2.3}Zn_{0.4}O_4$ after synthesis at different regimes.

Table 1. Results of X-Ray analysis

Concentration of zinc (X_{Zn})	Sample type	Phases	Phase concentration, %
0.2	A	$Li_{0.4}Fe_{2.4}Zn_{0.2}O_4$	100
	B	$Li_{0.4}Fe_{2.4}Zn_{0.2}O_4$	88
		$Li_{0.5}Fe_{2.5}O_4$	12
0.4	A	$Li_{0.375}Fe_{2.375}Zn_{0.25}O_4$	33
		$Li_{0.315}Fe_{2.315}Zn_{0.37}O_4$	64
		$Li_{0.295}Fe_{2.295}Zn_{0.41}O_4$	3
	B	$Li_{0.5}Fe_{2.5}O_4$	14
		$Li_{0.385}Fe_{2.385}Zn_{0.23}O_4$	48
		$Li_{0.295}Fe_{2.295}Zn_{0.41}O_4$	20
		$Li_{0.215}Fe_{2.215}Zn_{0.57}O_4$	18

The presence of mass jumps on TG curves at the Curie temperatures was observed for each of the magnetic phases in the sample (not shown in this paper). Thus, number of peaks on DTG(M) curves derived from TG(M) curves indicates a number of dominant magnetic phases in the samples. Figure 3 and 4 show DTG(M) curves for $Li_{0.4}Fe_{2.4}Zn_{0.2}O_4$ (Figure 3) and $Li_{0.3}Fe_{2.3}Zn_{0.4}O_4$ (Figure 4). It was shown that experimental dependence of DTG(M) are complex peaks for all types of samples (symbols in Figure 3 and 4), which were separated to the simpler peaks by the mathematical modeling (dotted lines in Figure 3 and 4). Mathematical modeling results of separation DTG(M) curves are shown in solid lines in figures and are in good agreement between the experimental and calculated curves.

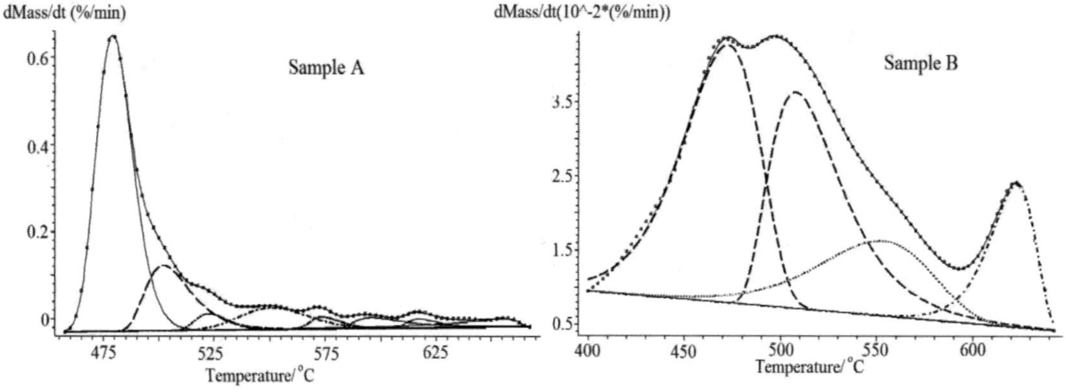

Figure 3. Experimental and calculated DTG(M) curves for lithium-substituted ferrospinel with zinc content X_{Zn}=0.2; symbols – experimental points, lines – calculated curves

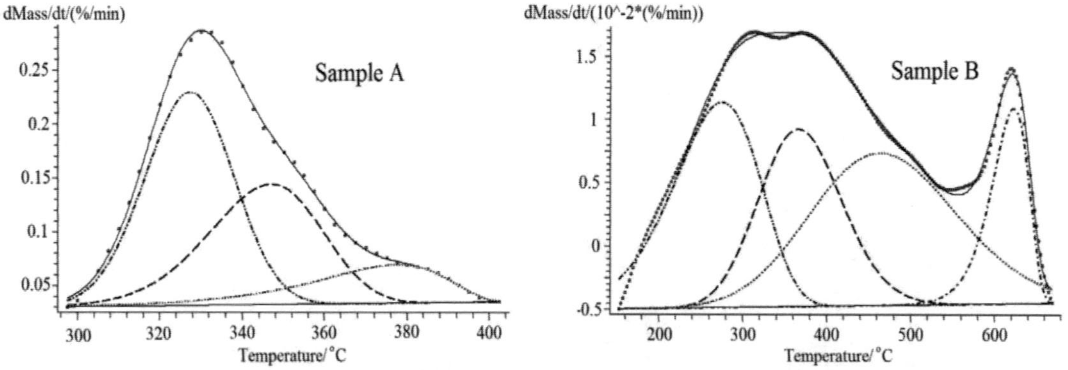

Figure 4. Experimental and calculated DTG(M) curves for lithium-substituted ferrospinels with zinc content X_{Zn}=0.4; symbols – experimental points, lines – calculated curves

Table 2 presents numeric values of peak parameters, which were determined via the modeling. Temperature position of each peak corresponds to Curie temperature of the magnetic phase with a zinc content determined from the concentration dependence of Curie temperatures for reference samples with zinc content X_{Zn}=0÷4 [5]. It can be seen that TG(M)/DTG(M) analysis allows to define the temperature of the magnetic phase transition more precisely, thereby it can more fully characterize the phase composition of sample. On the contrary, the result of XRD analysis is largely determined by the availability of complete catalogs of XRD patterns for reference samples. In the absence of non-magnetic or weakly magnetic phases in the samples, as confirmed by XRD analysis, a values of the peak areas obtained from DTG(M) analysis (see Table. 2) can be compared with quantitative content of ferrite phase.

Table 2. Results of DTG(M) analysis

Content X_{Zn}	Samples type	Number of peak	Peak position, °C	Peak area, %	X_{Zn}
0.2	A	1	479.0823	60.9	0.28
		2	502.066	17.5	0.18
		3	522.6296	2.938	0.15
		4	550.7789	8.307	0.11
		5	573.8409	2.042	0.09
		6	595.3677	3.282	0.05
		7	616.894	1.602	0
	B	1	472.1778	39.171	0.21
		2	507.5629	30.701	0.17
		3	552.4597	17.340	0.07
		4	622.5932	12.788	0
0.4	A	1	327.31	49.9	0.41
		2	346.84	36.2	0.38
		3	377.3	13.89	0.34
	B	1	274.69	28.83	0.48
		2	365.5	23.55	0.36
		3	462.51	35.54	0.22
		4	621.93	12.06	0

4. Conclusions

It was shown that thermomagnetometric analysis allows to define the temperature of the magnetic phase transition more precisely, thereby it can more fully characterize the phase composition of sample as compared to X-Ray diffraction analysis. However, the complex analysis using of both X-ray and thermomagnetometric methods will be the most optimal in case of the formation of non-magnetic or weakly magnetic phases in the samples.

The phase composition of products in synthesis of $Li_{0.4}Fe_{2.4}Zn_{0.2}O_4$ and $Li_{0.3}Fe_{2.3}Zn_{0.4}O_4$ lithium-substituted ferrites under different conditions including the operations of intermediate grindings and mixings was evaluated using thermomagnetometry and XRD methods. It was shown that the synthesis process of lithium-zinc ferrites passes through the formation of intermediate $Li_{0.5(1-x)}Zn_xFe_{2.5-0.5x}O_4$ phases, where $0 \le x \le 0.6$. Moreover, the degree of homogeneity of the synthesized ferrite at the same time-temperature parameters decreased with increasing concentration of zinc in initial mixture. Using the operations of grinding and mixing leads to increasing the homogeneity degree of ferrite with large zinc concentration.

5. Acknowledgements

Acknowledgements: This work was supported by The Ministry of Education and Science of the Russian Federation in part of the "Science" program.

References

[1] Smit J, Wijn HPJ 1959 *Ferrites: Physical properties of ferrimagnetic oxides in relation to their technical applications.* (Philips Technical Library, Eindhoven) p 369

[2] Baba P, Argentina G, Courtney W, Dionne G and Temme D 1972 Fabrication and properties of microwave lithium ferrites *IEEE Trans Magn.* Vol.**8** 83-94

[3] Jozef S, Martin ˇS, Gruskova A, Dosoudil R, Janc V and Degmova J 2013 Magnetic properties of selected substituted spinel ferrites *Journal of Magnetism and Magnetic Materials* Vol. **326** 251–256

[4] Mazen S and Abu-Elsaad N 2012 Structural and some magnetic properties of manganese-substituted lithium ferrites *Journal of Magnetism and Magnetic Materials* Vol. **324** 3366–73
[5] Surzhikov A P , Pritulov A M , Lysenko E N , Vlasov V A , Vasendina E A and Malyshev A V 2013 Analysis of the phase composition and homogeneity of ferrite lithium-substituted powders by the thermomagnetometry method *J. Therm. Anal. Calorim.* Vol.**112** 739-745
[6] Surzhikov A P, Lysenko E N, Vasendina E A, Sokolovskii A N, Vlasov V A and Pritulov A M 2011 Thermogravimetric investigation of the effect of annealing conditions on the soft ferrite phase homogeneity *J Therm Anal Calorim.* Vol.**104** 613-617
[7] Gallagher P K 1997 Thermomagnetometry *J. Therm. Anal. Calorim.* Vol. **49** 33–44
[8] Levin B E, Tret'yakov Y D and Letyuk L M 1979 *Physicochemical principles of preparation, properties, and applications of ferrites.*(Moscow: Metallurgiya, Russia) p 470

IOP Conf. Series: Materials Science and Engineering **66** (2014) 012038 doi:10.1088/1757-899X/66/1/012038

Numerical study of Pavlovskiy coal pulverized combustion in the furnace of BKZ-210-140 steam boiler

A S Zavorin[1] A V Gil[1] P S Khaustov[1] R B Tabakaev[1] and D A Buslov[1]
[1] Tomsk Polytechnic University, Tomsk, Russia

Email: AndGil@tpu.ru

Abstract. In this paper pulverized combustion of insufficiently investigated low-grade Pavlovskiy coal is simulated using the modern engineering software FIRE 3D. The object of study is a widespread in Russia BKZ-210-140 steam boiler. The results of computer simulation are represented with average temperatures in horizontal sections and oxygen concentration. Curves are plotted for three steam generating capacity loads of the boiler: 100 %, 70 % and 50 %.

Introduction

Different modifications of BKZ-210-140 steam boiler units are widespread in different regions of Russia. For example, the Vladivostok Central Heating and Power (CHP) Plant-2 has 14 boilers of this type. It's a drum-boiler with a U-shaped layout and dry ash removal. The combustion chamber with a volume of 1093 m^3 is furnished with four tangentially mounted straight-flow burners and four pulverizing fans. The steam generating capacity is 210 tons per hour; the superheated steam pressure is 14 MPa and its temperature is 570 °C; boiler's efficiency (gross) is 90.3 %.

BKZ-210-140 steam boiler is produced since 1959, but nowadays there is a decreasing of performance, especially that at modes with low load conditions, resulting from the used coals quality reduction.

The boilers at the Vladivostok CHP Plant-2 are designed for burning the Chikhezskiy lignite with the following characteristics: moisture = 41 %; ash content = 8.9 %; calorific value = 12260 kJ/kg. But presently, coals of the Pavlovskiy Deposit with a moisture of = 41.5 %; ash content = 19.6 %; calorific value = 91260 kJ/kg are used [1].

Quality reduction of combustible fuel increases the loads on the fuel-preparation system, decreases the stability of combustion process, and adversely affects the efficiency of power engineering equipment operation.

Because the used direct-fired pulverized-fuel system, the increased load on the pulverizing fans leads to a fuel and air flow reallocation along the burner devices which later affects the tangentially swirling flame formation. Flame displacement or non-intensive vortex formation leads to slagging of the heating surfaces and incomplete fuel burning. These processes reduce reliability and efficiency of the steam generating unit, especially at lowered loads.

Principles of study

Currently, the development of computer simulations has reached a level where it is possible to construct physical and mathematical models of processes occurring in the combustion chamber with accuracy sufficient for engineering practice. Models of multicomponent vortex motion, non-isothermal medium with polydisperse dust, radiation transfer, combustion and chemical reactions of gases, coal dust drying and pyrolysis processes and other physical and mathematical models allow to simulate the operation of real equipment. Modern computation equipment has enabled to implement these models in the form of engineering software available for personal computers.

The Euler-Lagrange method for describing aerothermal and chemical processes in gas-dispersion atmosphere based on the application package FIRE 3D [2] has been employed in this paper. Turbulent characteristics of gas are calculated with a k-ε turbulence model that also considers the effect of particle motion. Radiation heat transfer in the two-phase flow is represented within the framework of the P1 approximation of the spherical harmonics method [3] which displays positive results when used for pulverized-coal furnace.

The task of boundary conditions for equation systems is determined by the border type. For input boundaries homogeneous characteristic distributions are used. Mild boundary conditions are provided at outflow boundaries. Such output boundary position must be extensively moved far from the areas of possible flow recirculation. The furnace surfaces boundary conditions are used: conditions of attachment for velocity, boundary conditions of the first type for gas temperature, zero equality in the normal derivative of concentrations, gas component.

Calculation results

Fig. 1 presents numerical simulation results of average temperatures in horizontal sections and oxygen concentration at the three loads of 100 %, 70 % and 50 % of boiler steam generating capacity. It should also be noted that at load of 70% equipment operates with a single switched off burner.

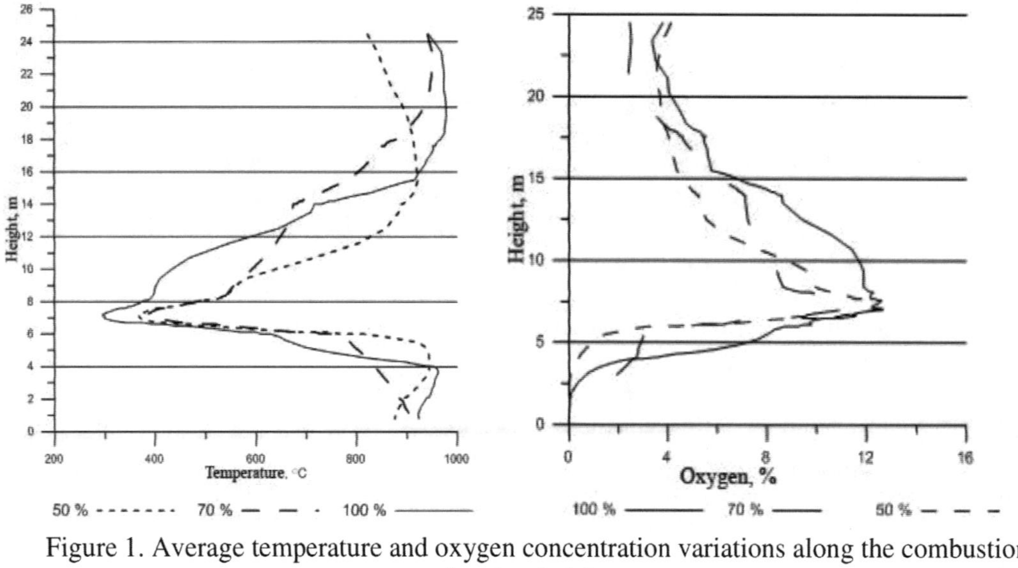

Figure 1. Average temperature and oxygen concentration variations along the combustion chamber height

Curve profiles are similar in all three cases. Comparing the numerical simulation results, it must be mentioned that temperature distributed more uniformly over the combustion chamber volume occurs when the boiler is operated at loads of 100 % and 50 % since flow movement is symmetric in relation to heating surfaces. Temperature values at a height of 7 m are minimal in all cases due to fuel combustion in straight-flow burners and fuel drying by recirculation gases. Temperatures of outflow from the combustion chamber are as follows: at 100 % – 970 °C, at 70 % – 940 °C, at 50 % – 850 °C.

Oxygen concentration in all researched variants has similar values, whereas the curve profile corresponds to the combustion process in the furnace chamber (see Fig. 1). The concentration value of the outflow from the combustion chamber is 3-4 %.

The concentration of volatiles is rather low at all operation modes and its average value equals 1% (see Fig. 2). Variations of the CO concentration correspond to changes of the temperature level in the combustion chamber and are minimal at its outflow.

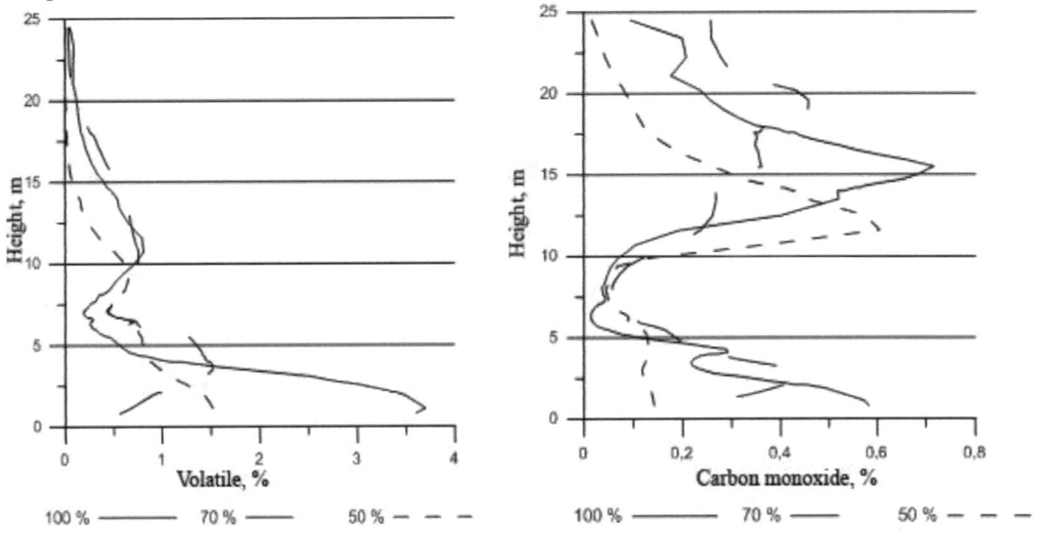

Figure 2. Variations of the concentration of volatiles and CO along the combustion chamber height

Summary

1. Wood wastes and the Kandinsky peat can be considered promising raw material for thermal technological processing both into fuel gas and peat briquettes.
2. The Sukhovsky peat may be used for power generation only after conducting a thorough investigation and feasibility study of its use in certain conditions.
3. Using the Karasevsky sapropel and Arkadevsky peat is considered unpromising due to low ash content of raw material and final products of thermal technological processing.

The reported study was partially supported by RFBR, research project №13-08-980713.

References

[1] *Teplovoj raschet kotlov (Normativny metod)*. Saint-Petersburg: Energiya, 1998. 256 p. [in Russian]
[2] Gil A V and Starchenko A V 2012 *Thermophysics and Aeromechanics* vol. **19**, Issue 3, pp. 503-519.
[3] Taylasheva T S 2009 *Bulletin of Tomsk Polytechnic University* **4** pp. 42-47.

Numerical research of heat and mass transfer at interaction of several hot particles with typical polymeric material

D Glushkov[1] and K Vershinina[1]

Tomsk Polytechnic University, Tomsk, Russia

E-mail: dmitriyog@tpu.ru

Abstract. The results of numerical investigation of heat and mass transfer at polymeric material ignition by several small-size hot particles are presented. The developed mathematical model considers the thermal conduction in polymeric material and hot particles, thermal decomposition of condensed substance, diffusion and thermal convection of combustible gases in air. It has been determined that the joint effect of several local energy sources on the ignition delay time is absent when the distance between hot particles exceeds their characteristic size.

Keywords: hot particles, polymeric material, thermal conduction, thermal decomposition, diffusion, thermal convection, ignition, mathematical modeling.

1. Introduction

Dozens of large-scale fires occur every day at power plants, residential and public buildings. These incidents are very dangerous for human health. The fires consequences are characterized by high material losses. The large proportion of fires happens due to the interaction between small-sized hot particles and combustible substances (solid and liquid fuels, wood, polymers, etc.) [1–3]. Polymeric material products are used recently as a construction and decorative elements in industrial, public and residential buildings. These products have shortcomings: variation of the strength characteristics, a high fire hazard and a toxicity of pyrolysis products emitted under heating.

Values of the main integral characteristics [4,5] were determined as a result of the numerical investigations of polymeric materials (polystyrene, polymethyl methacrylate) ignition by the single hot particle. The formulated theoretical results can be applied for developing recommendations and measurements to prevent ignitions of polymeric materials products. In practice, as a rule, the fires occur at interaction between a group of particles heated to high temperatures and a condensed substance. The heat and mass transfer characteristics may be different at ignition process in the systems "single hot particle – polymeric material" and "group of hot particles – polymeric material". It is known that at high concentrations of hot particles per unit of combustible material surface the ignition delay time is close to the same characteristic for materials heating by a massive plate with a constant temperature. Therefore, the more detailed representations about the characteristics of physical and chemical processes during the induction period at the interaction between the polymeric material and the stream of particles are necessary for developing the preventive measures.

The purpose of the present work is the numerical investigation of the joint effect of several local energy sources heated to high temperatures on the integral characteristics of the polymeric material ignition, taking into account diffusion and thermal convection of combustible gases in the outside air.

2. Problem Statement

Characteristics and conditions of heat and mass transfer processes depend on the distance between two neighboring particles at different quantity of the hot particles located on the surface of condensed substance. Therefore investigation of ignition process was carried out in the system "two hot particles – polymeric material – air" (Figure 1).

As a result of the polymeric material near-surface layer ($0<t<t_d$) heating by the energy of the hot particles thermal decomposition of polymeric material occurred and combustible gases emitted to the outside air. The gaseous products of polymeric material pyrolysis mixed with the oxidant due to the diffusion and thermal convection in the air. The gas mixture was heated moving along the side faces of the energy sources ($x=x_1$, $x=x_2$, $y_1<y<y_2$). The gas-phase ignition in the system (Figure 1) happened at the critical values of the combustible component concentration and the gas mixture temperature.

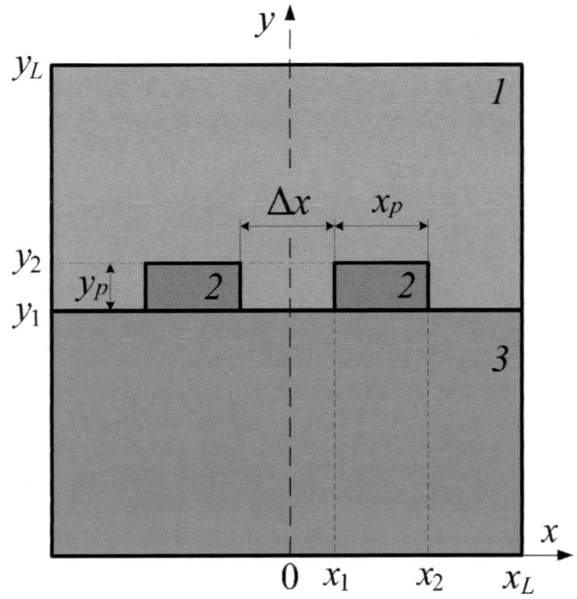

Figure 1. A scheme of the solution domain at t=0:
1 – air (gas mixture at $0<t<t_d$), 2 – hot particle, 3 – polymeric material

The assumptions were taken at the problem statement:

1. A contact between the energy sources and the polymeric material was ideal. A possible gas gap formation was not considered.
2. A possible burning out of polymeric material was not considered. Burning out of the near surface layer has an insignificant influence on the integral ignition characteristics under the local heating during short time period (less than 0.5 s).
3. As a result of polymeric material thermal decomposition only one gaseous substance with known thermal and kinetic characteristics was formed.

Ignition conditions [6]:

1. The energy released as a result of oxidation reaction of polymer material thermal decomposition products is more than the energy consumed by the hot particles to heating of a polymeric material and a gas mixture.
2. The temperature of a gas mixture in the zone of exothermic reaction exceeds the initial temperature of energy sources.

IOP Conf. Series: Materials Science and Engineering **66** (2014) 012039 doi:10.1088/1757-899X/66/1/012039

3. Mathematical Model and Solution Methods

To describe the interrelated physical and chemical processes in the system "two hot particles – polymeric material – air" (Figure 1) the mathematical model (non-stationary differential equations of mathematical physics in partial derivatives with corresponding initial and boundary conditions) was used [4,5]. It model takes into account heat and mass transfer at the conditions of chemical reaction.

At the solution of a system of differential equations the algorithm was used similar to the [4,5]. The system of equations with corresponding initial and boundary conditions was solved by the finite difference method. The alternating direction method was applied for solution the elliptical equations – continuity and motion. Difference analogues of energy and diffusion differential equations were solved by the locally one-dimensional method. The system of difference equations was solved by iteration method and sweep method at each iteration (for nonlinear equations) with using the implicit four-point difference scheme.

4. Results and Discussion

The numerical investigations of heat and mass transfer were carried out at interaction between hot steel particles in the form of parallelepiped with sizes $x_p=4\cdot10^{-3}$ m, $y_p=2\cdot10^{-3}$ m and polymethyl methacrylate (PMMA). An area with sizes $x_L=20\cdot10^{-3}$ m, $y_L=20\cdot10^{-3}$ m was selected in the polymeric material and air. The initial temperature of air and PMMA was $T_0=300$ K; the initial temperature of hot particles was varied in the range $T_p=700$–1500 K. The distance between two neighboring particles is characterized by a dimensionless value L ($L=\Delta x/x_p$). It varied in the range of $L=0.25$–2.

Table 1 shows the ignition delay time t_d for PMMA at different values of L for the initial temperature $T_p=900$ K of energy sources. It was established that in the system "two hot particles – polymeric material – air" the ignition delay time rose at increasing the distance between energy sources. At $L>1.5$ the duration of investigated process does not change ($t_d=0.0939$ s). This result is similar to the ignition delay time of PMMA by single hot particle (Table 2). Thus, the value of $L=1.5$ is the limit (at $T_p=900$ K). In this case the neighboring particles still have a joint effect on the intensity of heat and mass transfer processes.

Table 1. The dependence of PMMA ignition delay time on dimensionless value L at $T_p=900$ K

L	0.25	0.5	0.75	1	1.25	1.5	1.75	2
t_d, s	0.0874	0.0915	0.0928	0.0934	0.0937	0.0939	0.0939	0.0939

At variation of distance between two neighboring particles the maximum change of ignition delay time in the system (Figure 1) is 6.7%. The joint influence of the hot particles on the intensity of heat and mass transfer processes decreases at the increasing initial temperature of energy sources.

The dependences of PMMA ignition delay time on the initial temperature $t_d=f(T_p)$ of single particle and two particle at $L=0.25$ are given in Table 2. It is shown that at $T_p>1150$ K the values of ignition delay time are identical. This result can be explained by the fact that at increasing of T_p the heat content of energy sources rises and influence of neighboring particles on the polymeric material near-surface layer warming and gas mixture decreases. Therewith the temperature of the PMMA pyrolysis products increases, respectively, less time is required for warming up the mixture of combustible gases with an oxidant to the critical values. Ignition zones, as in the case of a single particle [4,5], are formed near the side faces of the local energy sources and shifted to the border "polymeric material – hot particle" ($x=x_1$, $x=x_2$, $y\rightarrow y_1$).

Table 2. Dependence of PMMA ignition delay time on initial temperature of hot particles

T_p, K	700	800	900	1000	1100	1200	1300	1400	1500
t_d*, s	0.2861	0.1506	0.0939	0.0685	0.0545	0.0462	0.0403	0.0358	0.0321
t_d**, s	0.2317	0.1371	0.0874	0.064	0.0519	0.0462	0.0403	0.0358	0.0321

* – single particle.
** – two particles at $L=0.25$.

Figure 2 shows the isotherms in the system (Figure 1) at the ignition moment at T_p=900 K for three different values of L.

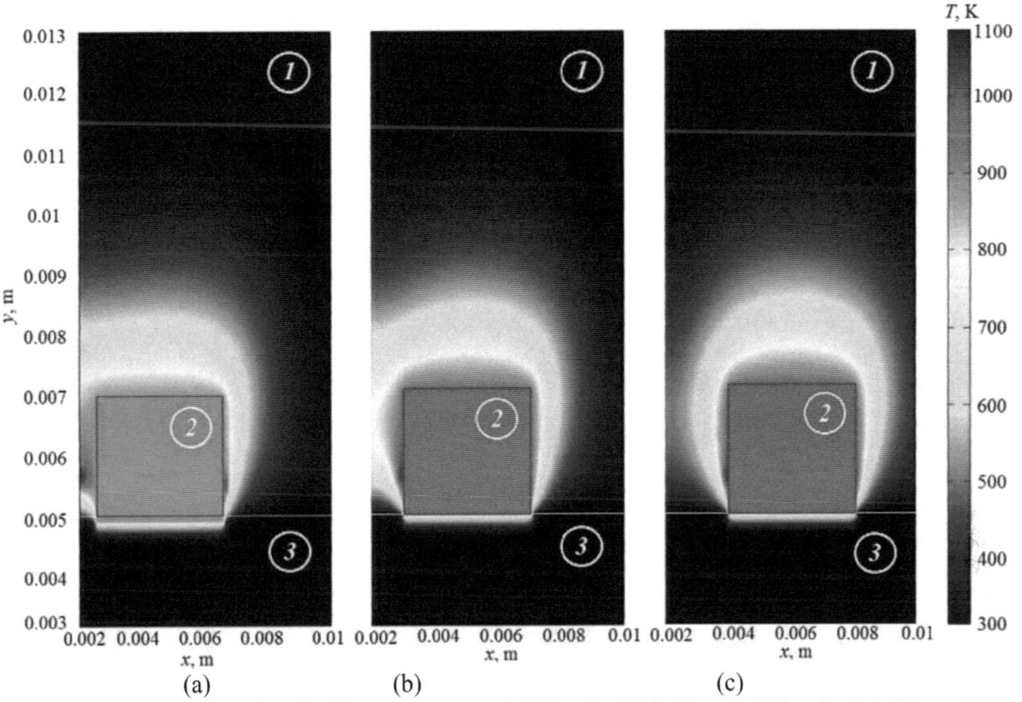

Figure 2. Isotherms (T, K) at ignition moment t_d=0.087 s, L=0.25 (a), t_d=0.092 s, L=0.5 (b), t_d=0.096 s, L=1 (c) at T_p=900 K:
1 – gas mixture, 2 – hot steel particle, 3 – PMMA

Two ignition modes were found for the system "single hot particle – polymeric material – air" [4,5]. They are characterized by the position of oxidation reaction zone in the gas area near the hot particle in dependence on the initial temperature of local energy source. In this work two ignition modes were revealed too. They are characterized by the distance between two neighboring particles. The one local ignition zone (Figure 2 (*a*)) was formed at small value of L=0.25. This zone was situated on the symmetry axis of solution domain (Figure 1). As a result of joint effect of energy sources the maximum temperature gradients and the greatest concentration of PMMA pyrolysis products is in this area compared with the area $x>x_2$. At increasing of L and another equal conditions (Figure 2 (*b*), (*c*)) two ignition zones were formed. Each of them was located in a small neighborhood of own energy source. At L>0.25 the concentration of polymeric material pyrolysis products is the same near the side faces of hot particles ($x=x_1$, $x=x_2$, $y_1<y<y_2$). However, the area of maximum temperature gradients was formed between the hot particles ($0<x<x_1$, $y_1<y<y_2$), where the ignition conditions were executed.

5. Conclusion

Numerical investigations carried out at the using of developed mathematical model of heat and mass transfer processes in the system "two hot particles – polymeric material – air" allowed to determine the maximum deviation of PMMA ignition delay time 12.3% (at L=0.25 and T_p=700 K) at polymeric material heating by a single hot particle and a group of hot particles. It means that analysis of the characteristics and laws of heat and mass transfer at the gas phase ignition of polymeric material by several small-sized particles may be accomplished at the using simpler mathematical model "hot single particle – polymeric material – air".

Acknowledgments
The reported study is supported by the Grant of the President of the Russian Federation (project MK-2391.2014.8) and subsidy within implementation of the Program of competitiveness increase of National research Tomsk polytechnic university.

References

[1] Wang Y and Zhang J 2012 Influences of specimen size and heating mode on the ignitability of polymeric materials in typical small-scale fire test conditions *Fire and Materials* **36** 231–240
[2] Bonati A, Merusi F, Polacco G, Filippi S, Giuliani F 2012 Ignitability and thermal stability of asphalt binders and mastics for flexible pavements in highway tunnels *Construction and Building Materials* **37** 660–668
[3] Liu Y, Ren Y, Dong E 2011 Development of enviromental friendly flame retardants for wood plastic composites (WPC) *Advanced Materials Research* 332–334 1880–1883
[4] Glushkov D and Strizhak P 2012 Heat and mass transfer at ignition of solid condensed substance with relatively low calorific power by a local energy source *Journal of Engineering Thermophysics*. **21** 69–77
[5] Glushkov D, Kuznetsov G, Strizhak P 2014 Theoretical assessment of ignition stability of a typical polymeric material by a local power source *Fire and Explosion Safety* **23** 10–19 [in Russian]
[6] Vilyunov V and Zarko V 1989 *Ignition of Solids* Elsevier Amsterdam

Modelling of heat and mass transfer to solve the problem of particle ignition water-coal fuel

VV Salomatov[1a], **SV Syrodoy**[2b] and **NY Gutareva**[2c].

[1]Novosibirsk, 630090, Russia, Institute of Thermophysics named after S.S. Kutateladze SB RAS
[2]Tomsk, 634050, Russia, National Research Tomsk Polytechnic University.

[a]vvs@itp.nscl.ru,[b]ssyrodoy@yandex.ru,[c]gutarevanadezhda@mail.ru,

Abstract. The problem of particle ignition coal-water fuel (WCF) has been solved numerically, located at a high flow environment. Two different approaches to describe the physics of the process have been applied. According to the results of numerical simulation of the effect of the scale has been set about ambient temperature, the particle size on the inertia of the studied processes. A comparison of ignition delay times has been obtained using both methods of describing heat and mass transfer processes preceding fire.

1. Introduction

Use of coal-water fuel (WCF) in the heat is associated with the solution of a number of technical and technological problems [1]. One of the problems is the burning particles of WCF. The presence of water in such a structure leads to an increase in fuel required to initiate the combustion reaction heats the coal and hence significant increase of ignition delay time [2]. Known experimental data [1,2] on the ignition characteristics of particles WCF have been obtained only for sufficiently narrow ranges of parameters describing the process, and a small group of fuels.

A set of simplified formulations has been developed in the absence, until recently, the general mathematical models, methods and results of solving heat and mass transfer in the structure of WCF under intense heat [3-5]. Only one stage of a multistage process heat and mass transfer has been considered in a particle flowing coal-water fuel when heated in a stream of high-temperature gases in solving each of the problems [3-5]. The results of mathematical modeling of the processes preceding ignition WCF, based on simplified models [3-5] are the basis for assessing the true values of the ignition delay time of the particles of coal-water fuel (t_{ign}). It is advisable to compare these values with those obtained (t_{ign}) solving the problem of heat and mass in the formulation [11], taking into account the simultaneous occurrence at specific intervals throughout the group of basic physical and chemical processes [3-5] in the structure of WCF when heated to high temperatures. The joint implementation of "stages" [3-5] warming-up, evaporation, thermal decomposition of the organic portion of the coal and the subsequent ("secondary") products of chemical reactions physicochemical transformations between them in the appropriate temperature ranges are determined by a large temperature gradient across the thickness of the particle and continuous WCF state of the change in time as the removal of fuel from the heating surface.

Four problems have been solved in this paper. Three productions of which are given in [3-5] not resolved approximately analytically and numerically (finite difference method). Results of the solution in the form of dependency characteristic times of the processes of heating, evaporation and thermal decomposition (preceding ignition) the ambient temperature are shown in Fig. 1-3. Figure 3 shows the main characteristics of the process — the ignition delay time $t_з$ (T_c).

2. Mathematical simulation

The mathematical model, the corresponding formulation of the problem [3-5] is described by the following system of differential equations:

- Energy equation for the initial (wet) of the fuel:

$$C_1\rho_1\frac{\partial T_1(r,t)}{\partial t}=\lambda_1\cdot\left[\frac{\partial^2 T_1(r,t)}{\partial r^2}+\frac{2}{r}\cdot\frac{\partial T_1(r,t)}{\partial r}\right]. \quad (1)$$

$t>0,\ 0<r<r_V,\ T\le T_V;$

- Energy equation for "dehydrated" part WCF:

$$C_2\rho_2\frac{\partial T_2(r,t)}{\partial t}=\lambda_2\cdot\left[\frac{\partial^2 T_2(r,t)}{\partial r^2}+\frac{2}{r}\cdot\frac{\partial T_2(r,t)}{\partial r}\right]+Q_i\cdot\rho_2\cdot\exp\left(-\frac{E}{R\cdot T(r,t)}\right). \quad (2)$$

$t>0,\ r_V<r<r_0,\ T\ge T_V;$

The system (1.2) has been solved with the following boundary conditions and closing ratios:

$0<r<r_0;\ T(r,0)=T_0;$

$$-\lambda_1\frac{\partial T_2(r_0,t)}{\partial r}=\alpha\cdot[T_c-T_2(r_0,t)]+\varepsilon\cdot\sigma\cdot[T_c^4-T^4{}_2(r_0,t)]. \quad (3)$$

$$\frac{\partial T_1(0,t)}{\partial r}=0. \quad (4)$$

The problem has been solved in a general setting heat and mass transfer in a particle of coal-water fuel in addition to problems [3-5]. The system of nonlinear no stationary differential equations with appropriate boundary conditions describing within the overall physical model the process under study is presented below:

- Energy equation for the initial (wet) of the fuel:

$$C_1\rho_1\frac{\partial T_1(r,t)}{\partial t}=\lambda_1\cdot\left[\frac{\partial^2 T_1(r,t)}{\partial r^2}+\frac{2}{r}\cdot\frac{\partial T_1(r,t)}{\partial r}\right]. \quad (5)$$

$t>0,\ 0<r<r_V,\ T\le T_V;$

- Energy equation for "dehydrated" part WCF:

$$C_2\rho_2\frac{\partial T_2(r,t)}{\partial t}=\lambda_2\cdot\left[\frac{\partial^2 T_2(r,t)}{\partial r^2}+\frac{2}{r}\cdot\frac{\partial T_2(r,t)}{\partial r}\right]-\sum Q_i\cdot W_i-u_s Cp_s\cdot\rho_s\frac{\partial T_2(r,T)}{\partial r}. \quad (6)$$

$t>0,\ r_V<r<r_0,\ T\ge T_V;$

- The equation of chemical kinetics for the organic portion of the fuel:

$$\frac{\partial\eta_2(r,t)}{\partial t}=(1-\eta_2(r,t))\cdot k_0\cdot\rho\cdot\exp\left(-\frac{E}{R\cdot T_1(r,t)}\right). \quad (7)$$

$t>0,\ r_V<r<r_0,\ T\ge T_V;$

- Filtration equation for water vapor and gaseous products of thermal decomposition:

$$\frac{\partial p_s(r,t)}{\partial t}=\frac{m\cdot\xi\cdot Z}{K_P}\left(\frac{\partial p_s^{\,2}(r,t)}{\partial r^2}+\frac{2}{r}\frac{\partial p_s(r,t)}{\partial r}\right). \quad (8)$$

$t>0,\ r_V<r<r_0,\ T\geq T_V;$

The boundary condition of the 4th kind has been used for the system (1-4) and (5-17) on the interface "original WCF — dry coal":

$$\lambda_1\left.\frac{\partial T_1(r,t)}{\partial r}\right|_{r=r_u-0}-\lambda_2\left.\frac{\partial T_2(r,t)}{\partial r}\right|_{r=r_u+0}=Q_V\cdot W_V. \quad (9)$$

$T_1(r_V;t)=T_2(r_V;t)=T_V$

The system of equations (1-9) has been solved with the following boundary conditions and closing ratios:

$0<r<r_0;\ T(r,0)=T_0;\ P(r,0)=P_0;\ \eta(r,0)=\eta_0;$

$$-\lambda_2\frac{\partial T_2(r_0,t)}{\partial r}=\alpha\cdot[T_c-T_2(r_0,t)]+\varepsilon\cdot\sigma\cdot[T_c^4-T^4{}_2(r_0,t)]+\sum_i Q_i\cdot W_i. \quad (10)$$

$$\frac{\partial T_1(0,t)}{\partial r}=0. \quad (11)$$

$$\frac{\partial P(r_V,t)}{\partial r}=\frac{K_P}{v}V_P(r_V;t). \quad (12)$$

$$p(r_0;t)=p_0 \quad (13)$$

Speed reversible reaction has been calculated from the following expression:

$$W_i=k_i\cdot c_i\cdot\rho_i\cdot\exp\left(-\frac{E}{R\cdot T(r,t)}\right). \quad (14)$$

Pyrolysis rate has been calculated using the following equation:

$$W_i=(1-\eta(r,t))\cdot k_i\cdot\rho_i\cdot\exp\left(-\frac{E}{R\cdot T(r,t)}\right). \quad (15)$$

ρ-density gaseous components has been determined from the equation of state:

$$\rho_i=\frac{P(r,t)\cdot\mu}{T(r_0;t)\cdot R}. \quad (16)$$

Filtration rate has been determined by a pair of Darcy's law:

$$u=-\frac{K_P}{v}\frac{\partial P_P}{\partial r}. \quad (17)$$

Notation, the initial temperature of T_0— particles, K; T_c— ambient temperature, K; r_V –radius evaporation front, m; r_0— outer radius of the particle, m; σ — constant blackbody radiation, $\frac{W}{m^2\cdot K^4}$; α-convective heat transfer coefficient, $\frac{W}{m^2\cdot K}$; λ_1— thermal conductivity of the original WCF $\frac{W}{m\cdot K}$; λ_2— dry thermal conductivity of the WCF, $\frac{W}{m\cdot K}$; C_1-heat capacity of the initial part of WCF, $\frac{J}{kg\cdot K}$; C_2-is the heat capacity of the initial part of the dry part of WCF, $\frac{J}{kg\cdot K}$; W_V— mass evaporation rate of water,

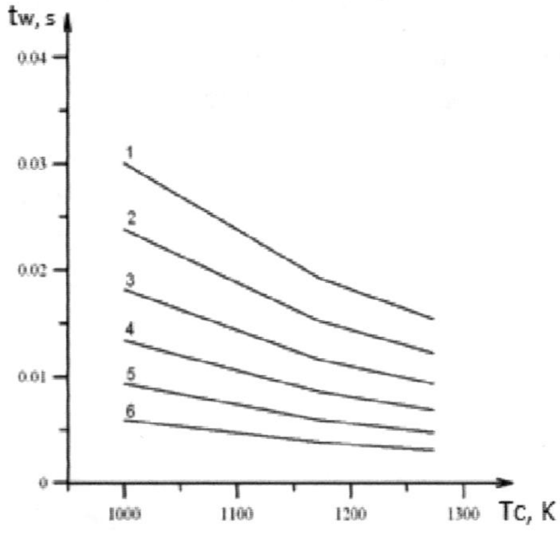

Figure.1. Dependence of the characteristic times for warm particle diameter WCF: 1-0.9 · 10^{-3}м; 2-0,8 · 10^{-3}м; 3-0.7 · 10^{-3}м; 4-0.6 · 10^{-3}м; 5-0.5 · 10^{-3}м; 6-0.4 · 10^{-3}м.

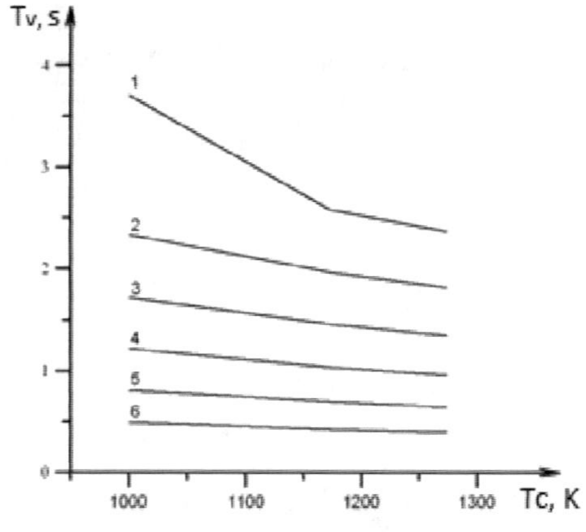

Figure.2. Dependence of the characteristic times for vaporization particle diameter WCF: 1-0.9 · 10^{-3}мм; 2-0,8 · 10^{-3}мм; 3-0.7 · 10^{-3}мм; 4-0.6 · 10^{-3}мм; 5-0.5 · 10^{-3}мм; 6-0.4 · 10^{-3}мм

$\frac{kg}{m^2 \cdot s}$; Q_V— thermal effect of the evaporation of water, $\frac{J}{kg}$; Q_i-thermal effect of the chemical reaction, $\frac{J}{kg}$; k_i-pre-exponential factor of the chemical reaction, $\frac{1}{s}$; E_i-activation energy of a chemical reaction, $\frac{J}{mol}$; u_S—filtration rate of water vapor, $\frac{m}{s}$; Cp_S-isobaric heat capacity of water vapor, $\frac{J}{kg \cdot K}$; ρ_S— vapor density, $\frac{kg}{m^3}$; p_S-vapor pressure in Pa; m-porosity; Z— degree of compressibility of water vapor; K_P-permeability porous structure, m^2; μ-molar mass of water, $\frac{kg}{mol}$; ν— kinematic viscosity of water vapor, $\frac{m^2}{s}$; C_i— concentration of the respective component of the chemical reaction; i— ordinal chemical reaction component.

Mass evaporation rate has been calculated from the expression:

$$W_V = W_0 \cdot \exp\left(\frac{Q_V \cdot \mu \cdot (T_1 - T_0)}{R \cdot T_0 \cdot T_1}\right).$$

Where: T_1— evaporation temperature at the boundary, K; T_0-fluid temperature corresponding to the point of freezing, K; W_0— rate of evaporation at temperature, T_0; μ-molar mass of water vapor, $\frac{kg}{mol}$; R — universal gas constant, $\frac{J}{mol \cdot K}$.

The system (1) — (5) has been solved by finite difference method. The iterative algorithm has been used for solving nonlinear problems of heat and mass, designed for the simulation of intense evaporation of liquids [6, 7] and thermal decomposition of polymeric materials [8] under the intense heat. Choosing grid parameters should be adjusted to the specifics of the problem being solved (high temperature gradients and large rate of change of T) by analogy with [9, 10]. According to the results of solving the problem of heat conduction for a spherical particle WCF (Fig.1.) It can be noted that the characteristic time of warming up in the case of particles with a diameter of 0.4 mm to 0.9 mm, less than $3 \cdot 10^{-2}$ c.

Comparison of these values with the experimental data [3.4] allows us to conclude that the process of inert (non-thermal decomposition and evaporation of organic carbon) warming-up time is not more than 2.1% of the total duration. Consequently, the time period during stepwise consideration can be analyzed independently. Dripping processes (Fig.2.) extend substantially longer (about 100 fold) as

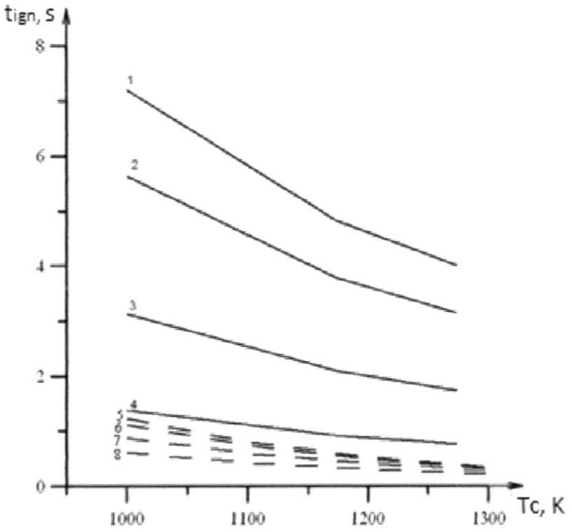

Figure.3. Dependence of the characteristic times of the ignition delay WCF particle diameter: 1-0.9 $\cdot$ 10^{-3}мм; 2-0,8 $\cdot$ 10^{-3}мм; 3-0.7 $\cdot$ 10^{-3}мм; 4-0.6 $\cdot$ 10^{-3}мм; - solution of the system (1-4). 5-0.9 $\cdot$ 10^{-3}мм; 6-0,8 $\cdot$ 10^{-3}мм; 7-0.7 $\cdot$ 10^{-3}мм; 8-0.6 $\cdot$ 10^{-3}мм; - solution of the system (5-17).

compared with the period of heating and the inert differ significantly from the total task time of ignition.

Figure 3 shows the results of numerical modeling of the test process as a dependency t_{ign} (T_c) for particles of different diameters WCF obtained during the solution of (1-4) and (5-17). Comparison of the ignition delay time, resulting from the application of two different approaches to the description of heat and mass transfer in the particle coal-water fuel (Fig. 3) shows quite significant differences t_{ign} values. Last process, can be most likely due to the combined influence of these processes of heat transfer on the formation of the temperature field in the particle. For example, the evaporation of water takes place at any temperature up to the reflux temperature. The speed of evaporation and accordingly, the intensity of the absorption of energy in a small neighborhood of the evaporation surface are exponentially dependent on temperature. When relatively low temperatures, this heat sink is not taken into account in solving the problem of heat conduction only, the temperature at each point of the particles will be considerably higher WCF compared with the real one. As a result, the delay time t_{ign} is less than the true value. Similar effects will occur stepwise modeling in the future with increasing temperature and as you move the water and evaporation fronts beginning of thermal decomposition of the organic portion of the coal particles deep into the WCF. Furthermore, filtering products of vaporization through the porous structure is formed after the completion of the processes of evaporation and thermal decomposition would result in both a reduction of the absolute values of the temperatures at each point, and a decrease gradient T.

In conclusion, it should be noted that despite the fairly significant differences ignition delay times calculated using two different approaches to the above description of the basic processes occurring in coal-water fuel particle at high temperature, the simplified model [3-5] provide a generally acceptable for the initial stage analysis of numerical estimates of the values of t_{ign}.

References

[1] Deljagin G.N. 1967. Problems in the theory of combustion of coal-water slurry in a stream of air. Conf. (Moscow).p. 45-55.

[2] Deljagin G.N. 1969. In. Kinetics and Aerodynamics combustion processes. (Moscow).p.111-127.

[3] Huang Z. et all. 1986. Proc. 8-th Int.. Symp. Coal Slurry Fuels Preparation and Utilization. (USA: Orlando.) V **1**. p. 343-358.

IOP Conf. Series: Materials Science and Engineering **66** (2014) 012040 doi:10.1088/1757-899X/66/1/012040

[4] Matthews K.J., Jones A.R. 1986. Proc. 8-th Int. Symp. Coal Slurry Fuels Preparation and Utilization. (USA: Orlando.) V **1**. p.. 388-407.
[5] Salomatov V.V., Kravchenko I.V. 2007. Combustion and plasma chemistry. V. **5** no **3** p. 178-188.
[6] Kuznetsov G.V., Stpizhak P.A. 2008. Journal of Engineering Thermophysics. V. **17**. no **3**. p. 244-252.
[7] Kuznetsov G.V., Stpizhak P.A. 2009. Journal of Engineering Thermophysics. V. **18**. no **1**. p. 72-79 .
[8] Strakhov V.L., Garashchenko A.N., Kuznetsov G.V., Rudzinskii V.P., 2001. Combustion, Explosion and Shock Waves. V. **37**, no **2**, p. 178-186.
[9] Kuznetsov G.V., Mamontov G.Ya., Taratushkina G.V., 2004. Combustion, Explosion and Shock Waves. T **40.** no **1**. p. 70-76.
[10] Kuznetsov, G.V., Mamontov G.Ya., Taratushkin, G.V. 2004. Chemical physics. V.**23.** no **5** p. 62-68.
[11] Syrodoy S.V., Salomatov V.V., Kuznetsov G.V., 2013. Polzunovskii Vestnik. V. **4** no. **3** p. 28-33.

Numerical investigation of conjugate heat transfer in a local working area in conditions of its radiant heating

A Nee[1,a] and T Nagornova[1,b]
[1]Tomsk Polytechnic University, Tomsk, Russia

E-mail: nee_alexander@mail.ru

Abstract. Mathematical modeling of unsteady conjugate heat transfer in a closed rectangular area with a local heat-conducting object in a gas cavity in conditions of radiative heat supply is conducted. Fields of temperatures and stream functions, illustrating the influence of Grashof number on the character of heat transfer are formulated. The dependence of average Nusselt number from time at different Grashof numbers is given. The influence of heat-conductive object on the intensity of heat transfer in under study solution domain is shown.

1.Introduction

Use of gas infrared radiators (GIR) [1] is the most prospective at the local heating of working area.

For investigation the temperature regimes of premises heated by GIR, a model [2], which allows the unsteady dimensional character of conjugate conductive - convective heat transfer was developed. However, in study [2], an assumption that the flow of radiant energy coming from GIR is only evenly distributed over the inner surface of the lower boundary was made. Also, the model [2] doesn't describe the effect of heat-conducting objects located in the working area on the intensity of heat transfer.

Experimental study [3] of radiant heating system has a number of significant defects, since it ignores, as a rule, the dimensional dynamics of heat transfer.

The aim of this study is the numerical investigation of unsteady radiant heating process of a typical heat-conducting manufacturing object in the gas cavity of a closed rectangular area in conjugate formulation.

2.Problem formulation

The boundary value problem of conjugate heat transfer in area consisting of six rectangular subdomains (fig. 1) is considered. Heat insulation conditions were adopted at external borders. At internal borders "air – enclosure structures", "air – heat supply object" are the fourth type boundary conditions.

Assumptions that thermal properties of air and enclosure structures don't depend on temperature are introduced. Flow regime is laminar. The air is considered as a Newtonian fluid, incompressible, satisfying the Boussinesq approximation and absolutely transparent for thermal radiation. Total radiant flux coming from the GIR can be represented as the sum of heat fluxes values of which were determined by zonal method as shown in [4].

Investigated heat transfer process is described by the unsteady Navier - Stokes equations and the energy for the air and the heat conduction equation for enclosure structures within the adopted model. Dimensionless equations of Navier - Stokes and energy in variables "vorticity Ω - stream function Ψ - temperature Θ " are as follows [2, 5]:

$$\frac{\partial \Omega}{\partial \tau}+U\frac{\partial \Omega}{\partial X}+V\frac{\partial \Omega}{\partial Y}=\frac{1}{\sqrt{\mathrm{Gr}}}\cdot\nabla^2\Omega+\frac{1}{2}\frac{\partial \Theta_1}{\partial X}, \tag{1}$$

IOP Conf. Series: Materials Science and Engineering **66** (2014) 012041 doi:10.1088/1757-899X/66/1/012041

$$\nabla^2\Psi = -2\cdot\Omega, \tag{2}$$

$$\frac{\partial\Theta_1}{\partial\tau} + U\frac{\partial\Theta_1}{\partial X} + V\frac{\partial\Theta_1}{\partial Y} = \frac{1}{\Pr\sqrt{\mathrm{Gr}}}\cdot\nabla^2\Theta_1, \tag{3}$$

$$\frac{\partial\Theta_2}{\partial Fo_2} = \nabla^2\Theta_2, \tag{4}$$

$$\frac{\partial\Theta_3}{\partial Fo_3} = \nabla^2\Theta_3, \tag{5}$$

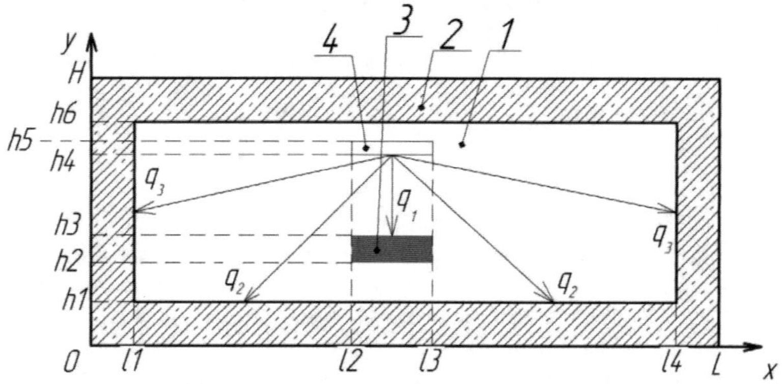

Figure 1. Solution domain: 1) air; 2) enclosure structures; 3) heat supply object; 4) gas infrared radiator (symbolic notation).

The initial conditions for the system of equations (1) – (5) are as follows:

$$\Psi(X,Y,0)=0,\ \Omega(X,Y,0)=0,\ U(X,Y,0)=0,\ V(X,Y,0)=0,$$
$$\Theta_1(X,Y,0)=\Theta_2(X,Y,0)=\Theta_3(X,Y,0)=0.$$

The boundary conditions at the outer boundaries of the solution domain are as follows:

$$X=0,\ X=1,\ 0<Y\le 1:\ \frac{\partial\Theta_2(X,Y,\tau)}{\partial X}=0,$$
$$Y=0,\ Y=1,\ 0<X\le 1:\ \frac{\partial\Theta_2(X,Y,\tau)}{\partial Y}=0.$$

at internal interfaces "heat supply object – air", "solid wall – air", parallel to the axis OX:

$$\Psi=0,\ \frac{\partial\Psi}{\partial Y}=0,\ \begin{cases}\Theta_i=\Theta_j,\\ \dfrac{\partial\Theta_i}{\partial Y}=\dfrac{\lambda_j}{\lambda_i}\cdot\dfrac{\partial\Theta_i}{\partial Y}+Ki_{q_k},\end{cases} \text{где} \left| \begin{matrix} i=\overline{1,3}\\ j=\overline{1,3}\\ k=1,2.\end{matrix}\right.$$

at internal interfaces "heat supply object – air", "solid wall - air", parallel to the axis OY:

$$\Psi=0,\ \frac{\partial\Psi}{\partial X}=0,\ \begin{cases}\Theta_i=\Theta_j,\\ \dfrac{\partial\Theta_i}{\partial X}=\dfrac{\lambda_j}{\lambda_i}\cdot\dfrac{\partial\Theta_i}{\partial X}+Ki_{q_3},\end{cases} \text{где} \left| \begin{matrix} i=\overline{1,3}\\ j=\overline{1,3}\end{matrix}\right.$$

where $Fo-$ Fourier number; $Gr-$ Grashof number; $Ki-$, Kirpichev number; $\Pr-$ Prandtl number; X,Y – dimensionless Cartesian coordinates; U, $V-$ dimensionless velocities along X,Y directions correspondingly; ∇ – dimensionless Laplace operator. Indexes: 1,2,3 - design elements; for Ki: - heat flow values. Indexes: 1, 2, 3 – element number; for Ki: q_1, q_2, q_3 – heat flux values.

Equations (1) – (5) with the corresponding initial and boundary conditions were solved by applying the finite difference method, as in [5]. Locally one scheme of A.A. Samarskiy was used for approximation of equations (1), (3) – (5) [6]. Approximation of Poisson's equation was done by the scheme of variable directions [6]. Woods condition [6] was used to determine the boundary conditions for the vortex velocity. One - dimensional difference analogues were solved by the sweep method [5]. In order, to evaluate the reliability of results of the computational modeling, the conservatism of the difference scheme was checked analogously to [7].

3.The results of the numerical simulation

Numerical investigation was performed for the following values of dimensionless criteria: $10^5 \le Gr \le 10^7$, Pr=0,71, Ki_1=42, Ki_2=25, Ki_3=5. The results of solving boundary value problem for different Gr are shown in figures 2 and 3.

Evolution of the air flow with increasing Gr is shown in figure 2 and 3. Lifting force is insignificant at low Grashof numbers and heat transfer in solution domain under study occurs mostly by heat conduction. Increasing Gr leads to a modification of convective cells on the left and right of the heat supply object. Boundary layer with a considerable temperature gradient at the surface y = h1, l1 <x <l4 is formed at $Gr = 10^7$ and ascending air is created at section X = 0,6. As the heat supply object is conductive, it absorbs radiant flux directed at its upper surface. Consequently, the air near the surface of the object is heated not as intense as at the lower base of the solution domain.

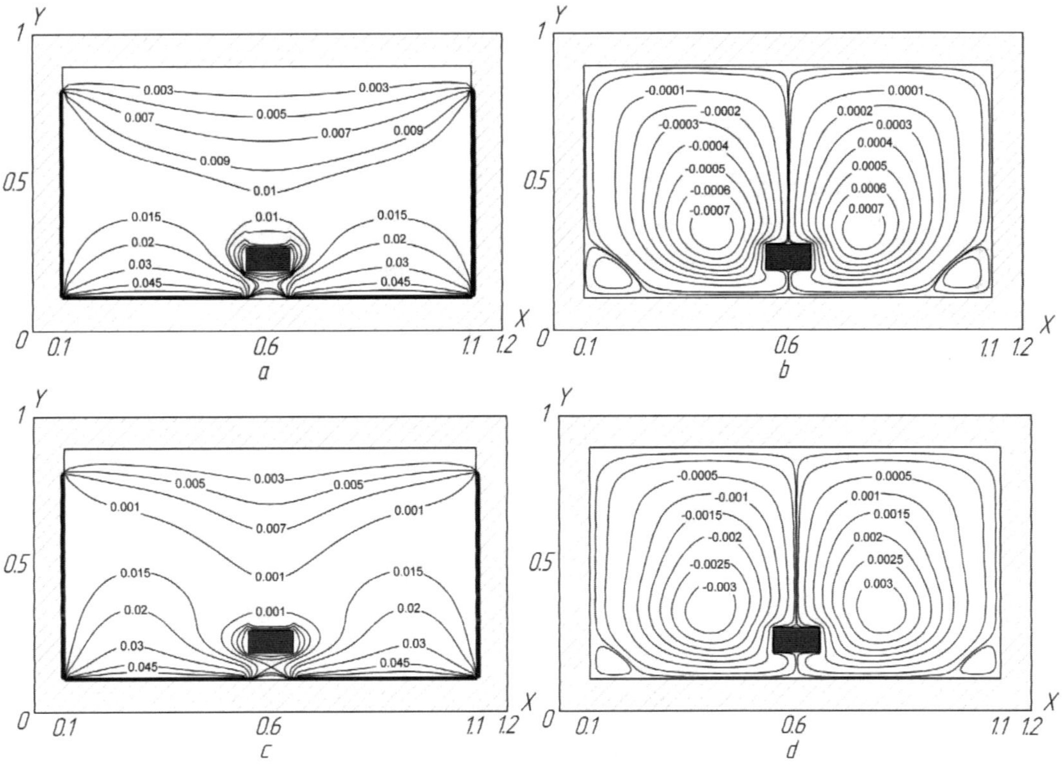

Figure 2. Fields of temperature (a,c) and the stream function (b,d) at при τ = 3600: $Gr = 10^5$ (c, d) $Gr = 10^6$ (c, d).

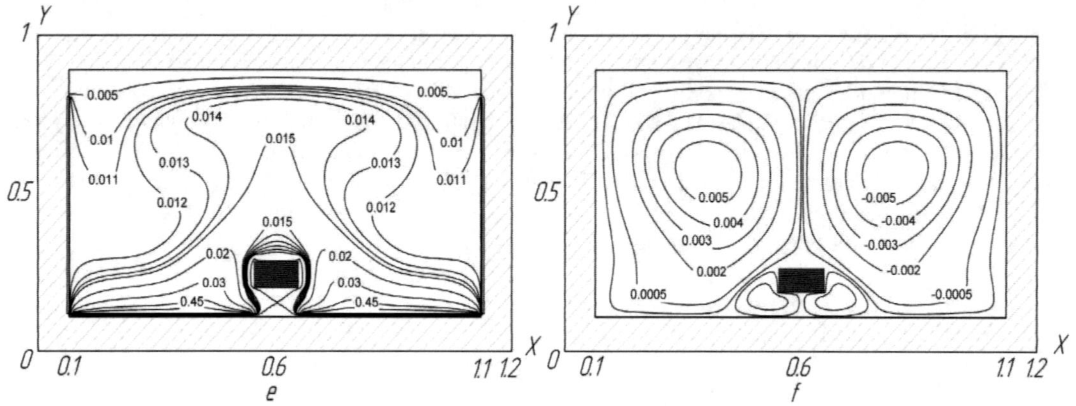

Figure 3. Fields of temperature (e) and the stream function (f) at при τ = 3600: $\mathrm{Gr} = 10^7$.

Analysis of the influence of the Grashof number on the average Nusselt number at the interface of "enclosure structure - air» (y = h1, 11 <x <l4) is carried out by analogy with [8]:

$$\overline{Nu} = \int_{0,1}^{1,1} \left|\frac{\partial\Theta}{\partial Y}\right|_{Y=0,12} dX.$$

Dependences $\overline{Nu}$ from τ for different Gr are presented in figure 4.

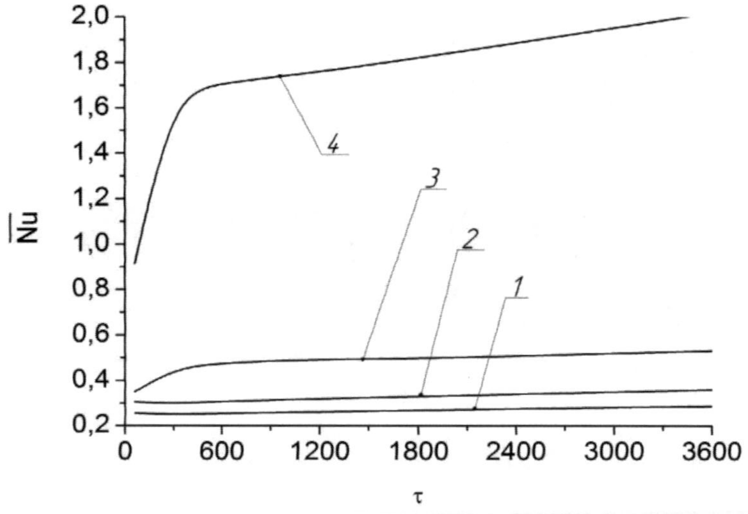

Figure 4. Dependences $\overline{Nu}$ from τ: 1 – $\mathrm{Gr} = 10^5$; 2 – $\mathrm{Gr} = 10^6$; 3 – $\mathrm{Gr} = 10^7$; 4 - $\mathrm{Gr} = 10^7$, heat-conducting object in gas cavity is absent.

$\overline{Nu}$ is increased with rising Gr. Average dimensionless heat transfer coefficient varies insignificantly with time at $\mathrm{Gr} = 10^5, 10^6$, because the heat exchange rate for such values of similarity criterion is low. Increasing the average Nusselt number occurs before $\tau \approx 600$ at $\mathrm{Gr} = 10^7$. Average dimensionless heat transfer coefficient varies insignificantly in the sequel. The presence of heat-conducting object in the gas cavity decreases $\overline{Nu}$ 5 times, and therefore significantly effects the heat transfer rate in the solution domain

4.Conclusion

The numerical analysis of the radiant heating process of manufacturing heat-conducting object in a closed rectangular area in conjugate formulation is done. It was established that Grashof number significantly effects both the movement of air masses and the temperature regime in the solution domain under study. The influence of heat-conducting object in a gas cavity on the intensity of heat transfer is shown.

This work is the part of the Research Works of the State publisher "Nauka" (code of the target-oriented scientific-technological program 2.1321.2014)

REFERENCES

[1] Roth K, Dieckmann J and Brodrick J 2007 Infrared Radiant Heaters, *ASHRAE J.* **4** pp 72 – 73.
[2] Kuznetsov G, Maksimov V, Nagornova T, Kurilenko N and Mamontov G 2013 Heat transfer under heating of a local region of a large production area by gas infrared radiators *J. of Eng. Phys. and Thermoph.* **86** pp 519 – 524.
[3] Kimball B, Conley M and Lewin K 2012 Performance and energy costs associated with scaling infrared heater arrays for warming field plots from 1 to 100 m *Theor. and App. Clim. J.* **108** pp 247 – 265.
[4] Bukhmirov V, Krupennikov S and Solnyshkova Yu 2010 Algorithm for solution of radiant heating premises systems *Vestn. Ivanovsk. Gos. Energ. Univ.* **Issue 4** pp 23–25.
[5] Kuznetsov G and Sheremet M 2009 Mathematical modeling of complex heat transfer in a rectangular enclosure *Thermoph. and Aer. J.* **16** pp 119 – 128.
[6] Paskonov V, Polezhaev V and Chudov L 1984 Numerical Simulation of the Heat and Mass Transfer Processes *Nauka*.
[7] Kuznetsov G and Strizhak P 2008 Heat and mass transfer at the ignition of liquid substance by a single "hot" particle *J. of Eng. Thermoph.* **16** pp 244 – 252.
[8] Kuznetsov G and Sheremet M 2008 New approach to the mathematical modeling of thermal regimes for electronic equipment *Rus. Microel. J.* **37** pp 131 – 138.

Peculiarities of producing decorative articles by the method of lost foam casting

Oleg Utyev[1], Anna Lukyanchenko[2]
Tomsk, 634050, Russia, National Research Tomsk Polytechnic University
E-mail: [1] utievv@mail.ru, [2] annlukyann@gmail.com

Abstract. This paper investigates the method of lost foam casting performed with consumable patterns. The dependence of the fluidity of the silumin on the polystyrene's characteristics, casting temperature and the pattern's parameters is revealed. Recommendations regarding the application of various polystyrene types for artistic casting are provided.

1. Introduction
The lost foam casting (LFC) method is one of the cutting edge techniques for producing castings, which appeared in the second half of the XXth century. Lost foam casting is a casting method with the use of patterns, which are not removed from the mold but remain in it and are gasified under the action of the heat energy of a metal being poured into the mold. The described technique was firstly patented in 1958 by the American architect F.H. Schroyer [1].

2. Theoretical background
The LFC method is consist of following main stages.

- Production of a pattern. For decorative castings a pattern is produced from the ready-to-use foamed polystyrene slabs by using high-speed milling or a device which has a heated nichrome wire with a current flowing through it at 12V. Separate parts of the pattern can be glued together with the easily burnable materials while the defects can be removed by the use of wax materials (paraffin, stearin, etc.).
- Assembly of the pattern blocks. For the LFC method, the bottom gating system, desirably with air gates, is used. The pattern blocks are assembled by gluing.
- Dressing of a pattern. A pattern is dressed with non-stick paints to prevent the burn-on of the casting surfaces and the deformation of the pattern.
- Molding of a pattern. When molding with a dry filler without a binder, we use perforated flasks the bottom of which is covered with a layer of sand. A pattern block is then affixed to the bottom of the flask. Vibration is applied to ensure better compaction of the sand.
- Pouring of a metal. The metal pouring rate affects the casting surface quality: it should not be too high or too low.

The downside of the lost foam casting method is a single use of a pattern [2].
The lost foam casting method is mostly applied in industrial production. It is used to simplify the molding process, increase productivity, enhance the casting quality, reduce the production costs, etc [3].

For artistic works of authorship, which are normally produced in a single copy, packaging polystyrene and a polystyrene commonly applied for heat insulation are used. The present paper provides an investigation of the polystyrene brands used to produce consumable patterns, as well as the optimum sections and lengths of the samples, proper casting temperature and method for the specified alloy.
As with any other method of casting, lost foam casting is characterized by a number of technological features. In this paper, the relationship between the cross-section and the length of the pattern is examined on the example of the AL12 (10-13% Si, the balance Al) and AL5 (1-1,5% Cu, 4,5-5,5% Si,

the balance Al) silumin alloys and initially two polystyrene brands (expanded polystyrene (EPS-25) and extruded polystyrene Penoplex 35). Further on, Penoplex 31 is added to extend the range of the investigation.

To determine the lowest possible cross section of a casting, test samples of the 150 mm length and the 5×5mm (S=25 mm2), 10×10mm (S=100 mm2), 15×15 mm (S=225 mm2) cross-sections have been made. The 100×100×20 mm test samples of each polystyrene brand are weighed and measured with the accuracy of ± 0.05 mm. By using the obtained data, density of the samples of each polystyrene brand is found: 15 kg/m3 for EPS-25, 29 kg/m3 for Penoplex 31, and 31 kg/m3 for Penoplex 35. Then, three samples of the different cross sections of each polystyrene brand are assembled into the pattern blocks. The AL12 silumin alloy is used for casting at a pouring temperature of 600, 700, 800, 900°C. The obtained samples have been measured at the gating system. The data are presented in the diagram (figure 1). Polystyrene EPS-25appeared to provide the best results allowing the production of the casting with the 100 mm2 section for the entire length of the pattern. The other polystyrene brands demonstrated excellent results at the 225 mm2 cross sectional area of the samples and a melt pouring temperature of above 800°C.

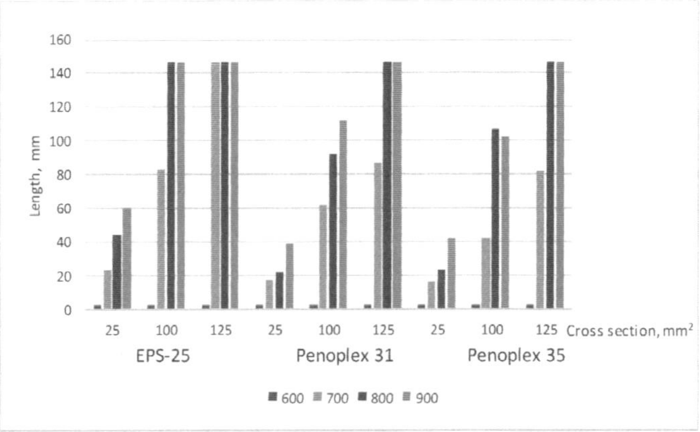

Figure 1. Dependence of length and cross section area of the samples on the silumin's pouring temperature

Based on the results analysis of the described investigation stage it can be stated that small cross sections of the patterns (≈25 mm^2) are acceptable when the product's length is limited (20-40mm). The best results of casting by the LFC method can be achieved by using the EPS-25 polystyrene and a pouring temperature of 800°C. Max results are achieved with the 100 and 225 mm^2 cross sectional area of the samples [4].

The next step is to study the dependence of the AL5 silumin's fluidity on pouring temperature using spiral test samples. To determine fluidity, different test samples cast in the molds of a special design are used. The most common varieties are different spiral samples. Molds designed for casting such samples have an extended spiral channel located in the horizontal plane. The cross section of the channel is constant on its entire length and typically has a trapezoid shape of the 8 mm height, 8 mm width at the top, and 7 mm width at the bottom. One of the spiral samples' designs is shown in figure 2. Fluidity is measured in millimeters along the length of the cast spiral. For the convenience of measurement, protrusions are made in the upper part of the channel; the distance between them is 50 mm [5].

The spiral test sample of the specified construction has been designed in the SolidWorks software and produced on a milling machine.

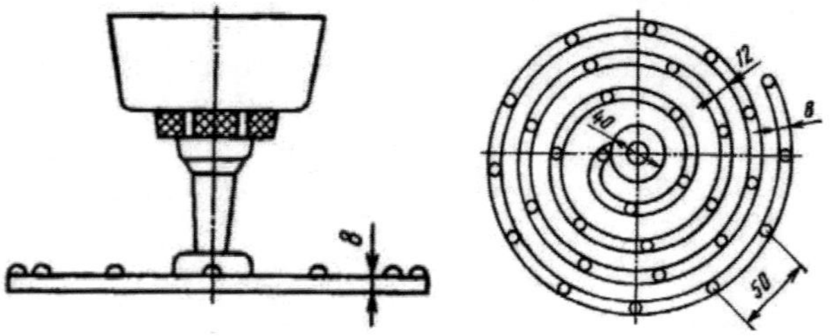

Figure 2. Spiral fluidity test sample

The polystyrene foam Penoplex 35 with the given average density characteristics is used for milling (table 1). The quality certificate data are verified under laboratory conditions.

Table 1.

Density characteristics	Density (kg/m^3)
Technical requirements of the 5767-016-56925804-2011 specification	25.0-35.0
Data in the certificate of quality	28.1
Data obtained in the laboratory	31

The selection of the polystyrene brand and its treatment are characterized by the following features:

- EPS-25 is a soft, relatively coarse-grained and easily delaminated material. Hence, it is hardly suitable for milling treatment;
- the properties of Penoplex 31 are similar to those of Penoplex 35 [6], although Penoplex 31 is more dense, easily machined by milling equipment, which makes it possible to obtain small parts without layering of the polystyrene foam.

In the course of the investigation spiral test samples are cast by the LFC method from the AL5 silumin alloys. The metal fluidity along a curved trajectory is analysed (figure 3).

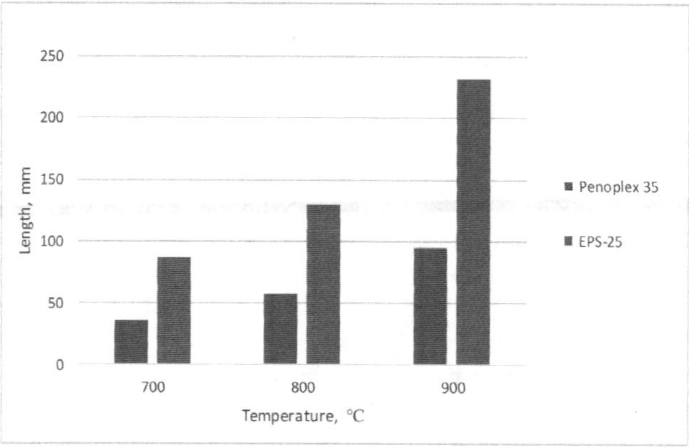

Figure 3. Dependence of the spiral test samples' length on pouring temperature

It should be noted that the surface quality vary when using different polystyrene brands for the production of the patterns (figure 4). The patterns are not coated with any non-stick coatings or coatings aimed at improving the casting surface. The length of the samples cast into consumable patterns which are made of a less dense, coarse-grained EPS-25 polystyrene foam is maximal. However, their surface quality is much worse compared to that of the samples cast into the patterns which are made of a denser, fine-grained Penoplex 35. Therefore, it is preferable to use Penoplex 35 as it will reduce costs of the follow-up mechanical treatment of an article. In the case when the cross section and the length of a pattern do not allow the use of Penoplex 35, surface of the EPS-25 polystyrene foam pattern should be treated with paraffin.

a) b)

Figure 4. Appearance of the sample surface:
a) EPS-25 pattern; b) Penoplex 35 pattern

3. Results

For the production of bulky castings of a complex geometrical shape it is better to use the LFC method at a pouring temperature of 800°C. The best mold filling is achieved by using the EPS-25 polystyrene foam patterns.

For the production of articles with long curved elements it is reasonable to use EPS-25 polystyrene foam since it ensures the maximum mold flow but requires manual treatment. Penoplex 35 allows the production of articles of a sophisticated configuration and with small parts even by using metal-cutting equipment.

References

[1] Ivanov V N 2007 *Spetsial'nye vidy lit'ya* (Special casting methods) (Moscow: MSIU) [in Russian]
[2] Shulyak V S 2007 *Lit'ye po gazifitsirovannym modelyam* (Lost foam casting.) (Saint-Petersburg: NPO "Professional") [in Russian]
[3] Grigoryev V M 2002 *Lit'ye po vyzhigaemym modelyam* (Cavityless casting.):nCoursebook: (Khabarovsk State Technical University Publishing) p 58 [in Russian]
[4] Soboleva L I 2012 Investigation of the dependence of silumin fluidity on the temperature of the alloy and properties of polystyrene foam *Proc. Int. Conf. of Students on mod. tech. and technol.* **263** pp. 79-80
[5] GOST 16438-70 *Formy peschanaya i metallicheskaya dlya polucheniya prob zhidkotekuchesti metallov* (Sand and metal form for samples metal fluidity) [in Russian]
[6] Utyev O M, Lukyanchenko A A 2013 Poisk tehnologicheskogo sozdaniya skul'ptury pri lit'ye po gazifitsiruemym modelyam (Technological solution to the creation of a sculpture by lost foam casting) *Proc. of the Academy of Tech. Aest. and des. №2.* **41** 11 p [in Russian]

IOP Conf. Series: Materials Science and Engineering **66** (2014) 012043 doi:10.1088/1757-899X/66/1/012043

Alternative technologies of manufacture and decoration of wood balusters

D Krauinsh[1], N Zhukova[1], M Volkova[1], A Kukhta[2]
1 Tomsk Polytechnic University, Tomsk, Russia
2 Tomsk State University, Tomsk, Russia

E-mail: dima_krauinjsh@tpu.ru

Abstract. Despite the fact that balusters made of stone, metal, gypsum can be found in modern interior, wood is the most popular material for balusters still. The aim of research is to research new forms of designer balusters and technologies of their production. Segmentation allows plane and curved surfaces to be produced independent from each other, and increases the artistic (aesthetic) expressiveness of products.

1. Introduction

Woodworking - most important sphere of modern business in Russia. With the process of development of low-rise building (houses, townhouses) there is a constant demand production of wooden stairs, columns, balusters, etc. And, as a consequence, the search of new technologies for manufacture don't lose its relevance. Alternative technology and the decoration of wood balusters, considered in this article, may be useful in the manufacture of products (stairs, pillars, balusters) for low-rise building (houses, townhouses), as well as in the manufacture of wooden furniture (wardrobes, chairs, tables). These same technologies are applicable for the production of souvenir products (boxes, vases, etc.) for design of ornithological art objects, etc. The authors of this article, having sufficient experience in various areas of woodworking, offering an alternative method of making and decorating products, obtained by transforming the bodies of revolution in the segment structure.

Baluster is support column of railing, made in the form of a pillar shape. Wooden baluster is made by turning traditionally. The basic function of balusters is security, second function is decorative of stairs. During the manufacturing process balusters given a variety of forms that provide visual harmonization of interior space. Note that a typical baluster manufactured from the block , a cut of square from 40x40 mm to 80x80 mm, with a total length of 900 mm, and processed profile (graceful form) about 600 mm (Figure 1).

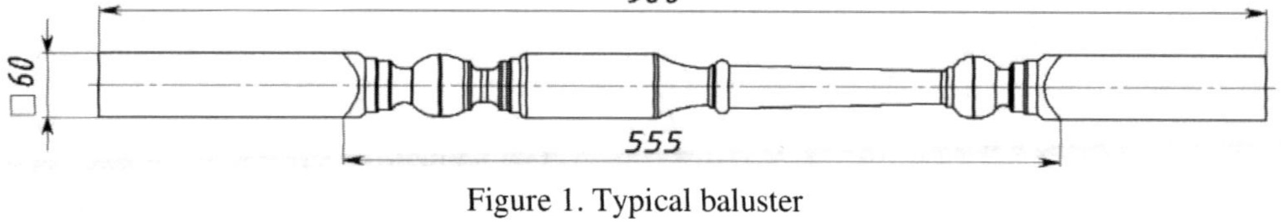

Figure 1. Typical baluster

2. Technology

Consider some possible unifying factors manufacturing of balusters as revolution bodies. The main factor is manufacturing of balusters on turning lathes. It can be as Universal Turning Machine (UTM) for wood and copying turning lathe for wood [1].

Turning on the lathe faster than milling. Skilled turner manufacture a typical turned baluster (Figure 1) for 15-25 minutes (including grinding). Total cost typical turned balusters for consumer is small.

Milling let you transform traditional turned forms in segment structure in various profile, as well as create additional independent of each other surfaces which constructor - designer can use to create an artistic shape of the product (balusters) [2, 3].

During milling on a single lathe milling machine with CNC processing time balusters be about 8:00, and its total cost to the consumer will increase significantly compared to the cost of a typical turning balusters. Processing surround woodwork largely determined by the finishing tool is weakly dependent on the depth of treatment, but depends on the total area of cultivated layer material. For "conditional" reduce processing time is used multi-place processing. But for such equipment will require additional space in the workplace.

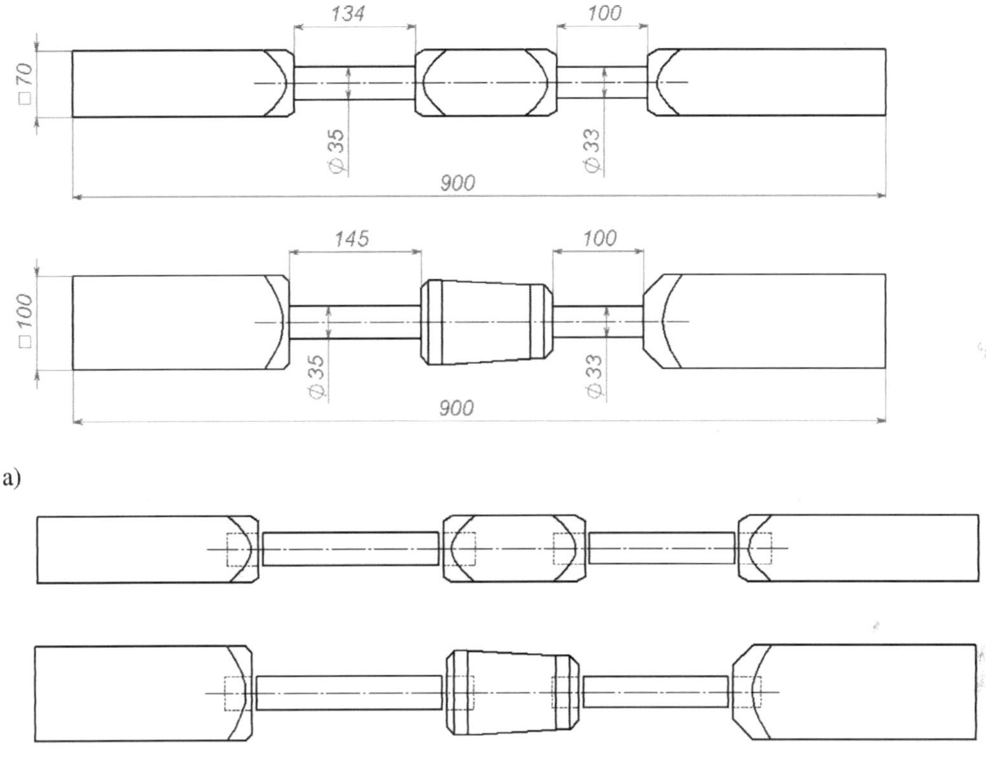

a)

b)

Figure 2. Workpiece of balusters

Authors of the article offered an alternative method of processing of balusters that significantly transforms the form standard balusters, giving it a form of milling and turning products. This method is based on the introduction of mass production techniques in processing that reduces the final cost of the treated this way balusters.

The production of balusters must have room, a set the necessary equipment and experienced designer.

Skilled constructor - designer must develop a design carved balusters so that the workpiece can be included laths - segments that can be obtained in different ways.

In the proposed method, instead of bulky specialized equipment is used as the main universal milling machine gantry type. Processing zone machine is 150 to 150mm, or 400 to 600mm, or 600 to 900mm. On this machine can be refined individual segments or made various multiplace matrix from gypsum for the mass production of various decorative elements.

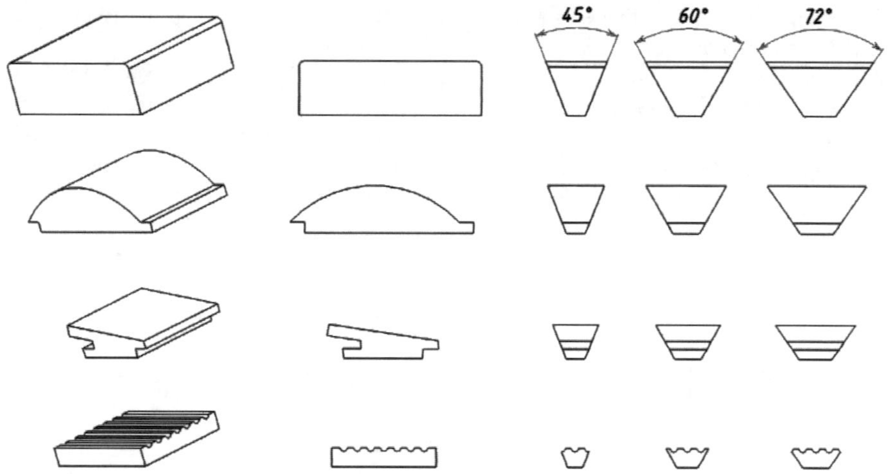

Figure 3. Typical element profiles segment inserts
(Segments with different angles: 45, 60 and 72 degrees)

Work piece of balusters (eg, Figure 2a) possible paste over a different segments (Figure 3) or other variant - to make the segment insertion (Figure 4), and then connect them with the rest of the balusters by the central rods (Figure 2b).

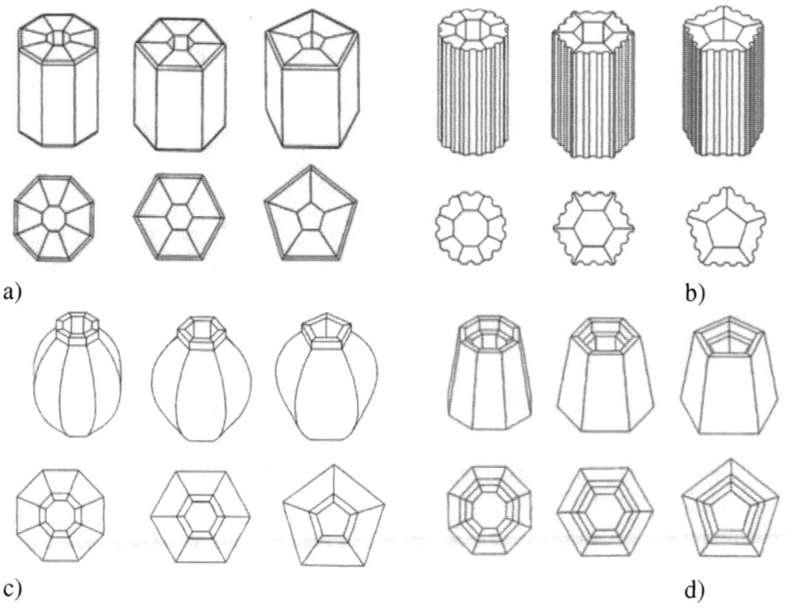

Figure 4. Segmental insert

What is the insert, and how production it?

For organize a competitive of technological process should seek to mass-produced segments of inserts, received by the productive equipment. For example, machines for quadripartite processing of balusters are used for making standard linear products including the desired tilting angles. Typical options linear profiles (Batten, block house, siding, door lining) are shown in Figure 3. For further processing of products and to obtain the desired segments a workpiece is divided on elements measured length or angular shaped segments. It should be noted that the top row of segments in Figure 3 can be obtained directly at the quadripartite machine processing. This simplifies the manufacturing process. Figure 4a shows the options segment inserts from the floorboard. Figure 4b shows the options segment of the finished panel inserts. Figure 4c shows the options segment inserts a block house. Figure 4d shows the options segment inserts of wood siding.

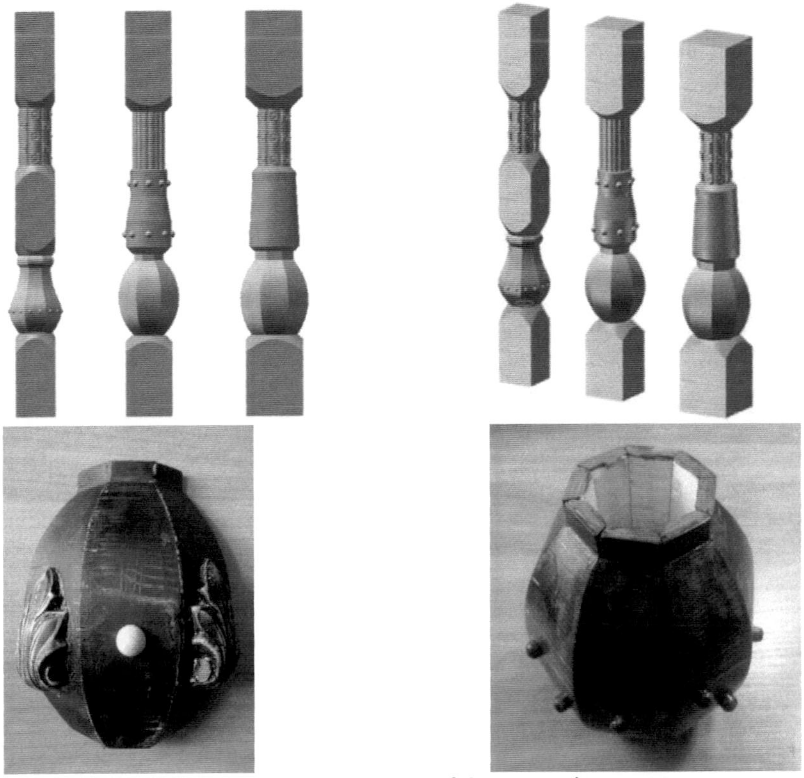

Figure 5. Result of the processing

Note also that the flat linear products can be obtained on the machines for rolling pattern embossed roller hot way, for example: «BORGORE 800 / I» Italian producer Renzo Borgonovo, or «DM-45" or «PC-TR1D». Typical maximum depth of relief on such machines is 3mm. This avoids time-consuming (time machine) milling process ща segments.

If the segment insertion modify simple technological operations such as drilling, gluing some ornament or a decorative element of polyacrylic type sealant «Parkett-202" [4, 5], we obtain the result shown in Figure 5.

3. Conclusion

As seen from the results, segmentation has allowed do more independent from each other plane and curved surfaces. Combining this surfaces in different combinations, or adding on segment surface other decorative elements can get the products presented in Figure 5. Described technology can increase the diversity of the range of products and their artistic (aesthetic) expressiveness and minimizing manufacturing costs. Equipment features allow you to post manufacture on smaller areas, what reduces costs [6]. Technology that described in this article allows the use of standard blanks (elements) and their subsequent decoration. This increases the variation forms and reducing it cost.

IOP Conf. Series: Materials Science and Engineering **66** (2014) 012043 doi:10.1088/1757-899X/66/1/012043

References

[1] Winchester K 1994 Wood turning. *Series: The art of woodworking* Time-Life Books 148
[2] El'tsov A and Scuba D 2012 Algorithms and methods for the transformation of industrial products in the design based on examples *Young scientist* **10** pp. 52-57 [in Russian]
[3] Kukhta M 2013 Effect of inherent models of organization of experience on the formation of visual images *Bulletin of Tomsk Polytechnic University* **323** (6) pp 223-226 [in Russian]
[4] Krauinsh D 2013 A method for producing a decorative stucco in product design with a curved surface *Proceedings of the Academy of Industrial art and design* **2** pp. 5-7 [in Russian]
[5] Chernich M 2011 Classification of wood art. Features engraved ornament decoration *Design. Materials. Technology.* **4** (19) pp.30-35 [in Russian]
[6] Kukhta M 2013 The unfluence of inborn experience organization models on visual images' shaping *Bulletin of Tomsk Polytechnic University* **323** (6) pp 227-230 [in Russian]

MTT'2014 IOP Publishing
IOP Conf. Series: Materials Science and Engineering **66** (2014) 012044 doi:10.1088/1757-899X/66/1/012044

Welding technologies in art processing of metal

M Kukhta [1,a], **A Sokolov** [1,b], **E Pelevin** [1,c]

[1]*Tomsk Polytechnic University, Tomsk, Russia*

[a]eukuh@mail.tomsknet.ru, [b]iscanderaga@rambler.ru, [c]eapelevin@gmail.com

Abstract. The article presents a comparative analysis of modern welding techniques which can be applied in the artistic machining of metals. Features of designing of art objects are defined and methods for their manufacture are offered, including stages of prototyping and full-scale modeling. Factors influencing the shaping of metal art objects are revealed. Practical application of the proposed recommendations is shown in the example of manufacturing of openwork metal mannequin.

1. Introduction

Art processing of metal involves manufacturing of objects representing not only material but also artistic value. Art objects are exclusive products which reflect the different tastes and edges of personality realization in a rapidly changing world. Art object may not have practical importance, but at the same time it has artistic value, the achievement of which is provided by high technology level of artistic processing of materials [1].

The purpose of this research is to study possibilities of application of welding technology in artistic processing of metal. To achieve this objective, several tasks should be solved sequentially.

It is necessary to reveal shaping features and technologies for creating metal art objects. Moreover we need to review the specifics of various types of welding and highlight their strengths and weaknesses. And also we must justify the need of prototyping and full-scale modeling.

Currently poorly studied issue about the specifics of welding technologies and their impact on the shaping of art objects, which are distinguished by high degree of esthetic expression. Urgency of work is due to the need to research the application of welding technology in the production of metal art objects.

2. Research

The development of various kinds of welding takes place in conditions when all is focused on the strength and durability of the resulting structures, while the requirements of esthetics were secondary. In the production of art objects main quality indicator is the high artistic expressiveness of the image and that requires careful selection of welding technology [2]. The main factors affecting the quality of welding in the manufacture of metal art objects are as follows: mobility of equipment, access to the weld area, the thickness of the welded parts, material weldability. All these factors affect the appearance of welds which is the main quality criteria of products.

In the art objects there are many parts in different attitude positions. Complex shapes require the free access to the weld zone, as well as mobility of welding equipment. Now let's analyze the most common welding methods and allocate their advantages and disadvantages.

Manual arc welding with coated electrodes (MMA) is currently one of the most common methods of manufacturing, installation and repair of various welded structures. The advantages of the method are the simplicity of the process and mobility of equipment. This method allows working in different attitude positions; it is possible to weld a large range of materials and thicknesses. These advantages provided broad application of MMA in creation of artistic objects.

The disadvantages of MMA are the following: weld quality depends on the skill of the welder; difficulties in welding of thin metal (less than 2mm) [3]; surface contamination of the metal and slag splashes that causes the need of additional surfaces cleaning; shape distortion of metal that must be considered in the design [4].

The gas welding technology is easier than MMA in many ways. In addition this method has one unique feature that can be used in welding of artistic objects. It allows correcting poorly fitted parts or even adjusting the composition during the welding process. It is also possible to work with small workpieces using low-power torches. Gas welding allows welding steel, cast iron, copper, brass, lead, bronze and aluminum alloys [3].

Disadvantages of gas welding as follows: high risk of warping of metal; spray and deposits, affecting the weld appearance; heightened danger of work with gas equipment.

The feature of gas metal arc welding (MIG/MAG) is continuous automatic supply of wire to weld zone. This allows making many welds without stopping. To protect the weld pool the special gas (active, inert or mixed) is used. Using the gas mixture ($Ar + CO_2$) minimizes the risk of warping and burn-through thin parts. The main drawback is the high gas emission in the welding zone.

This welding technology can be recommended for the production of large-scale art objects with many of easily accessible welds.

In gas tungsten arc welding (TIG welding) weld zone is protected by an inert gas and nonconsumable electrode is used. Main advantages are: high aesthetic quality of weld; no spatter and warpage; practically any metals can be welded; welding of very thin details.

However, the performance of this method is very low and time consuming preparation of metal before welding is needed. All this makes the method expensive and unsuitable for the manufacture of large structures.

Electric resistance welding is performed by short-term heating of the parts with passing electric current and applied plastic deformation [4]. For production of art objects it is reasonable to use spot welding. With proper welding mode there are no spatter and warpage.

Also there are portable welding devices and compact handheld devices for micro welding. However this equipment not very mobile and free access to weld zone is necessary. So in many cases, this method is not applicable.

The most efficient way is to use different types of welding for specific details. For remote weld areas can be used MMA, and for easily accessible zones gas metal arc welding will be most productive.

Outcome of the analysis can be represented as table 1.

As an example of welding in the manufacture of art objects let's examine production process of forged mannequin. [5]

In figure 1 presented final sketch of mannequin "Falcon" for men's clothing store [6]. The image is based on outline of a male figure which implies a comfortable arrangement of costume, hats, scarves and umbrellas. Unlike standard mannequins, this object is an artistic interior decoration, even without garments on it [7].

Table 1. Factors providing the quality of welding.

Welding method / Features	Manual arc welding	Gas welding	MIG/MAG welding	TIG welding	Electric resistance welding
Mobility of equipment	high	average	average	average	high
Accessibility of the weld zone	good	poor	average	average	poor
Minimum thickness of the welded parts, mm*	1,5	0,3	0,5	0,3	0,1
Range of welded metals	wide	average	wide	wide	average
Prime cost	low	low	average	high	average
Complexity of performance	high	high	average	average	low

* **Note:** The minimum thickness of the welded parts also depends on the skill of the welder.

Since sketch does not consider operational requirements, as well as structural changes that may occur during manufacturing, modeling phase is required. It includes: clarification of the artistic image (proportions, decorative elements, composition) to select the most balanced option; refinement of structural elements and their connections to each other to determine the technological features of the process of fabrication and assembly; verification of object perception from different angles to evaluate the aesthetic qualities of the object in different points of interior space.

Figure 1. Mannequin "Falcon": sketch (left) and layout (right).

The modeling process involves creating a layout that reflects the options of artistic solutions of object image, and full-scale modeling – production of a sample of the future product in full scale, that allows to reveal manufacturing features of the product.

Layout, made from simple materials, is designed to test the technical and artistic solutions. Mannequin layout is shown in figure 1. During creation fo layout main objectives are: to elaborate artistic image of future products; to work out the connection of separate parts and units of construction.

Full-scale model of the manikin considered technological features identified in step of prototyping. Volumetric part of the layout required to solve the problem of small decorative details connection.

It was decided to abandon the volumetric sample and perform planar version of object (Figure *2)*, which did not affect the aesthetic and artistic quality of the product. [8]

Figure 2. Welds in the construction of the manikin

As a material for a full-scale model was used steel bars as it has sufficient flexibility, strength and good weldability. For manufacture of parts was applied cold forging. For connection of details was selected manual arc welding, as satisfying the manufacturing requirements. [9]

Final steps are: cleaning, polishing and painting of mannequin.

3. Results and discussion

Based on the current research, we can recommend the manual arc welding as the most universal type of welding for connecting parts of metal art objects of any complexity.

Layout allows us to offer different variations of the artistic image and clarify technological methods of manufacturing, as well as proportions, decorative elements and specifics of the composition.

Manufacturing of full-scale model allows to obtain a sample of future products and to develop manufacturing technologies, to refine manufacturing operations and to make recommendations.

While designing the art objects stages of prototyping and full-scale modeling turn out to be important and necessary because they affect the process of shaping and ensure the quality of future products.

References

[1] Sokolov A and Kukhta M 2013 Peculiarities of creation and perception of objects of art-design *Design. Theory and practice* vol **13** pp 82-89
[2] Kukhta M 2013 Semantic capacity things in design. *Proceedings of Academy of technical aesthetics and design* **№1** pp 31-33
[3] Hicks J 1999 Welded Joint Design. New York: Industrial Press p 148
[4] Howard B Helzer C 2005 Modern Welding Technology Upper Saddle River, New Jersey: Pearson Education p 720
[5] Orlov D and other 1986 Technology and equipment for contact welding 2-e Izd M: machine-building p 352
[6] Kukhta M, Sokolov A and Sokur H 2011 Mannequins for storefronts made by technology of

artistic forging *Design. The materials. The technology* № **2(17)** pp 57-59
[7] Sokolov A and Kukhta M 2012 Peculiarities of design and technology of metal mannequins *Design. Theory and practice*. Vol **10** pp 112-128
[8] Kukhta M, Sokolov A and Pelevin E 2014 Peculiarities of application of welding technologies in the manufacture of art objects. *Design. Theory and practice.* in Vol **15** pp 70 – 77
[9] Kukhta M 2013 The unfluence of inborn experience organization models on visual images' shaping *Bulletin of Tomsk Polytechnic University* vol **323** №. 6 pp 227-230 (in Russian)

Advertisements and information objects' positioning technologies based on the territorial zoning of the city of Tomsk

R Solovyev[1, a], **V Seryakov**[1,b]

[1] Tomsk Polytechnic University, Tomsk, Russia

[a]sol.arrome@gmail.com, [b]seryakov@tpu.ru

Abstract. The main results obtained in the analysis of existing advertising facilities in the city of Tomsk were considered as a basis for media planning while choosing the most effective advertising medium and its location depending on the area of territorial and functional purpose of the city.

1. Introduction

Nowadays in the contexts of the dense urban construction and abundance of advertising on the streets it has become impossible to use the standard principles of media planning without violating the aesthetic component of advertising and the image of the city as a whole. It is essential to create a visually comfortable advertising look without losing the perceptual efficiency of the outdoor advertising.

The territory of any big city is divided into three main zones - historical heritage, special city significance and public city significance areas, which generally imposes certain restrictions on outdoor advertising positioning, taking into account the city's architecture and block density. The historical heritage zone includes: architectural and cultural monuments and adjacent territories, nature reserves, parks and squares. Positioning of outdoor advertising in this area is unacceptable. The only exception is the area along the road where it's permitted to place a man- proportionate advertising objects that do not cover the view of the building of historical and cultural heritage. The zone of the special city significance includes: central highways, squares on the main streets and avenues, squares at the railway stations, the area around the government buildings. It is prohibited to place certain types of advertising objects infringing architectural and spatial environment of the city in this area. The public significance zone includes the remaining area of the city: exit, entrance to the city, sleeping areas.

2. Positioning of the advertising constructions.

The cost of outdoor advertising and its effectiveness depends on the location area. Correspondence of the streets in Tomsk with the functional zones has enabled to identify the main categories of advertising objects positioning (Fig.1) [1].

Under Category №1 mainly fall information constructions, as well as advertising objects, not exceeding 10 square meters. Rationale: the inadmissibility to overlap the main scenery spots, historically-formed landscapes, architectural ensembles.

With Category №2 it is recommended not to use a large-scale advertising, billboards 6x3 m or more. The main advertising construction is city format with the advertising space no more than 1,2x1,8 m. It's also possible to use city board with the space of 3,7x2,7m. Rationale: retaining of "sight corridors" of cultural heritage within the pedestrian zone [2].

Category №3. Prohibition of positioning large-format advertising constructions with the space of the advertising field over 10 sq.m. Rationale: retaining of "sight corridors" of cultural heritage, ensembles of residential compounds and neighborhoods.

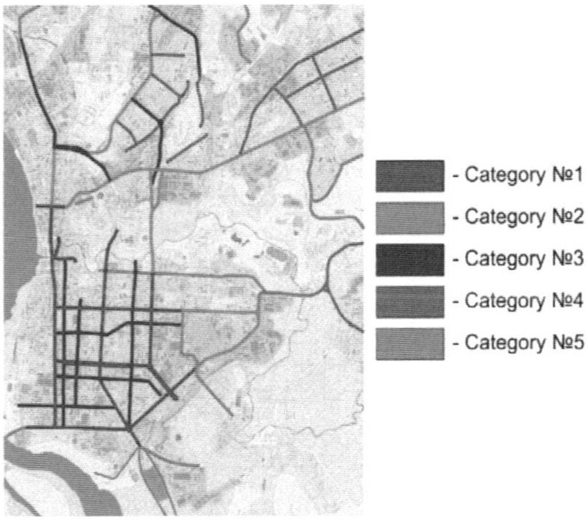

Figure 1. The city zoning scheme by categories of advertising objects positioning.

Category №4. Prohibition of positioning large-format advertising structures with the space of the advertising field over 18 sq.m. Rationale: This category includes the territory of roadsides, mostly passing through the dwelling zones.

Category №5. All types of advertising constructions are permitted except for banner constrictions. Rationale: This category included the territory of roadsides, passing mainly through industrial development zone.

Positioning of the advertising construction in any given area within the city depends on a man's perception of this construction in the context of the historical environment of the city, the distance to the object and the formation of the free pedestrian traffic. Thus, a distance to the outdoor advertising objects ensuring comfortable movement for the people and their perception of the historical character of Tomsk has been determined (Fig.2).

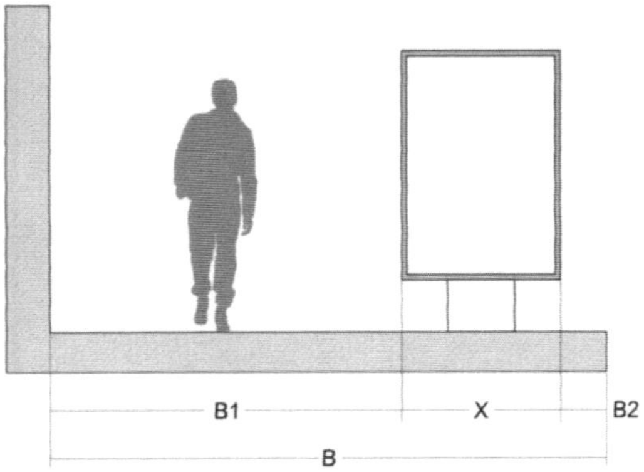

Figure 2. Dependence of the size of the advertising construction on the intensity of pedestrian traffic and All-Union State Standard (GOST).

The calculation was performed using the formula:

$$X = B - (B1 + B2),$$

where x- permissible width of the information field; B - total width of the sidewalk; B1 – the minimum permissible width of the passage between the curb and the advertising construction, determined by existing regulatory documents; B2 - the distance from the curb to the advertising construction.

In accordance with the requirements of the Construction Standards and Regulations (SNIP) II -60-75 «Planning and development of cities, towns and rural settlements", the width of the footways of the streets and roads of different categories should not be less than that specified in Table 1.

Table 1. Correspondence of streets and the size of the sidewalks.

Category	Streets and roads	Width of the sidewalk, m	Width of advertising object, m	Advertising objects				
				City Format	Stand/ Pillar	City board	Bill board	Super board
	Main streets			Outside of sidewalks				
1,2	City streets	4,5	1,5	v	-	v	-	-
3	Streets of district value	3	0,8-1,4	v	v	v	-	-
	Streets and local roads			Outside of sidewalks				
4	Residential streets	2,5	-	-	-	-	-	-
5	Industrial and municipal roads of warehouse districts	1,5	4-12	v	-	v	v	v
5	Township streets	1,5	6-12	-	-	-	v	v

According to All-Union State Standard (GOST) P 52044-2003 «Outdoor advertising on highways and areas of urban and rural settlements»:

"6.1. Outdoor signs should not limit the visibility of traffic management facilities, reduce the size of engineering structures, and should not be placed:

- Alongside of the highway or the road at a distance of less than 10 m < * > from the edge of the road bed of the highway (the curb stone) outside urban areas and at a distance of less than 5 m < * > - in the settlements/population centers;

< * > The distance to the nearest edge of the outdoor advertising.

- Alongside of the highway or the road at a distance of less than the height of the outdoor advertising, if the top spot is located at a height of over 10 meters or less than 5 m above the traffic way".

In the above calculations we used the average value. For city-format the distance to the edge of the road is 1 m, for Pillars, city boards, bill boards the distance is 2-2.5 m.

The calculation of the perception of outdoor advertising contributed to identifying the type of advertising construction relevant to its territorial location. The apparent sizes of objects, including signs, are defined in angular values.

The distance to the object is calculated by the formula:

$$L=\frac{H}{2\,tg\alpha/2};$$

where α – sight angle; H – height of the advertising construction; L – distance to the object.

The L value must be calculated in terms of optimal (α = 30 °), admissible (α = 45 °) values of the visual field. The distance of the object's perception will determine the following parameters - the height of the advertising construction and its location. Based on ergonomic data there was defined a distance of viewing the main advertising objects in Tomsk, at the angle of α = 30 ° (Table 2).

IOP Conf. Series: Materials Science and Engineering **66** (2014) 012045 doi:10.1088/1757-899X/66/1/012045

Table 2. Location of advertising objects depending on their perception.

Advertising objects	Construction height, m (mean value)	Distance to the outdoor advertising objects, m	Functional zone
City Format	2	3,5	Historical heritage zone
Stand/Pillar	3	5	Zone of the special city significance
City board	6	10	Zone of the special city significance
Super board	16	28	Public significance zone

In the course of the study there were revealed a number of violations in placing advertising constructions in the central part of the city of Tomsk in the area of the historical heritage, which is not consistent with the norms of advertising objects positioning in the urban environment and has a detrimental effect on the perception of the historical look of the city (Fig. 3) [3, 4].

Figure 3. Outdoor advertising objects impeding the perception of the historical part of the city

3. The illumination of advertising constructions

The important factor in placing information and advertising constructions in the urban environment is to provide high quality lighting for efficient visual perception and to preserve the aesthetic look. Creating a harmonious lighting requires consideration of the psychophysiological abilities of the observer and the surrounding urban environment, architectural and functional peculiarities, including the auxiliary lighting of the architectural structures, alleys or residential districts. Taking into consideration all the parameters for creating a visually comfortable look of the modern and informative urban environment requires calculations based on capacity of human vision, reactions of the visual apparatus depending on the energy parameters and spectral quality of the light in the perception of emitting sources at night [5].

Light emission is characterized by photometric parameters: luminous flux, luminous intensity, brightness, illumination. These parameters are taken into account in calculation of parameters of the interaction of various techniques ensuring the lightening of the city objects and advertising constructions, depending on their location in different functional areas of the city. The modern light sources, commonly used in everyday life and advertisement, with their calculated photometric parameters to compare are shown in Table 3.

Table 3. Photometric parameters of the modern light sources

Types of light sources	Luminous flux, Lm	Luminous intensity, Cd	Illumination to the area $1m^2$, Lx	Brightness, Cd/ m
Incandescent lamp 100 W	1350	100	1350	$2*10^6$
Halogen incandescent lamp 70 W	1170	90	1170	$1*10^6$
Fluorescent Lamp 200 W	11400	900	11400	$10*10^6$
Light-emitting diode 3.5 W	540	40	540	$2*10^5$
Light-emitting lamp 10 W	860	60	860	$6*10^5$
LED strip 1m , 240W	1200	90	1200	$1.2*10^6$

If the parameters of the light source are available, it's possible to make all necessary calculations to create a balanced sign lightning. At that, it's important to keep in mind the fact that the upper threshold of the eyes sensibility, delectability of the light emission on the black background, which can be compared with the night-time, is 10^6 Cd/m^2 and is characterized by pain, which roughly corresponds to the direct radiation of the filament lamp of 100 W at a distance (1.5 ÷ 2) m.

When performing calculations in the outdoor advertising design, one must consider what area of the city it will be located. The historical heritage zone covers architectural structures of historical significance, illuminated at night. As a rule, they are illuminated by the point light sources directed to the front of the building, floodlighting is used rarely. Therefore, when calculating the lighting it's necessary to consider not a luminous flux of the light sources, but the reflected light flux on the basis of the reflection coefficient of the material ρ.

$$\rho = \frac{\Phi_o}{\Phi},$$

where Φ_o – reflected light flux, Φ – luminous flux of the source.

Hence it follows that if the light wood facade is illuminated by halogen lamps of 1170 lumens, the reflection coefficient in this case is 50%. while luminous flux directed to the observer will be 585 lumens. If a city-format sign lightened by energy-efficient fluorescent lamps of 40 W with a luminous flux of 2000 lm, is placed against the building, even if taking into account the transmission coefficient of the thick glass closed by the advertising banner, this city format will completely distract and disturb the aesthetic and artistic perception of architectural structures with its brightness. A great example of this effect are Christmas decorations as shown in Fig. 4.

Powerful LEDs completely neutralize the effect of illumination of the building and distract the attention. In this case, they do not break, but substitute the functional purpose. During the Christmas holidays, this is a good example of how bright LED advertising can draw attention away from the architectural structures of historical significance [6].

City-format can also be bright enough, Fig. 5. As already discussed, even the most energy-efficient fluorescent lamp that backlights a city format creates a luminous flux of 2000 lm.

This level of lighting will be expedient in the special city significance area, highways and railway stations. At the same time on the highways and public roads advertising lighting should not be brighter than road and automobile lighting. Advertising constructions should not create a luminous flux more than 800 lm. If an advertisement is placed in the shopping areas, the brightness of its backlight with the abundance of advertising structures should be comfortable for perception and be no more than $(10^4 \div 10^5)$ Cd/ m .

Figure 4. Architectural lightning of the building, illumination, Tomsk, 9 Frunze Prospect

Figure 5. City Format, fluorescent lamps based light

The illumination of advertising constructions should be harmonious and balanced. This is the reason why it is necessary to make calculations using different lighting equipment within one zone for the purpose of presenting an excellent advertising that provides complete and accurate information and is the most effective at any time of day [7].

Thus, the effectiveness of the advertising construction depends on its size, the size of the advertising field and the physical location as well as on the level of illumination considering the external lights (lighting of buildings, street lights, highway lighting) [8, 9]. It's also important to consider the potential audience of a particular location of the advertising object. For example, doubling advertising on the city-formats is more susceptible and effective for the pedestrians in the area of historical heritage, for automobilists in the special city and public areas city and super boards are more effective.

References

[1] Berdyshev S 2010 *Effective Outdoor Advertising* 132
[2] Jerome Jewler A, Bonnie L. Drewniany 2002 *Creative Strategy in Advertising* 384
[3] Seryakov V 2010 Conceptual setting of the exhibitions facilities. *Modern techniques and technologies (MTT)* **3**
[4] Seryakov V and Tuzovskaya A 2013 The Meaning of Public Transportation Stops in the Modern Communication Environment. *Modern techniques and technologies (MTT)* **3**
[5] Kukhta M, Solovyev R 2014 Complex approach to design quality evaluation. *Design. Materials.Technologies (DMT)*
[6] Kukhta M, Solovyev R 2013 Demonstration Platform with an Interactive Control *Works of the Academy of Industrial Aesthetics and Design* **1** pp 17-21
[7] Kukhta M, Solovyev R 2013 Platform with an Interactive Control in the Design of Shop-Windows of the Industrial and Museum Complexes. *Design. Materials. Technologies (DMT)* **3** pp 16-19
[8] Kukhta M, Seryakov V 2010 The Specifics of Industrial Display Design's Forming and Structuring. *Design. Materials. Technologies (DMT)* **1** pp 18-24
[9] Kukhta M, Seryakov V, Sokolov A 2012 The Design of Expositions and Shop-windows: theory and practice. *Monograph, Saarbrucken: Palmarium Academic Publishing* 184

IOP Conf. Series: Materials Science and Engineering **66** (2014) 012046 doi:10.1088/1757-899X/66/1/012046

The influence of glass and metal properties on the peculiarities of an item of art's shaping in ethnostyle

M Kukhta [1,a], O Kazmina [1,b], A Sokolov [1,c] N Arventyeva [1,d], A Soroka [1,e], O Homushku [2,f], S Zaitseva [2,j], M Sergeyeva [3,h]

[1] *Tomsk Polytechnic University, Tomsk, Russia*
[2] Tuvan State University, *Russia*

[a]eukuh@mail.tomsknet.ru, [b]kazmina@tpu.ru , iscanderaga@rambler.ru[c], goodboy2007@mail.ru[d], anutka-soroka@mail.ru[e], hom17@mail.ru[f], mashaist@mail.ru[j], zsayana@bk.ru[h]

Abstract. Modern esthetics dictates the domination of glass and metal not only in architecture. The topicality of artistic material processing technologies' synthesis is demanded in different spheres of interior, environmental and jeweller's design. The objective of the work is the comparative analysis of different methods of joining glass and metal and practical recommendations on items' production on their basis on the pattern of items of art taking into consideration these materials' properties. The influence of technological and material conditionality on an item's shaping is revealed in solving the problems of a design object's visual-tactile sensing.

1. Introduction

The history of creating glass and metal items of art starts from stained-glass windows' making. The technology of coloured pieces of glass' joining together by means of small H-shaped leaden stripes was worked out in the VI th century. Scientists of different countries are still interested in stained-glass windows' production technique of the medieval period [1, 2]. The possibilities of glass and metal's joining together in the modern world are not restricted to stained-glass windows and are realized in a great variety of spheres of interior and environmental design, the design of art-objects and jeweller's technique. The production combining a metal foundation and a glass surface (tables, decorative panel) is widely spread in interior design, and glass and metal synthesis is seen in street lamps and stained-glass window compositions in environmental design.

The peculiarities of glass and metal items of art's shaping are described in this article. On the one hand, the topicality of the work is conditioned by the necessity of harmonious and esthetically expressive objects' creating. On the other hand, it is conditioned by technological factors' influence on such products' shaping. One must take into consideration the properties of items combining glass and metal when choosing the technology of their production.

The objective of the work is the comparative analysis of different methods of combining glass and metal and practical recommendations on items' producing on their basis on the pattern of items of art taking into consideration the properties of these materials.

Technological and material conditionality are chosen for items of art's esthetical characteristics in the work [4]. Technological conditionality characterizes the correspondence of an item's form to the materials and technology. Form's plastic must correspond to materials' properties. Thus, form's rigid graphic is possible and justified for metal items, and soft outlines and big radiuses of details' junction are preferable for glass items, which are shaped in a plastic state. Form's plastic is also determined by the method of an item's producing. Glass items produced by the method of free shaping have more soft and fluid forms than pressed. Metal items produced by casting method are characterized by contour softness as compared to items produced by cutting method.

Material conditionality characterizes the correspondence of a material to its function and displays the level of decorative qualities (colour, transparency, facture, etc.). The oxides of different metals (chromium, vanadium, copper, nickel, cobalt, iron, manganese, titanium, cerium), selenium, and sulphur as well as sulphureous chemical combinations of lead, copper, cadmium, and iron are used for coloured glass' production. Using these combinations, one can get different tints and colours: green, violet, milky, red. The surface of an item is covered by paints, or special methods of surface treatment (chemical, electrochemical or heat treatment) are used for the diversity of colour. Facture as a property of material characterizing form exterior structure is estimated through visual-tactile perception – roughness, smoothness, relief, lustre, dullness, etc. [5]. Modern technologies allow one to produce most diverse factures on glass and metal's surface.

2. Methods of Joining Glass and Metal together, Taking into Consideration their Properties

There are two main methods of joining glass and metal together, which may be conditionally subdivided into "cold" and "hot". The cold method is glass processing in a solid state and joining it together with metal by special glues or pastes. A variation of this method is a mechanical joining of glass and metal with the help of screws, through resiliency of the mechanical construction as well as combined technologies. The advantage of the cold method is a relative simplicity in the technological aspect and the fact that there are practically no restrictions on the dimensions of an object. Agglutination in interior design has a great preference over other technologies of joining due to the simplicity of the equipment and the absence of restrictions on the dimensions of an object. It is also applied when glass is laid on metal and when metal is laid on glass. A glue joint is absolutely invisible, that is why it is the most preferable one for the production of objects, the main function of which is esthetic. The imperfection is the necessity of obligatory preparation of glued surfaces and the absence of distortion of the junction's geometrical form in the process of production and exploitation of an item. Besides, a relatively low thermal capacity and small elasticity are typical of glue joints, which lowers exploitation characteristics.

The method of hot joining is devoid of these faults. When applying this method, glass is in a molten or softened state. Processes of physico-chemical nature are the basis of glass and metal's soldering. Moistening of metal with a liquid glass is made by its spreading over the metal's surface and also due to surface tension forces. The technological problem of items' producing by this method is thermal stress in glass and metal soldered joints. Residual stresses resulting from the interaction of heterogeneous materials with different properties can lead to cracking of an item.

The values of thermal expansion coefficient (TEC) characteristic of some glasses and metals are given in table 1. According to the correlations of glass and metal's TEC soldered joints are subdivided into coordinated and non-coordinated. In coordinated soldered joints the difference between glass and metal's TEC doesn't exceed 10 % in the whole interval of temperatures to which the joint can heat or cool in the process of production or exploitation. Non-coordinated joints are those which TEC values differ greatly. The comparative characteristics of technological and decorative properties of the values given in the table shows that such widely spread metals as aluminium, iron, copper as well as alloys on their basis differ greatly in thermal expansion coefficient from glass. Glasses of different structure also have a large dispersion of coefficients' values. If one views the most widely spread silicic glasses, which are mostly used in glass items' production, one can see that an average TEC value is at an interval of 85 – 90 × 10^{-7}, deg^{-1} . In that case the choice of metals suitable for soldering with glass is very limited from the point of view of the joint's coordination.

Table 1. Characteristics of glass and metals' technological and decorative properties

material		technological		decorative		
		TEC × 10^{-6} deg^{-1}	temperature of melting(softening)	colour	facture	light-reflecting capacity and transparency
metal	Al	24	660	mono-	depends on the	intransparency
	Cu	16	1083	chroma-	value of surface	dull lustre
	Fe	12	1538	ticism	microdistortions	
glass	of silicate	9	≈600	diversity	smooth, soft	transparency of
	of quartz	1	≈1400	of colour	relief	different degree
	leaden	10	≈565	and tints		

3. Results and Discussion

Two items of art made of glass and metal are chosen as the objects of investigation. The item in the form of a table is produced according to the cold method by means of gluing a metallic semi-manufacture and a tempered glass together. Metallic semi-manufacture's producing has different levels of mechanization and automation depending upon the technologies, which may be conditionally subdivided into plane and volumetrical. Plane technologies are those where two-dimensional details are worked up; in the case of volumetrical technologies – three-dimensional details. Plane are the most productive ones, wholly automatic. The basis of volumetrical technologies are different kinds of casting. In that case a metallic half-finished product is made according to the technology of artistic forging with the use of soldering for joining the elements of the construction together.

First of all, a glass upper surface of a table is cut according to the dimensions, and then it is thermally treated by a special mode of operation to temper it. It is well-known that a tempered glass differs from untempered not only in an improved mechanical and thermic resistance, but also in its safety. The safety of such glass is determined by the fact that it breaks to pieces with blunt edges when hitting.

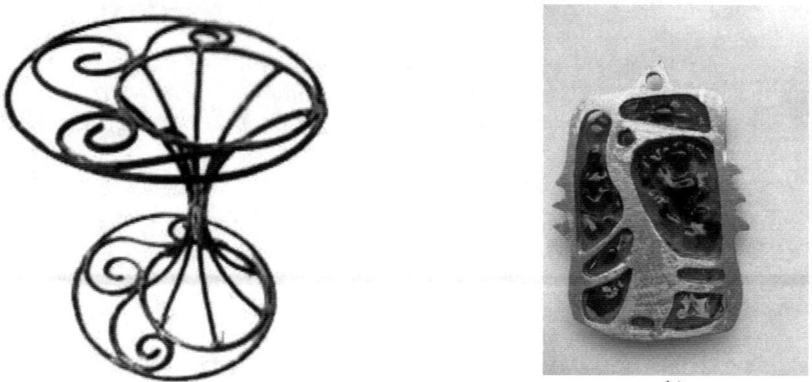

(a) b)

Figure 1. Glass and metal items of art: (a) table (Denisov M.), (b) decorative panel (Soroka A.)

The second item of art made by one of the authors is the most complicated one in the technological aspect of joining glass with metal. A combination of two methods is applied here. A metallic foundation is made according to the technology of casting by means of pouring of a liquid brass with further crystallization. Casting to smelted models is a kind of artistic casting which is applied for producing thin-walled, complicated in their configuration mouldings. The background of the pendant is made by fusing technology, which foresees conglomerating of glass of one and the same composition, but of different colour in that case into one volumetrical composition. Such properties as the temperature of glass softening, the temperature of an item's virtual deformation and thermal expansion coefficient become the most important ones in glass conglomerating. Some characteristics of the glass used in the work are given in table 2. The calculations are made with the help of SciGlass 7. Calculated values of glass thermic properties allowed us to determine the temperature conditions of its treatment. Such characteristics as glass length and crystallization index point at glass behavior in the process of heating and cooling of an item. Index 20 corresponds to glasses with low crystallization capacity and glass suitableness to mechanized moulding. Glass length of 808 °C corresponds to slowlyhardening glasses. The temperature of softening defined by Littletone's method (thread lengthening under the influence of its own mass when heating) corresponds to glass viscosity of $10^{6,6}$ Pa·s. The values of refractive index and light scattering coefficient are characteristic of glasses of silicate composition.

Table 2. Composition and properties of the glass used in the work

Glass composition, oxide content, mass fractions %							
SiO_2	Al_2O_3	Fe_2O_3	CaO	MgO	Na_2O	K_2O	SO_3
72.18	1.09	0.04	8.85	3.60	13.62	0.26	0.35

Glass properties							
TEC up to 400 °C, 10^{-6}, deg^{-1}	optical		glass "length" °C	crystallization index	temperature, °C		
	refractive index	light scattering coefficient			of fusion	of virtual deformation	Littletone's
9,14	1,516	0,00871	808	20	605	634	726

The technological process of soldering glass and metal includes the following main operations: preparation of the materials' surface to soldering; annealing of a soldered joint; control of an item's quality. Glass and brass used for the panel's production have TEC values of $9{,}14 \cdot 10^{-6}$ and $18{,}9 \cdot 10^{-6}$, deg^{-1}correspondingly. This tells about a non-coordinated joint's making and the possibility of dangerous stresses' formation. This problem is solved by means of a differentiated annealing of an item. For this purpose the difference in temperature is supported, so that thermal pressing of metal and glass would remain approximately the same during the whole process of cooling of an item combining glass and metal.

Glass items of art's producing by the method of conglomerating allows one to get different decorative effects. For example, one can make large stained-glass windows by means of fitting glass panels produced by agglomeration method on a metal or combined frame. Agglomeration method can be combined with glass sheet bending in fireproof ceramic moulds. An interesting decorative effect is achieved if one fills hollows in the mould by crushed glass of the same composition as a sheet which is laid upon it. Two items made by the method of fusing and conglomerating of glass on a metal foundation are represented in figure 2. The decorative panel representing a strength joint of several layers of coloured glass is produced by means of partial fusing of one piece of glass into another under a temperature of 750 °C preserving its configuration. A glass with both a reflecting facture and a smooth transparent surface is chosen for obtaining a volumetrical image. The main means of artistic expression in that case are local colour stains of glasses. The substrate of coloured glass let us to add an emotional content of colour palette consisting of turquoise, dark blue and sky blue tints to cast metal's graphics.

According to the depth of an item's decorating one can single out three main groups: deep, surface and intermediate (Table 3). The criterion of classification is the depth of the factor's spreading over an item's thickness. In deep technologies the decorative factor (colour) is spread over the whole thickness of an item or its element – in the case of fusing. In surface technologies colour is concentrated in a thin surface layer. Surface technologies are presented mainly by decol and paintings [7]. Intermediate technologies unite different kinds of mechanical treatment and chemical etching as well as their combinations [8]. The decorative factor in that case is the relief and facture of the surface.

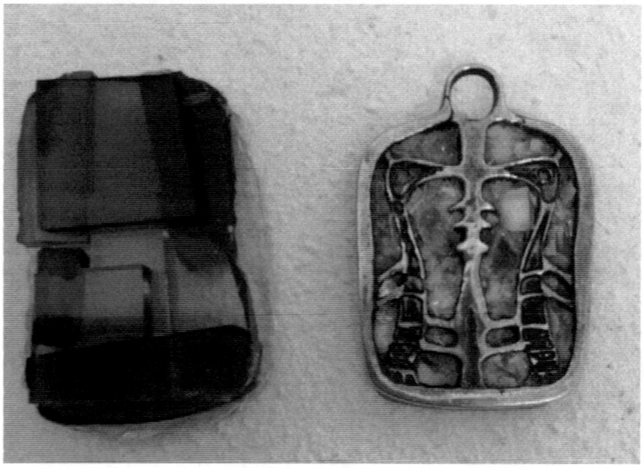

Figure 2. Glass items of art produced by: fusing – on the left, agglomeration – on the right (Soroka A.)

Table 3. Comparative characteristics of a decorative effect in glass items of art

Group	Peculiarities of a decorative factor	Automation	Decorative effect	Level of effect
deep	is spread over the whole material's or layer's thickness	low	purity of colour	high
			detailed drafting of the pattern	middle
inter-mediate	controlled depth of a decorative factor	middle	relief of the surface	high
surface	is concentrated in a thin surface layer	high	purity of colour	middle
			detailed drafting of the pattern	high

Thus, different techniques of production of items of art combining glass and metal exist nowadays. One can single out three main groups of these items. The fist one is glued items of sheet glass on a metal foundation. The second – items of volumetrical pieces of glass joined together by means of conglomerating and soldering. Combined techniques of joining refer to the third group. The influence of technological and material conditionality on items of art's shaping is revealed first of all in solving the problems of a design object's visual-tactile sensing.

4. Resume

The comparative analysis of different methods of glass and metal's joining allows us to give the following practical recommendations on items of art's production on their basis.

1. When producing items of art by means of hot joining of glass and metal, taking into consideration the fact that thermal expansion coefficients of glass and metal differ more than in 10 %, it is worthwhile using a differentiated annealing. This allows one to avoid residual stresses, which lead to defects in a finished item.

2. Glue joining of glass and metal is a relatively simple one in a technological aspect, it doesn't require choosing materials according to TEC values and is suitable for large objects. However, an

additional technological operation of glued surfaces' preparation and the absence of a joint's geometrical form's distortion are obligatory in production and exploitation of items made by a cold method.

3. The combined method, which foresees soldering of glass and metal and agglomeration of glasses of different colour, allows us to produce different decorative effects of colour spreading over an item's thickness and facture. The combined method can be supplemented with other different methods, mechanical and chemical decorative treatment of glass, which allows one to improve esthetic qualities of a design object's visual-tactile sensing.

References

[1] Kunicki-Goldfinger J, Freestone I, McDonald I, Hobot J, Gilderdale-Scott H, Ayers T 2014 Technology, production and chronology of red window glass in the medieval period – rediscovery of a lost technology J. Arch. Sci. vol 41 pp 89-105

[2] Vilarigues M, Fernandes P, Alves L C, Silva R C 2009 Stained glasses under the nuclear microprobe: A window into history *Nuclear Instruments and Methods in Physics Research Section B: Beam Interactions with Materials and Atoms*. vol **267**, issues 12-13, 15 pp 2260-2264

[3] Kukhta M S, Sokolov A P, Danila K G 2012 Analysis of shaping in decorative lamps' design *Design. Materials. Technology* №. **1(21)** pp 10-15 (in Russian)

[4] Stepanov S A 2006 *Universal Quality Management* (SPb: Publishing house of SPbSETU "LETI") (in Russian)

[5] Kukhta M S 2013 The unfluence of inborn experience organization models on visual images' shaping *Bulletin of Tomsk Polytechnic University* vol **323** №. 6 pp 227-230 (in Russian)

[6] Sokolov A P 2013 Hammered tables' shaping, conforming to a style and technologies *Works of Academy of technical aesthetic and design* № **2** pp 12-16 (in Russian)

[7] Boiko J A 2010 Print as a means of ceramics' decorating *Design. Theory and Practice* № **4** pp 36-52 (in Russian)

[8] Galanin S I Vologdina V O 2010 Decorative etching of coloured glass insertions in the items of jeweller and art *Design. Theory and Practice* № **5** pp 38-50 (in Russian)

IOP Conf. Series: Materials Science and Engineering **66** (2014) 012047 doi:10.1088/1757-899X/66/1/012047

Triad "Metal - Enamel - Glass"

T Mukhina[1,a], S Petrova[1,b], V Toporova[1,c] and T Fedyaeva[1,d]
[1]Saint-Petersburg, 190008, Russia, Saint-Petersburg State Marine Technical University

[a]email, muhodaw@mail.ru, [b]email, psg1@yandex.ru, [c]email, nikanorka.05@mail.ru, [d]email, Fedun07@mail.ru

Abstract. This article shows how to change the color of metal and glass. Both these materials are self-sufficient, but sometimes used together. For example, enameling. In this case, the adhesion between metal substrate and stekloobraznae enamel layer, which was conducted on a stretching and a bend, was tested.

Metal and glass are unique materials, which are well combined with each other according to their decorative and durable features. Speaking about metal we usually mean monochromium sculptures. But this is wrong to consider.

In sculpture color doesn't play a role as in painting, but it gives a feeling of form, space and light and shade, it is used for expressing ideas and notions. Sculptor can create colour including natural colour of material or change it by means of a toning, paints, golding, patination. All chemical compounds, used in patination of copper alloys can be combined the following groups: nitrates, sulfurous, sulfates, carbonic, acelates, chlorides, chlorates, ammonia in different concentrations, acid, oxides, chromates and separated chemical compositions which are impossible to combine in groups.

In patination the most common chlorides compositions are ammonia and sodium chlorides. They are both paint copper alloys in yellow, orange, light-red, pale blue, blue colours and also in green, brown and their shades. Besides using a sodium chloride can produce grey and shades of black and with ammonia chloride - violet colour. Chloride mercury gives blue and green colours. The range of chloride zinc extends from brown to dark-red. Stannous chloride and potassium chloride are used for painting in brown colour. Chlorates compositions are mostly limited by the brown spectrum, but in addition to them berthollet's salt gives violet and potassium chlorate - dark-red colours on the metals.

Sulfide ammonic shows the most various colours of sulfurous compositions - yellow, orange, dark red, red-brown, blue black. The following is giposulfid sodium, which forms golden or grey films on the surface of metals. Arsenic trisulfide antimony and calcium sulphides show the brown colour and it's shades and giposulfid potassium - light red. Copper sulphate is the leader among sulfate compositions, it's spectrum extends from golden to brown through red, plus green, blue, violet and grey colours. Using sulphates compositions on the base of gidrosulfite except golden colour can get the following spectrum of colours: from light to dark red, from pale blue to violet. Sulfate potassium gives black, brown, blue and violet colours and their shades. The brown films appear on the patinated things because of the sulfates of nickel, iron and ammonia.

Chromates are used rather seldom in patination of copper alloys. Using a potassium dichromate we can get golden and blue colours, but sodium bichromate and dichromate paint metals in golden and yellow.

In patination sufficient amount of acids are used. In particular acetic acid is popular. It gives yellow, orange, red, brown, green, pale blue colours and their shades. The range of sulfuric acid colours extends from golden to brown plus blue colour. Citric acid suits for painting in red, pale blue, blue, violet colours. Other acids are used seldom, because they can help to get rare colours in patination, for example grey colour which is produced by nitric, hydrochloric, arsenic acids.

Among carbonic compositions the most popular are copper and ammonia carbonate. Metal can be painted in golden, yellow, orange, blue, brown, black colours and their shades and also in yellow, orange, blue and green respectively.

Acelates compositions have sufficient spectrum of colours. In particular copper acelate can paint metal from golden to brown plus blue, green and plumbum acelate - in golden, dark-red and also including spectrum from pale blue to violet.

Ammonia vapors are used for producing black, light-red, pale blue and violet colours. But 25% ammonia solution - blue, grey and colours for restoration.

Patination with nitrates compositions is very rare, because the range of colours is limited mainly by the green colour (nitrate sodium). In addition to the green colour nitrate copper gives the brown colour and it's shades and also the range of violet colours.

No one from the oxides give a wide spectrum of colours. Crocus is limited by the red-brown gamma, borax - blue-green, zinc oxide - blue, iron oxide - brown.

Besides the above mentioned chemical compositions there are other compositions which are not combined in groups but they are popular and in some cases even more popular, for example caustic soda. By means of it we can get all "natural" colours of patina and also the range of colours from golden to red-brown, from blue to violet. Potassium permanganate gives painting from golden to brown through red. Sulfuric liver suits rather well for restoration of colors and also for red, blue and grey colours. Lactose, chalk, sodium dichromate are limited in golden gamma. Sodium arsenate is used for producing golden and light-red films. Sulphuricsodium salt, tartur emetic, granular tin, iron filings give grey colour. Except grey colour copper oxychloride gives violet colour, tartar in blue, acetic plumbum - violet and dark-red colour. Segnetova salt is used for dark-red and blue colours but soot - yellow and orange.

Chemical compositions possessing various spectrum of colours are widespread in patination of metals. Their use is profitable and practical. But don't forget that some chemical composition which is not included into more or less widespread group and is difficalt of access can give the colour or shade suitable and necessary far the artist in patination. Then it's study and "hunting" for a new technique will lead to something unique.

However, the sculptor should always remember that the colour sculpture and environment must be in harmony with each other and taken separately whichever colour is chosen.

The decoration of metals can be made by brighter colours than patination for example, smelting of the stained glass. Such technique is used for the long-lived decoration and it is called an enameling. In this process adhesion plays a great role between metal and enameling layers.

However, this technique hasn't been enough studied in enameling of the artistic things. There is no accurate method of measuring adhesion in the joint of enamel-metal. The pattern of the enameled metal was put to the widespead test on tensible strength and it was testroyed in the result of bending, tursting, shock.

In the worst case of removing enamel the light and bright surface of metal remains; on the surface of metal the dark coupling layer with the remains of enamel is formed.

The thicker enamel covering more often possesses the best coupling and resistance to shock, but however in reality the opposite is true [6].

The strength of enamel is 15-20 times more in the case of pressing than under stretching[6,7]. The comparative experiments connected with stretching were carried out showing the quality coupling enamel base and copper layer between enamel DKZ № 10 and new glass bases T-5, T-6, which are widespread silicon-sodium-potassium systems.

The experiment was carried out by means of testing machine (CDM-2,5) with driving gear. The aim was to define the quality of coupling between glass forming bases T-5, T-6 and enamel DKZ №10 on the surface of copper (mark M1) using the comparative method.
Special patterns from copper marc M1, thickness 2mm were produced.
Then one composition (T-5, T-6, DKZ № 10) was piled up on each pattern for comparative analysis. Every pattern was put to stretching until the complete destruction of the enamel layer happened.
In the case of the good coupling the priming covering separates from metal in the form of the small needles, the surface under the separated enamel layer is dull, moreover the undamaged layer of the prime coating remains partially on it. In the case of the bad coupling the prime coating separates from the metal with large layers, in addition to that the brilliant surface of the in oxidizing metal appears [8]. The destruction of the patterns with the glassforming bases T-5 and T-6 took place according to the same scheme.
The remains of the enamel and oxiding layer we seen in the place of exfoliation. The destruction of the enamel pattern DKZ №10 took place accorting to small lengthening. The type of destruction demonstrates the insufficient adhesion.
Thus the conducted research confirmed the formation of the transient layer between the studied bases T-5, T-6 and the substrate of copper(mark M1). One can assume in the base of the observed formation of the transient layer the diffusion processes between metal and enamel take place.
And in addition the comparative experiment of resistance to wear was carried out. For this experiment the layers of copper (mark M1), will thickness 0,3mm,leght 35mm and with 10 mm were prepared 0,5 g of enamel or glassforming base were piled up on the surface of each layer. After glazing and complete cooling the patterns were put to bending. In the character of pattern showing the bending the spherical object with diameter 39 mm was chosen.
The results of the experiment showed that all tested glassforming bases and enamel possess of sufficient adhesion for their use in the enameled artistic technique.
Besides the glass can be used as an independent material.
"The real ancient painting is a mosaic," - Girlyandayo said [9]. Mosaic – a special kind of art, the aim of which is to construct the topical image from monochromatic or achromatic elements, coupled with each other and the base with the astringent substance which has came from ancient times and hasn't lost its interest up to our days.
For quite a long period of formation four main techniques of mosaic art were formed: opus barbaricum (set of pebbles), opus tesselatum (set of tesserae – stones of the same shape), opus sectile (different form of elements), opus vermicalatum.
Not only the technique of mosaic composition was improved, but also the material.
From natural material – pebbles, came to the artificial, more perfect material – smalt with richer palette to be well processed by traditional mosaic tools.
Smalt is a special kind of silicate glass. Due to the introduction of different combinations of decolour substances, dyes smalt can be transparent, translucent or opaque and dull, blotchy and wiry, with gold and silver coloured and colourless cantarell (upper thin layer of glass). The longer the art existed, the wider the smalt palette became and the more the amount of shades took part in combination.
The technological parameters influence the characteristic of smalt. The process of preparation of smalt consists of several stages: glass alloy, "deaf" white glass – "whitewash"; schmelz – basic non-ferrous alloys produced by melting "whitewash" with dyes, boiling and melting smalt.
The continuous development of the mosaic art makes us to improve material. Despite the perfect features of smalt it has essential disadvantages, namely its cost and its inability to find quickly the right shades.
After work in the mosaic workshops a lot of fragments of smalt unsuitable for further use remain. This material is well melted in a muffle furnace.
The most suitable are white glazes because the colour is predictable even after addition of the dyes.
The lead arsenic white glazes suitable for capricious pink red colours.
Tin lead are for the most of others.

IOP Conf. Series: Materials Science and Engineering **66** (2014) 012047 doi:10.1088/1757-899X/66/1/012047

Not all white glazes and colourful ones retain their colour specifications and derocative properties after melting.
Green glazes lose their colour, becoming transparent but this disadvantage can be eliminated. But it is difficult to make dissemination of the transparent glaze into the thiokness of the damp colour. During the second melting the dirty shade is formed. Therefore for melting and subsequent painting it is necessary to use uniformly coloured white.
There is such a kind of glass as a glaze and enamel. The introduction of these types of glass can improve the properties of smalt, reduce or even remove the dirty shade during the second melting, to give transparency or to create blotches.
To melt the glazes it is necessary to reduce their metallic grounds in a mortar, sieve and remove unnecessary metul particles using a magnet. Melting itself can be done in a traditional way in crucibles and then to pour the melt into the form with mica molding the material at once.
Smalt like only other glass has no melting point. According to the experiments conducted in secondary melting smalt begins to soften at about 750°C.
However not all dyes can be dissolved in the melt, they require higher temperature. For example, the cobalt requires 880°C, manganese and copper begin to dissolve, but the copper and cobalt have no time to react with each other to produce a new colour.
It should be remembered that too high temperature is undesirable. At 950°C melt begins to buble and as a result it affects the internal and external structure of glazes. However the diffusion goes well, the due (in this case copper oxide) is distributed over the entire thickness of the melt, forming dark and light areas of bright blue.
The slow cooling of the melt with furnace will allow the material to acquire the proper strength and keep it from cracking.
Such a method of melting smalts with due and sometimes with each other for creating the unusual effects can take place. It allows you to expand the palette smalts, to get a large number of variants from some kinds. This is particularly important for small workshaps.
Along with coloured smalts the gold and silver with colourless and coloured cantarell glazes with a metallic sheen have been used since ancient times. It is possible to prepare gold and silver smalts of different shades. They are prepared by sintering between two panes of glass or transparent coloured gold leaf or silver. However there is a new method of creating smalts with silver shade. It consists of spraying thin films on glass in a vacuum.
With this method it is possible to sputter metals, alloys and chemical compounds (oxides, nitrides, borides, carbides, etc.). Application of chemical compounds is preferable, because they are stable compound structure and they are not subjected to future changes and they are similar with glass. The given material can be processed with the same tools as conventional smalt. The characteristics of strength are rather high and a layer of sprayed substance remains on the glass surface even after treatment of the material on the grinding machine. However, palette is not so diverse, but the gold and silver smalt is quite possible to replace this material.
As in the case of conventional smalt the colour will be changed depending on the chemical compositions.
For example, Ti-Si-N has a golden colour, TiCN-bronze-purple colour, Ti_2N-golden straw, TiZn-gold-silver, TiMo-light grey, TiAl-golden-purple.
All titanium nitride coatings are golden-yellow shades. TiO_2 coatings have blue shades [10]. Not only the chemical combinations, but the thickness, of the film, the saturation of colours and their variations with various parameters including the installation of equipment, performing sputtering are of great importance for coating.
The most important of these parameters are pressure of the reactive gas, the distance from the evaporator and the substrate, the operating temperature of the arch current, the deposition material on the substrate and others.

IOP Conf. Series: Materials Science and Engineering **66** (2014) 012047 doi:10.1088/1757-899X/66/1/012047

For a long period of the glass use in the mosaic, it has under gone great change. The use of new technological methods for two types of smalts allows to get new materials on the base of already existing ones.
Moreover this technique is not harmful for the environment because the equipment meets the environmental requirements.

REFERENCES

[1] Odnoralov N V 1989 Sculpture and decorative trim of art metal wares *Textbook Allowance Moscow Art* p 208
[2] Gutov L A, Nikitin M K 1994 Directory of artistic metalwork *St. Petersburg Polytechnic* p 435
[3] Ref. ed. Azhogin F F, Belenky M A, Gall I E 1987 Electroplating *et al. Moscow Metallurgy* p 736
[4] Ref. ed. Belenky M A, Ivanov A F 1985 Metallic electrodeposition coatings *Moscow Metallurgy* p 288
[5] Ref. ed. Virbilis S per. from pol. ed. Ivanov A F 1990 Electroplating for masters *Moscow Metallurgy* p 208
[6] Shraider L L 1981 The corrosion
[7] Petrosold A, Peshmann G 1990 The enamel and enamelling *Moscow Metallurgy*
[8] Vargin V V 1972 Enamelling of metal products *Leningrad The machine building industry*
[9] Vinner A V 1953 Materials and equipment mosaic art *Moscow ART*
[10] Panfilov Y 2007 Deposition of thin films in vacuum *Technology in the electronics industry* **№ 3**
[11] Solntsev Y P, Pryahin E P 2004 Materials science *St. Petersburg HIMIZDAT.*

Phase analysis study of the copper-aluminum contact pair obtained by plasma dynamic method

A Sivkov[1,a], A Saygash[1,b], J Kolganova[1,c], I Shanenkov[1,d]
[1]Tomsk Polytechnic University, Tomsk, Russia

[a]sivkovaa@mail.ru, [b]s_nast@mail.ru, [c]julia_kolganova@mail.ru, [d]swordi@list.ru

Abstract. We obtained the contact pair copper-aluminum copper coating by using a magneto plasma accelerator. The process is realized during supersonic copper plasma jet flowing into the chamber filled with the air atmosphere. Copper jet is carried out of the accelerating channel towards the aluminum target. Plasma jet is generated by coaxial magnetoplasma accelerator (CMPA) based on copper electrode system. The CMPA is supplied from the pulsed capacitive energy storage with the maximum value of stored energy of 360 kJ. The obtained copper-aluminum contact pairs have been analyzed by X-ray diffractometry and Nano hardness tester. The copper coating on the aluminum surface is uniform with thickness about 100 µm. Also in this paper it is shown that transitional contact resistance of copper-aluminum contact pair is at 2,5 times less than a direct connection of copper and aluminum (test contact pair).

1. Introduction

In the place of joining dissimilar metals such as copper and aluminum, the contact is oxidized that leads to deterioration of its quality, the reduction of the useful life and the increase of the transitional contact resistance.

To join copper and aluminum contacts such ways as the using of bimetallic gaskets, copper-based powdered lubricants and copper-coating deposition on the aluminum surfaces are used [1-3]. However all of them have advantages and disadvantages. Bimetallic gaskets allow to join cooper and aluminum contact but the resistance is increased due to the high quantity of boundaries. The method of cold gas-dynamic spraying is the most widely-used but the low adhesion of coatings negatively impacts on the long life of such contacts [4].

It was shown that it is possible to obtain a copper-aluminum contact pair by using the pulsed jet of the copper electro erosion plasma [5]. The jet is generated by the coaxial magnetoplasma accelerator with copper electrodes. This method allows to obtain copper-aluminum contact pairs with the high adhesion and the stability to dynamic loads [6].

To investigate the influence of the coating phase composition on the contact pair parameters such as hardness and transitional contact resistance, a series of experiments has been carried out.

2. Experimental

The coaxial magnetoplasma accelerator (figure 1) has been used as a source of copper plasma that influences on the aluminum surface. The power supply has been provided from the capacity energy storage.

Based on the terms of previous studies [7], optimal energy and constructional parameters for copper coatings deposition are as follows: a charging voltage U = 3,5 kV, a charging capacitance C = 12 mF, the charging energy W_c = 73,5 kJ, the diameter of the accelerating channel (AC) d_{ac} = 12 mm, the distance between end of the AC and an aluminum target l=270 mm, the process is realized in the air atmosphere. The coating, obtained with these conditions, is uniform and the coating area is about of 130 cm2.

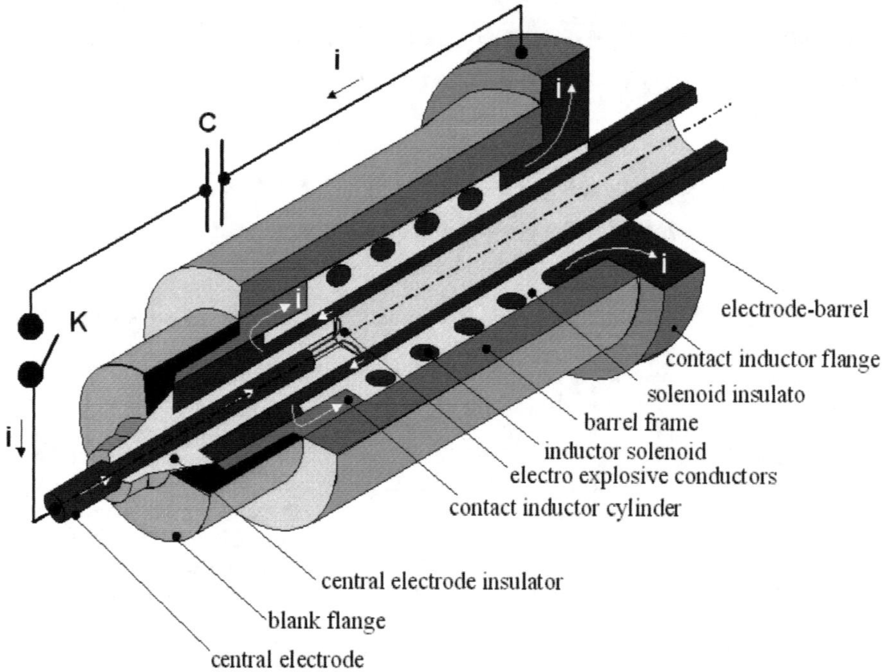

Figure 1. Coaxial magnetoplasma accelerator.

To investigate the phase composition of the copper-coated aluminum samples, the method of X-ray diffractometry (XRD) has been used by using the diffractometer Shimadzu XRD7000 with a Cu Kα radiation. The analysis of XRD patterns has been carried out using the software "PowderCell 2.4" and databases PDF2 and PDF4+. The hardness of samples has been investigated using the NANO Hardness Tester NHT-S-AX-000X and Vickers hardness. The coating microstructure has been investigated with the optical metallographic microscope Olympus GX-71. The value of the transitional contact resistance has been measured by the volt-ampere method.

3. Results and discussion

The typical XRD pattern of obtained copper-aluminum contact pairs is shown in figure 2. It has been found the content of the main crystalline phase of copper Cu (Space Group (SG): F 4/m – 3 2/m) varies in the range ~61÷83% by depending on the coating conditions. Besides, such crystalline phases as copper oxide CuO (SG: F 4/m – 3 2/m) containing up to ~25%, aluminum Al (SG: F 4/m – 3 2/m) up to ~20%, aluminum nitride AlN (SG: F 4/m – 3 2/m) up to ~20% have been identified.

The aluminum phase presence in the coating is caused by its receipt from the substrate surface due to factors such as melting, hydrodynamic mixing and diffusion in the liquid phase at a high speed jet interaction with the surface. The crystalline phases of CuO and AlN are formed as a result of the process of the copper deposition in the air atmosphere. The absence of the aluminum oxide is the positive factor of the obtained copper-aluminum contact pair, because its presence deteriorates the contact quality.

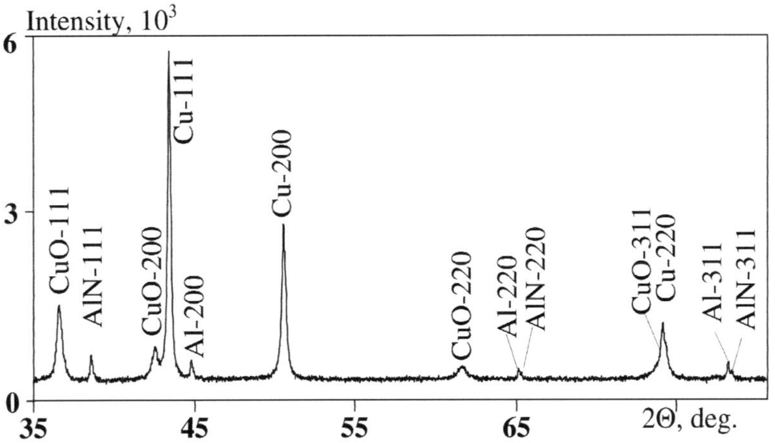

Figure 2. XRD pattern of the copper coating on the aluminum substrate obtained in the air atmosphere.

The comparative data analysis of the coating material phase structure and the nanohardness is shown in figure 3. There is also a correlation between the nanohardness of the copper surface and the presence of a harder phase of the aluminum nitride (AlN) and a less hard aluminum (Al) phase. The nanohardness increases like a linear function with the growth of the aluminum nitride presence in the above limits. An elevated presence of the aluminum phase leads to a significant reduction of the nanohardness.

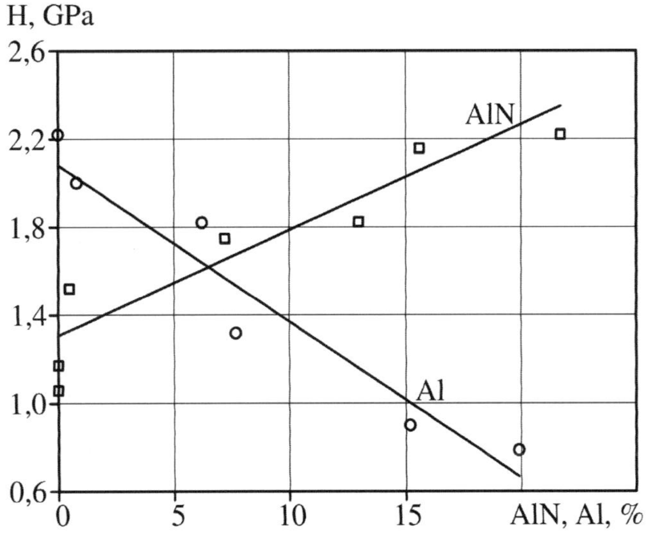

Figure 3. The dependence of nanohardness in the boundary layer on the phase concentrations.

It is well-known the transition resistance of the contact pair depends on the pressure load. The higher the value of pressure load is, the lower the transition resistance is [6-7]. However the dependence of the transition contact resistance on the nanohardness has been investigated (figure 4). It has been found, the transition resistance (Rres) reduces almost linearly with the growth of the nanohardness of the coating material, boundary and frontier layers of the substrate material.

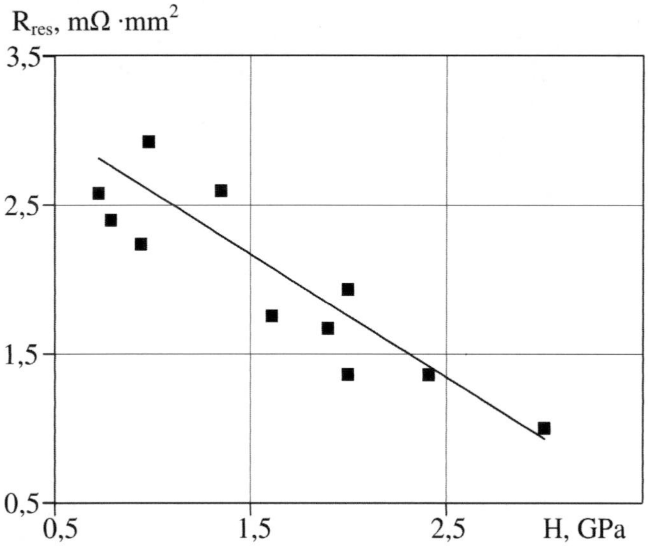

Figure 4. Dependence of the contact resistivity (R_{res}) on the nanohardness (H) of the coating.

The microstructure of the copper coating and the boundary layer has been analyzed by optical microscopy. Figure 5 shows the optical micrographs of cross-sections slice of aluminum sample with copper coating. Using the optimal energy and constructional parameters the uniform coating with the thickness of about 100 μm has been obtained. The cover adjoin tightly to the substrate replicating everything unevenness scope pairing. The boundary layer between aluminium and copper is absent according to micrographs. It is the positive feature of the method. The copper particles deeply penetrate into the substrate due to the hydrodynamic intermixing in the synthesis process. The hydrodynamic intermixing is the key factor of the strong adhesion between copper and aluminum. The transitional resistance of such contact pair is at 2,5 times less than a direct connection of copper and aluminum [8].

Figure 5. Optical micrographs of the cross-sectional slice of the aluminum sample.

4. Conclusion

It is shown the possibility of obtaining of the copper-aluminum contact pair with uniform copper coating with the thickness of about 100 μm by using the coaxial magnetoplasma accelerator. The phase composition of coating has been investigated and it has been found the presence of such crystalline phases as copper, aluminum, copper oxide and aluminum nitride. The dependence of the nanohardness on the aluminum nitride and the pure aluminum for the samples obtained by plasma dynamic method has been shown. The increase of the AlN phase leads to the increase of coating nanohardness. The increase of the nanohardness directly impacts on the reduction of the transitional contact resistance. Using the optimal energy and constructional parameters of the system based on the coaxial magnetoplasma accelerator, it is possible to reduce the transitional contact resistance of the copper-aluminum contact pair at 2,5 times.

References

[1] James J Licari 2003 *Coating materials for electronic applications (Elsevier Inc.)* p 912
[2] Peter M Martin 2010 *Deposition technologies (Elsevier Inc.)* p 312
[3] Zenin B S 2008 *Modern technology, and modified surface coating* p 75
[4] Christine Blanc 2010 *Corrosion Science* **52** 991-995
[5] Sivkov A A 2014 Patent No 137443 *Coaxial magnetoplasma accelerator*
[6] Sivkov A A, Korolkov L V, Saygash A S 2003 *Electrical engineering* **8** 41–46.
[7] Sivkov A A, Saigash A S, Kolganova Yu L 2013 *Russian Electrical Engineering* **84** 418-421
[8] Sivkov A A, Saigash A S, Kolganova Yu L 2012 *Proceedings of the higher educational institutions: Physics* **55** 138-141

MTT'2014 IOP Publishing
IOP Conf. Series: Materials Science and Engineering **66** (2014) 012049 doi:10.1088/1757-899X/66/1/012049

Antireflection and protective films for silicon solar cells

O V Semenova[1], V A Yuzova[2], T N Patrusheva[3], F F Merkushev[4], M Y Railko[5], S A Podorozhnyak[6]

[1] Senior Production Editor, 85 Kujbysheva str., apt. 259, Krasnoyarsk, Russia
[2] Production Assistant, 10 Borisova str., apt. 211, Krasnoyarsk, Russia
[3] Production Assistant, 5 Parizhskoy Kommuny str., apt. 40, Krasnoyarsk, Russia
[4] Production Assistant, 10 Urvantseva str., apt. 143, Krasnoyarsk, Russia
[5] Production Assistant, 85 Kujbysheva str., apt. 259, Krasnoyarsk, Russia
[6] Production Assistant, 12 A Yushkova str., apt. 54, Krasnoyarsk, Russia
E-mail: fedor-murkushev@mail.ru

Abstract. The optical properties of porous silicon structures coated with nanodiamond films have been investigated for use as antireflective and protective coatings of silicon solar cells.

Keywords: silicon solar cells, antireflective and protective coatings, porous silicon structures coated with the nanodiamond films.

1. Introduction

Currently, the most efficient renewable solar energy sources are semiconductor solar cells including silicon solar cells. Silicon is the basic material in the micro- and nanoelectronics, and photovoltaic industry. Because of the high refractive index of a single crystal silicon (n = 3,5), a significant portion of a solar radiation is reflected from the surface of the photovoltaic converter (reflectance value may be greater than 35 %) and, as a consequence, this does not contribute to the carrier pair generation process. Obviously, this leads to an efficiency reduction of these converters. Therefore, the creation of the antireflective coatings and the search for the materials for their production is really important. One of these materials is the porous silicon (por-S), formed on the surface of a monocrystalline silicon (c-Si). Seeing that the porous silicon is degradable, it is reasonable to use the protective films and coatings that do not negate the effectiveness of the solar energy conversion.

The coatings antireflect the light of a visible spectrum are applied on the protective glasses or directly onto the front surface of the solar cells [1-4]. The SiO_2 [1], TiO_2 [2], silicon nitride (SiN_x) [3] and boron nitride (BN_x) films which are good anti-reflective coatings for solar cells because of their spectral stability, high strength and sufficiently wide bandgap [4] are used as the materials with low reflection.

The promising protective materials are diamond-like carbon films (DLC) [5]. DLC effectively absorb the ultraviolet radiation, have a low infrared (IR) absorption, they are transparent to visible light and have the chemical inertness. They not only have the clarifying properties [6-7], but also provide a significant hardening of the functional layers of the porous silicon, significantly increasing the limit of fragility and fracture of toughness at the contact effects [7], and increase the radiation resistance of silicon also [8].

Thus, it should be assumed that the usage of the porous silicon structures with the diamond-like coatings can be one of the methods to increase the efficiency and protection of the silicon solar cells simultaneously.

In this paper we investigate the possibility of creating such structures.

2. Methods of the experiment

A quite simple technology of electrochemical anodization is used to create the porous silicon. The porous silicon samples was obtained in accordance with a manufacturing route previously developed [9] using the electrochemical anodization technique in the presence of the light sources with the different wavelengths (Fig. 1).

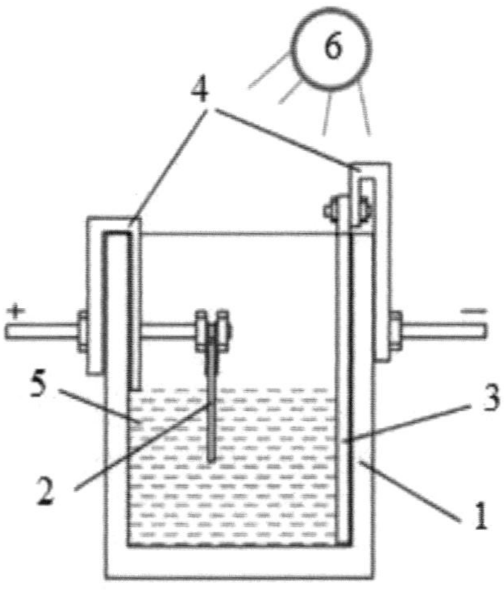

Fig. 1. Electrochemical anodizing device model:
1 – fluoroplastic bath; 2 – anode (silicon sample); 3 – cathode (nickel plate);
4 – electrode holdings; 5 – aqueous solution of a hydrofluoric acid; 6 – light source

Electrochemical treatment of the silicon samples was held in the aqueous solutions of hydrofluoric acid (HF: H_2O = 1:1). The cylindrical nickel plate was used as a cathode. The power supply have a range of operating voltages and currents (0.1-100) V and (0.01-5) A respectively was used at the experiment. The silicon substrates – the single crystal silicon wafers KEF-10 (100) before a process of the porous silicon samples producing had being treated in a peroxide- ammonia solution (NH_4OH: H_2O_2: H_2O = 1: 1: 4) at a temperature of 70–90 ° C for 15 min. A nanodiamond (ND) obtained as a result of an explosion synthesis followed by the chemical purification was used as an initial material.

In this research, the nanodiamond films were being obtained from the organic suspensions [10]. The nanodiamond suspensions was prepared as a result of stirring the ND powder in the low-boiling organic solvents (hexane, heptane, octane, toluene, benzene, isopropanol) on application of ultrasound. The transition of the smallest particles in the organic phase occured spontaneously as a result of the simple mixing of the liquids. More intensive transition was initiated by the ultrasonic treatment during which a disintegration of the particles aggregates and a content of the nanocrystalline fractions was increased. The nanodiamond formed the stable transparent suspensions in organic liquids, particularly in toluene, at a solids concentration of 0.01-0.05 g / liter.

The ND film deposition on the silica porous structure was carried out with a dipping method. The presence of the functional groups on the surface of the ND and its high surface energy gives the possibility of the nanoparticles adsorption on a substrate. The optical properties of the ND coatings are controlled by layer thickness and their number depending on the concentration of the suspension.

The investigations of the received samples morphology were carried out with a use of an optical interference microscope MII-4 and the optical properties were investigated with a use of UV-VIS spectrometer Specord M400.

3. The results and their discussion

The research results showed that the electrochemical treatment of the n-type silicon under the additional illumination (λ = 450 nm) allows to produce the porous silicon layers with high uniformity and the pores equally spaced on a whole surface of the c-Si substrate, as well as a sufficiently high density of the porous structure (Fig. 2). It was found that the blue light (λ = 450 nm) influences on the simultaneous growth of density and reduction in sizes of the pores (the diameter of the pores is less than 0,5 μm). This contributes to the extension of the porous structures surface area.

The experimental data confirmed the results of this research [11] that showed that the electrochemical anodization technique gives the opportunity to produce structures with the different optical and dielectric properties (refractive index and dielectric coefficient) by varying the programs parameters and conditions of the electrochemical anodization. The increase of current density and temperature of electrolyte results in the increase of pore diameter and change of the optical and dielectric properties.

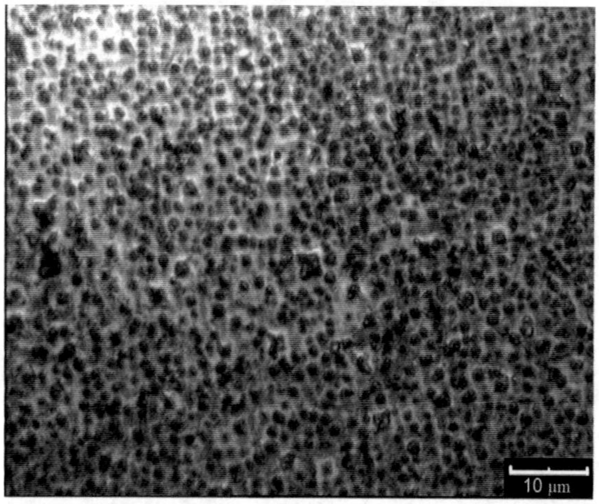

Fig. 2. Pores diameter dependence on current density and temperature of anodization

Introduction of various structural and technological decisions (change of a cathode form, the pretreatment of Si substrate using supplementary illumination of different intensities and wavelength during the anodizing process, drying of the samples under the influence of the light sources, the exposure time on the air and heat treatment) allows to produce the por-Si films of different thickness, reflectance and refractive indexes and the stability of the optical parameters. In order to obtain thin Si-por film with minimum reflectance (10 %) the following modes of an electrochemical process were determined: the voltage U = 10 V, current density J = 100 mA/cm^2, the anodization time is 40 min under the influence of wavelength radiation of a visible range λ = 450 nm.

Morphology, pore dimensions, chemical composition of the porous silicon surface and, as a consequence, the refractive index depends on the shape of the cathode and the lighting of the samples during of their producing. Usage of a cylindrical cathode increases the pores equitability on the plate surface and the reproducibility of the anodization process in comparison with the cathode in the form of a straight plate. In turn, joint use of the cylindrical cathode and additional lighting with λ = 450 nm allows to produce the thin pores and ordered structure of the porous silicon.

The results of the spectroscopic studies confirmed that the structures of the porous silicon with ND coatings have the low reflectance (2-8 %) in the visible region of the spectrum compared with the samples of the original single crystal silicon (40-80 %) (Fig. 3a and 3b) . The reflection coefficient depends on the number of ND layers (Fig. 3b). The possibility of varying the optical properties due to changes of the number of ND layers and film thickness during suppression allow to form coatings which meet the requirements of the optimal antireflection. The por-Si samples with 10 ND layers have a minimum reflectance coefficient in the range of 600–700 nm, that corresponds to the maximum response characteristics of the silicon solar cell batteries in the area of 600 nm.

The repeated spectroscopic analysis of the samples showed the stability of the optical properties after its expiration for a year.

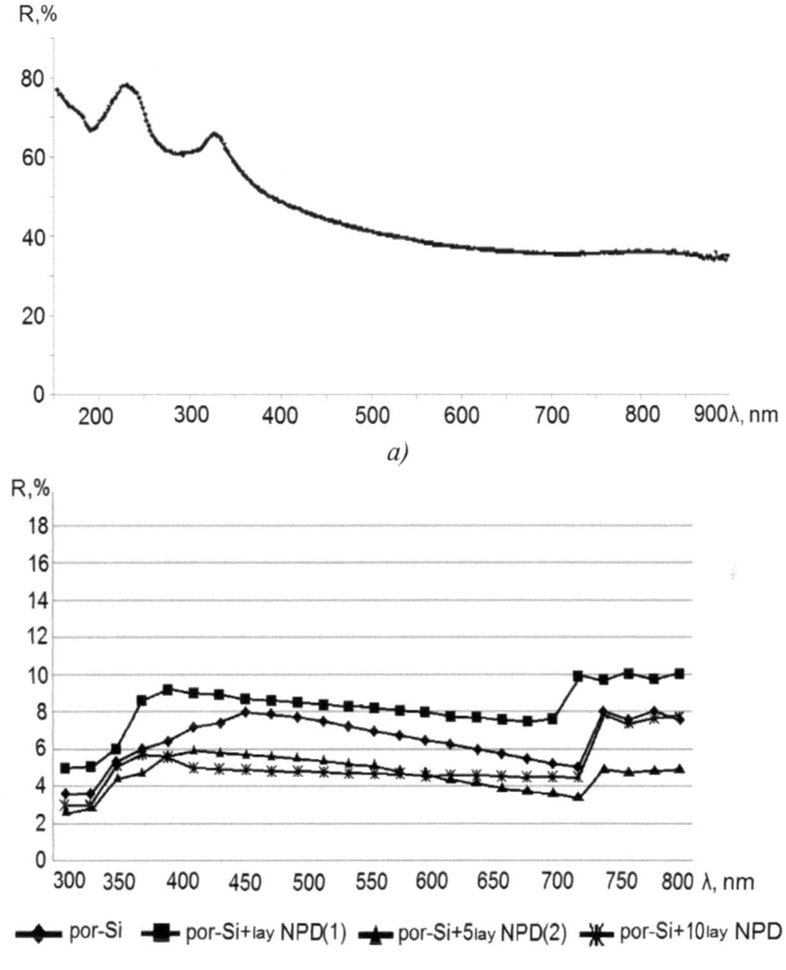

b)
Fig. 3. The reflectance spectra of the sc-Si original sample (a) and por-S with the NPD films (b)

According to these researches, the ND films produced as a result of a developed method on the porous silicon does not increase its reflection coefficient. It is shown that depending on the thickness of the ND coating the reflection coefficient changes in the range of 1–4 %, which allows to form the coatings which meet the requirements of the optimal reflectance for the solar cells.

In addition, due to their nature nanodiamond coatings have a protective function and stabilize the properties of the porous silicon at the time. It was found that the optical properties of the porous silicon, produced under the optimal conditions, and protected with the nanodiamond coating, did not change during 12 months.

4. Conclusion

The samples of the porous silicon with the equally spaced pores were produced during the process of the electrochemical anodization. It is shown that the porous silicon films effectively reduce the reflection coefficient in comparison with the monocrystalline silicon. The ND coatings were applied on the porous silicon surfaces in order to stabilize the properties and eliminate the degradation processes.

It was established that the ND films produced from the organic slurries practically do not change the reflection coefficient of the porous silicon. It is known that the diamond-like films applied on a working side of the silicon solar cells can increase their resistance to ultraviolet radiation and protons. This point to a perceptiveness of using the porous silicon structures with the nanodiamond coatings as a antireflection and protective coating for the silicon solar cells.

REFERENCES

1. Nemcova A A, 2009, *Dissertation author's abstract on scientific degree candidate of technical sciences*, (St. Petersburg) 17. [in Russian]
2. Patrusheva T N, Shelovanova G N, Snezhko N Y, Polyushkevich A V, Khol'kin A I, 2011, Alternative Energy and Ecology **3**, 8-15. [in Russian]
3. Prasad B, Bhattacharya S, Saxena A K, Reddy S R, Bhogra R K, 2010, *Solar Energy Materials & Solar Cells* **94** 1329–1332.
4. Alemu A., Freundlich A, Badi N, Boney C, Bensaoula A 2010, *Solar Energy Materials & Solar Cells* **94** 921–923.
5. Cluj N I, Litovchenko V G, Lukyanov A N, Neselovskya L V, Sarikov A V, Dyskin V G, Haziyev W H, Settarova Z S, Turusov M N, 2006, *J. of Physics tehnicheskly* **5** 122-126. [in Russian]
6. Sizov F F, Cluj N I, Lukyanov A N, Savkina R K, Smirnov A B, Evmenova A Z, 2008, *Letters Technical Physics* **9** 32-39. [in Russian]
7. Savenko V I, Belyanin A F, Perevozchikov B N, 1997, *Physics and chemistry of materials processing* **2** 59-64. [in Russian]
8. Won Seok Choi, Kyunghae Kim, Junsin Yi, Byungyou Hong. 2008, *J. Materials Letters* **62** 577-580.
9. Sakun E A, Polyushkevich A V, Kharlashin P A, Semenova O V, Koretz A Ya, 2010, *Engineering & Technologies* **3** 430-443. [in Russian]
10. Lyamkin A I, Petrov E A, Ershov A P, Sakovich G V, Staver A M, Titov A M, 1988, *DAN USSR* **3** 611. [in Russian]
11. Patrusheva T N, Koretz A Ya, Mironov E M, 2005, *J. of Pure and Appl. Phys.* **43**, 115–118. [in Russian]

IOP Conf. Series: Materials Science and Engineering **66** (2014) 012050 doi:10.1088/1757-899X/66/1/012050

Influence of nanopowders and pore-forming additives on sintering of alumina-zirconia ceramics

T A Khabas[1], **L V Maletina**[1], **K S Kamyshnaya**[1]

[1]National Research Tomsk Polytechnic University, Tomsk, 634050, Russia
E-mail: ksene4ka.01@mail.ru

Abstract. This work reviews the influence of nanopowders and pore forming additives on sintering and on properties of fine-pored alumina-zirconia ceramics. The optimum ratio of additives for ceramics' sintering, as well as the content of additives to obtain microporosity have been determined.

1. Introduction

At present, zirconium and aluminum oxide-based ceramics is widely used in machinery and medicine. Aluminum oxide-based ceramic materials are characterized by high durability and high resistance to abrasion [1]. Zirconium dioxide is also considered to be a promising material due to its high mechanical strength and fracture toughness [2, 3]. Nevertheless, each of these oxides has its limitations, such as high cost of zirconium dioxide-based ceramics, low fracture toughness of aluminum-based ceramics, and high temperatures of sintering. At this time ceramic materials based on the mixture of these oxides are being intensively developed. These ceramics appear to be rather promising due to their favorable physical and chemical properties [4, 5]. Medicine is a potential area of application for these ceramic materials. Implant materials should have specific porosity [6]. The optimal pore size of the tissue ingrowth into an implant varies from 5 to 10 μm and from 50 to 100 μm [7]. Introduction of various additives into ceramic mass allows to modify the temperature of sintering, porosity and strength of the obtained ceramics. Usually, the impact of small quantities of pore-forming additives (up to 1.5 mas. %) is generally disregarded, as their introduction to materials can lead to mixed results.

2. Experimental Part

This research was aimed at studying the influence of nanopowders and pore-forming additives on sintering and on properties of fine-pored alumina-zirconia ceramics. During the experiment the initial charge mixture was made of micron powders of aluminum oxide (produced by Nabalox, Germany, average particle size 1.20 μm) and of zirconium dioxide (produced by the Chepetsk Mechanical Factory in Glasov, average particle size 0.23 μm) with mass ratio 70:30. Specific surface area of powders was determined by Surface Area and Pore Size Analyzer Quantachrome Nova-2200-e. Formation of the porous structure with channels of the necessary size was achieved by introduction of a burnable fibrous material. To improve sample strength, a pore-forming additives in the form of synthetic fiber (thread length – 1-2 mm) or polyvinyl alcohol (PVA) granules with nanopowder of one of the oxides were added in the obtained charge mixture. The average size of the zirconium dioxide nanopowder particles was 0.065 μm, of the aluminum oxide – 0.070 μm. The choice of nanopowders of similar oxides as additives guarantees lack of impurities in the sintered material.

Oxides nanopowders were studied using analytical Field Emission SEM JEOL JSM-7500FA. To achieve even distribution of the additives in the micron powder, prepared mixture was blended in the Mini-Mill (Pulverisette 23). Cylinder shaped samples were molded from prepared charge mixture using uniaxial compaction method under specific pressure of 600 MPa. Samples were heat treated at

1550° C and 1580°C in air, rate of temperature rise ~ 2°C per a minute, curing at finishing temperature for 2 hour. Behavior of the synthetic fiber during heating was analyzed using thermal analyzer NETZSCH STA 449 F3 Jupiter®. After thermal treatment shrinkage, density, porosity and water absorption of the samples were defined using hydrostatic weighing with vacuum blowing method. For tensile strength determination at three-point bending, the samples were prepared in the form of beams with size 50 × 5 × 5 mm. For determination of strength at compression, the samples were prepared in the form of cubes, with size 5 × 5 × 5 mm.

3. Results and Discussion

Line charts showing relationship between shrinkage/total porosity and quantity/additive type

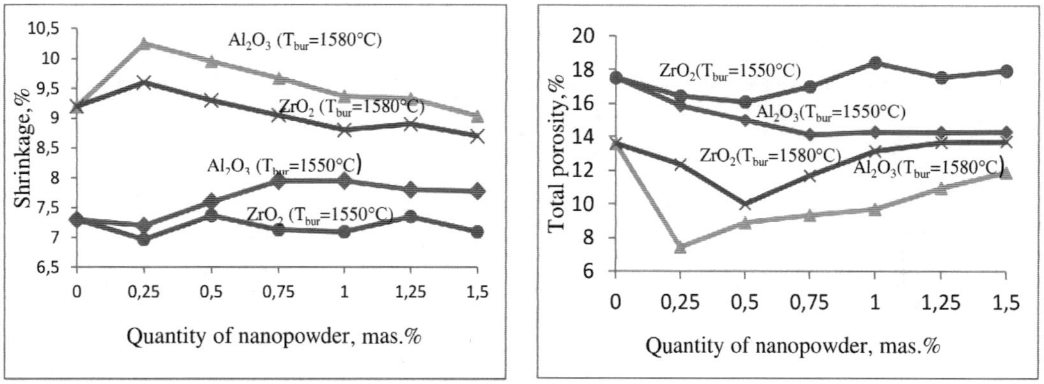

Figure 1. Relationship between shrinkage/total porosity and the quantity of Al_2O_3 and ZrO_2 nanopowder at the firing temperatures of 1550° C and 1580° C (types of nanopowder additive and firing temperatures are specified on the graphs).

Nanopowders in the content range under study at 1550°C carry out a function of a sintering additive, which appears from the increase in shrinkage of the samples and from the reduction of the total porosity when introducing nano-Al_2O_3 to 1 mas.% and nano-ZrO_2 to 0.75 mas.% (Fig.1). Sinterability at 1580°C is limited by the nanopowders' content at 0.25 mas. % of Al_2O_3 and 0.5 mas.% of ZrO_2. Porosity of the samples tends to increase with the increase of nanopowders' content greater than these values. It must be noted that properties of the initial powders greatly influence properties of the obtained samples. Therefore, pictures of the initial nanopowders were taken and a specific surface area of the initial powders was determined.

Microphotos (Fig. 2, Fig. 3) demonstrate that zirconium dioxide nanopowder has sphere structured hollow particles (0.065 μm), which later will influence the samples' porosity. Aluminum oxide nanopowder has spherical particles (0.070 μm) and "loose" structure. On the one hand, it complicates pressing process; on the other hand, it provides closer packaging of the particles in a raw material (density increases by 10%).

Figure 2. Microphotography of ZrO_2 nanometric.

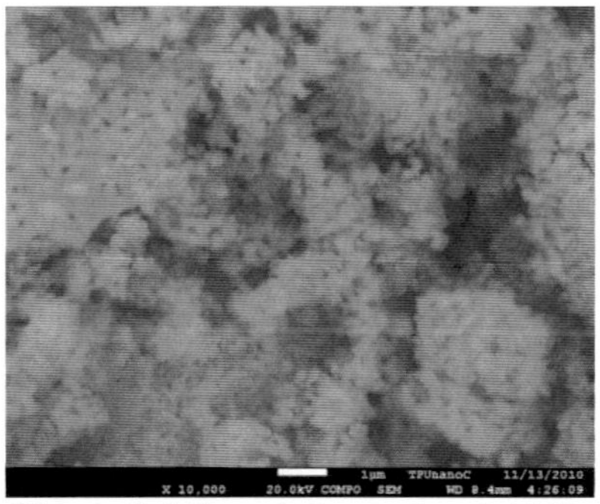

Figure 3. Microphotography of Al_2O_3 nanometric.

Not only PVA, but also small amounts (up to 0.1 mas.%) of synthetic fiber increase sinterability: they contribute to density of sintered materials, increase of shrinkage and decrease of porosity. (Fig. 4).

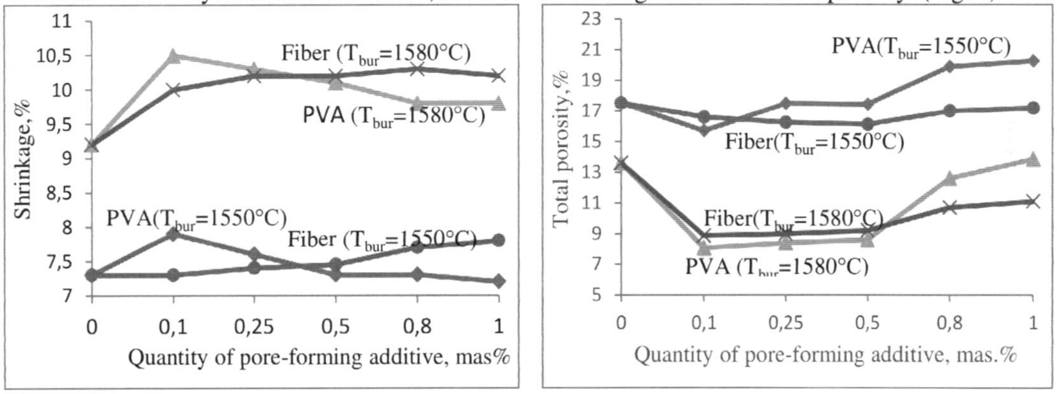

Figure 4. Relationship between shrinkage /total porosity and quantity/ type of pore-forming additive at firing temperatures of 1550° C and 1580° C (additive type and burning temperature are specified in the graphs)

With the increase of polyvinyl alcohol and fiber content in the sinter mixture, the role of PVA and fiber is progressively changing. At 0.5-1.0 % these additives contribute to the increase of porosity and to the formation of fine-pored structure. It was noted that with firing temperature increase by 30°C, linear shrinkage increases by 2.0-2.5%; total porosity increases by 4.0% (for samples without pore-forming additives), and by 10.0 % (for samples with 1 mas.% PVA).

Differential thermal analysis of the synthetic fiber was conducted to determine the role of the fibrous material. This material has to perform the pore forming function, i.e. to burn out completely when heated in air.

Thermogram (Fig. 5) shows that fiber starts melting at low temperatures 219.4 and 306.4 °C, that results in two small endothermic effects (circles in Fig.5); fiber starts oxidation at 359°C, and at 564.2° C the mass loss is 99%. This reaction is exothermic and it is accompanied by two effects at temperatures 375.5° C and 492.8° C through primary oxidation, formation of capsules and their additional oxidation. Thus, it is arguable that in the initial interval of the thermal treatment fiber melts, creating a small quantity of the melt that contributes to attraction of solid particles, so, up to a certain content in the charge mixture, it promotes material densification.

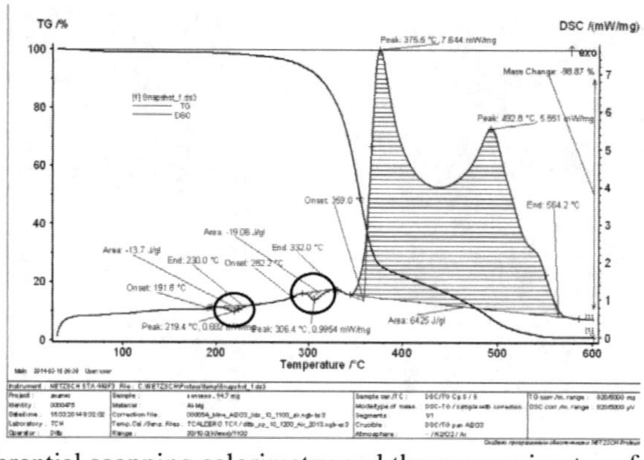

Figure 5. Differential scanning calorimetry and thermogravimetry of synthetic fiber.

At mass content greater than 0.5 mas.%, fiber creates porous structure while burning out. However, it is necessary to take into account, that fiber content greater than 1 mas. % complicates the pressing process and leads to formation of hairline cracks in samples. Microphotos of the samples confirm received results (Fig. 6, Fig. 7).

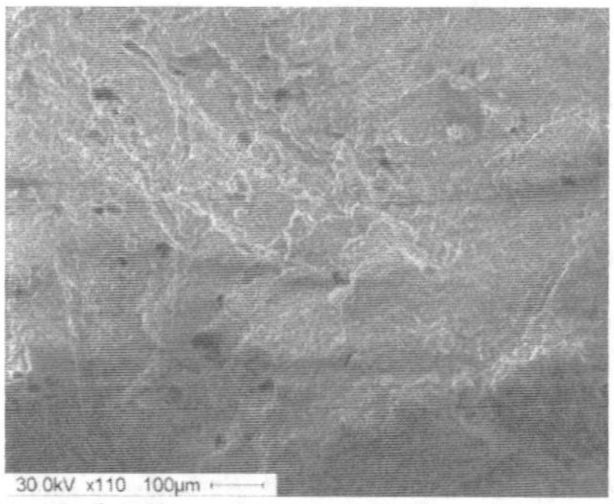

Figure 6. Microphotograph of a sample with support 1 mas.% of fiber (T_{bur}=1580°C)

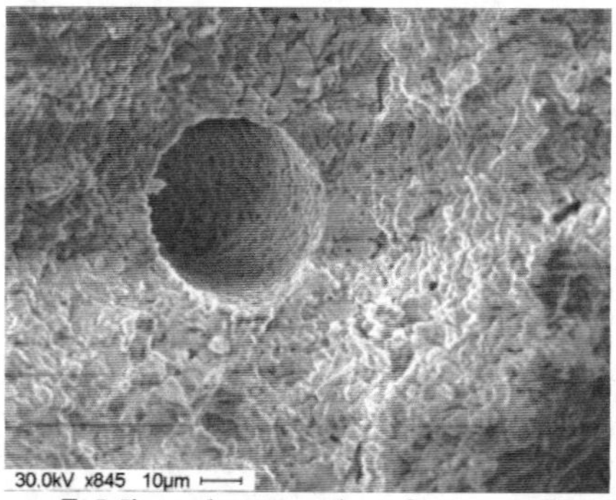

Figure 7. Microphotography of a pore after fiber burnout

Microphotos demonstrate that the increase in the amount of fiber leads to the increase in porosity of samples. It is noted that pores created by fiber are cylindrical in shape with diameter about 30 µm. The combined effect of nanopowder and fiber on the properties of sintered ceramics was also researched.
At 1580° C, the presence of fiber (0.1 mas.%) contributes to sintering of the ceramics, containing nano-ZrO_2 at 0.25 mas.% as an additive (without the fiber this additive activates sintering at 0.5 mas.%) (Fig.8). With introduction of nano-Al_2O_3 additive, sintering takes place when its contents is not greater than 0.5 mas.% (while without introduction of the fiber sintering occurs at the contents not greater than 0.25 mas.%). At 1550° C, sintering is limited by nanopowders' contents of 0.25 - 0.75 mas.%. Thus, when comparing shrinkage of samples with both nanopowder and fiber additives as opposed to samples with the nanopowder additive only, it can be concluded that the introduction of fiber in the amount of 0.1 mas.% insignificantly improves sintering of samples (shrinkage of samples is approximately the same), but it increases the content range of nanopowders, which leads to improvement of sintering. This fact has technological importance, since it allows adjusting additive amounts at a larger range.

IOP Conf. Series: Materials Science and Engineering **66** (2014) 012050 doi:10.1088/1757-899X/66/1/012050

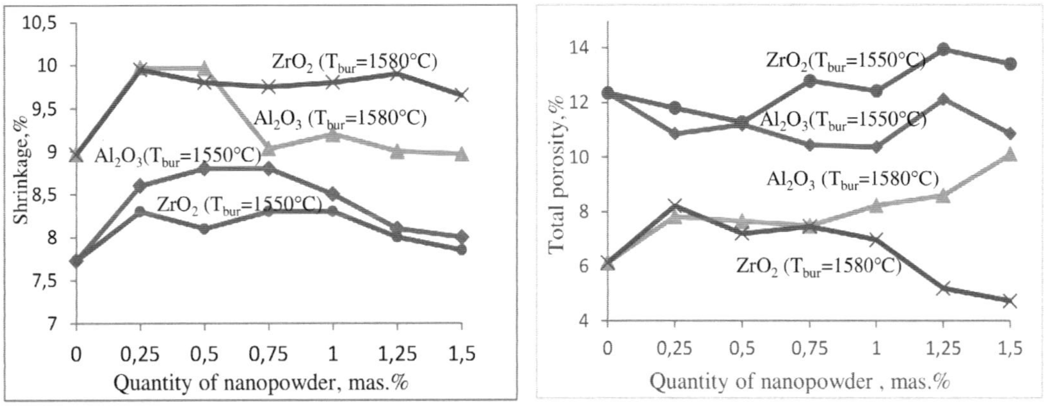

Figure 8. Relationship between shrinkage/total porosity and quantity of Al_2O_3 / ZrO_2 nanopowder at 0.1 mas.% fiber at firing temperatures of 1550°C and 1580°C.

Comparison of micropictures (SEM) shows that in samples without additives, separate agglomerates of grains and pores between grains are visible (Fig. 9).

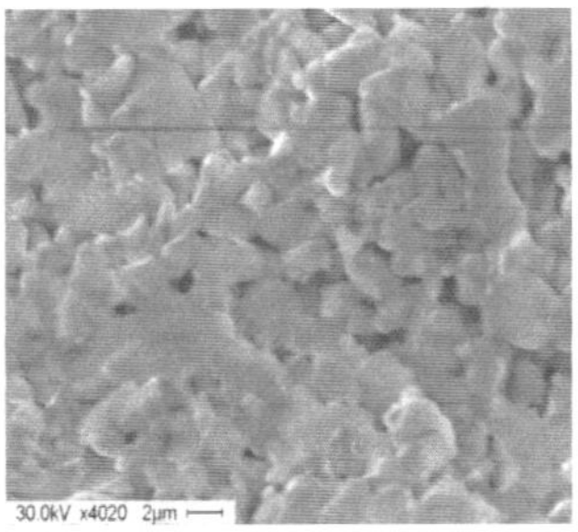

Figure 9. Photo of sample consist of 70%-Al_2O_3 and 30%-ZrO_2 (T_{bur}=1580°C)

Figure 10. Photo of sample with 1.5 mas.% of Al_2O_3 nanopowder (T_{bur}=1580°C)

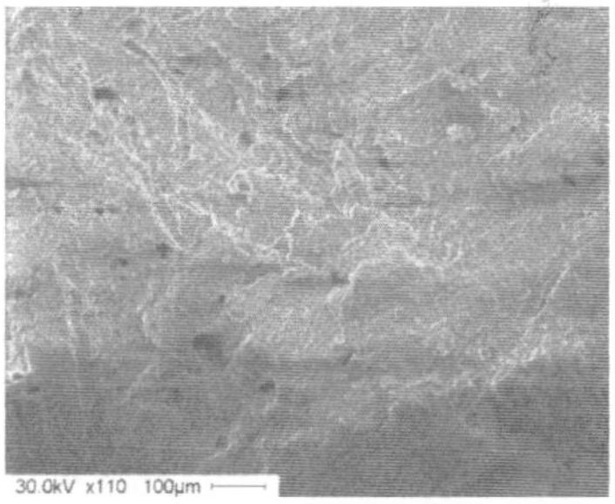

Figure 11. Photo of sample with 1.5 mas.% of Al_2O_3 nanopowder and 0.1 mas.% of fiber (T_{bur}=1580°C)

Nanopowder additive activates firing and creates small number of pores (Fig.10), also confirmed by the data demonstrated in Fig 1. Adding fiber to this structure, on the one hand, promotes proximity of particles at sintering, on the other hand, creates porous structure due to burn out (Fig.11). Samples with fiber and nanopowders have compact structure with large number of pores (Fig.11).

Tensile strength of samples on compression exceeds their flexural strength (Fig.12) approximately two times. The data obtained revealed that synthetic fiber as a pore-forming additive increases strength of the samples both at compression and at bending. PVA as a pore-forming additive increases strength of samples only at its content of no greater than 0.1 mas.%. Further increase of PVA content leads to decrease in strength properties of the ceramics, which is explained by the increase in the total porosity of the obtained samples. Increase in nanopowder content leads to increase of ceramics' strength. It is noted, that when combining nanopowder and synthetic fiber additives, a possibility of stronger ceramics can be achieved, compared to using a nanopowder or fiber additive alone. It can be explained by the fact that pores created by burning out of the pore-forming agent are strengthened by the powder, which makes the sample more durable.

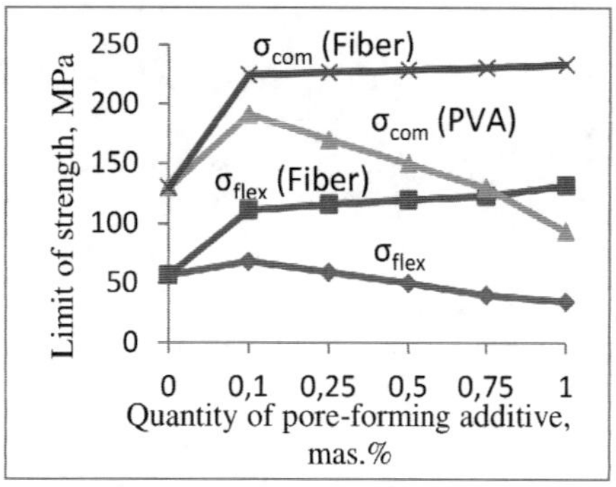

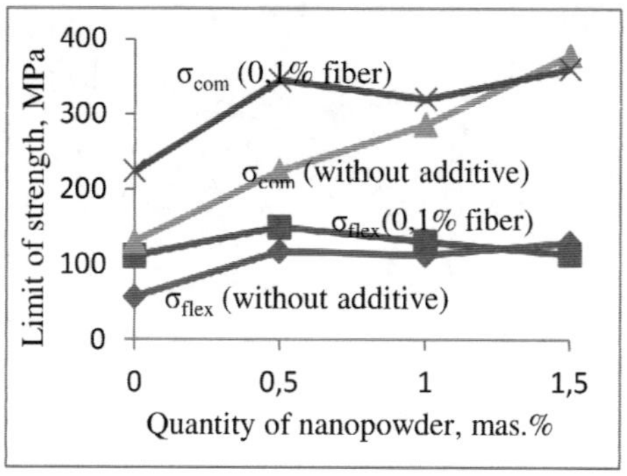

Figure 12. Relationship between the ceramics' strength (compression and flexure) fired at 1580 °C and type/ quantity of the additives.

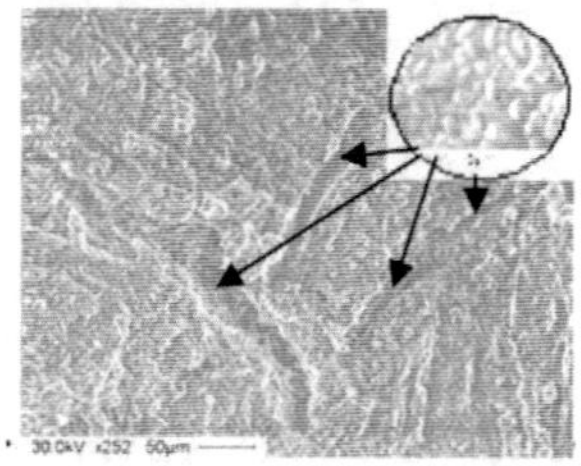

Figure13. Chanel micropores in ceramics. Separately shown: wall of the channel pore.

Microphoto of such samples show clusters of both loose and highly sintered oxide nanoparticles on walls of channel pores (Fig. 13).

4. Conclusion

Results of this research indicate that introduction of small amounts of additives of different nature and varying their ratio can significantly change properties of obtained ceramics (increased durability, porosity, degree of sintering). The nature of pore-forming substances has a great influence on ceramics' sintering and strength properties. Therefore, even small amounts of additives as well as processes occurring in the material at low heat treatment temperatures must be taken into account. Burning out, synthetic fiber as pore-forming additive in amounts greater than 0.5 mas.% produces pores with sizes about 30 μm.

References

[1] Burdick J A and Mauck R L 2011 A Review of the Past and Future Trends *Biomaterials for Tissue Engineering Applications* (New York: Springer-Verlag) pp. 1-5

[2] Zadorozhnaya O Yu, Bogaev O V, Tiunova A A, Khabas T A, Nepochatov Yu K 2014 Effect of adding nanosize ZrO_2 and compaction processing parameters on properties of composites based on Al_2O_3 and ZrO_2 *Refractories and industrial ceramics* **vol 54** pp 392-96

[3] Dubok V A 2000 *Powder metallurgy and metal ceramics* pp 381-94

[4] Kannan S, Balamurugan A, Rajeswari S, Subbaiyan M 2002 *Corrosion Reviews* p 342

[5] De Aza P N, De Aza A H, De Aza S 2005 Crystalline bioceramic materials. *Bol Soc Esp Ceram Vidr* pp135–45

[6] Chipsov V I, Sviridova I K, Sergeeva N S 2008 Research invitro of matrix quality of surface native porous granulate calcium phosphate ceramic materials *Cellular technologies in biology and medicine* **vol 2**. pp. 68–72

[7] Kulkov S N, Sharkeev Yu N, Rudenski G E 2013 RU Patent 134793

The possibility of boron carbide coating formation by using a coaxial magnetoplasma accelerator

A Sivkov[1], I Rachmatullin[1] and A Makarova[1]
1 Tomsk Polytechnic University, Tomsk, Russia

E-mail:sivkov@, rahmatullin_i@bk.ru, mafania1991@gmail.com

Abstract. A coaxial magnetoplasma accelerator can generate a dense and high velocity plasma jet by applying a pulsed high-current arc-discharge. The results of the experiment investigations of plasmodynamic synthesis in the B-C system have been shown while hyper speed jet boron carbide electric–discharged plasma steams onto copper substrate. The boron carbide coatings were formed on the copper substrate without a binder material. The formation of the crystalline boron carbide coating on the copper substrate was analyzed through X-ray diffractometry, transmission electron microscopy and scanning electron microscopy.

1. Introduction

Despite the superlative properties and the wide variety of potential applications of boron carbide, relatively little effort has been put. Boron carbide is one of the hardest materials. It has a high melting point, high resistance to chemical agents, low density and lightweight. It can be applied in many areas as abrasive powders and coatings, in nuclear applications as neutron radiation absorbent, ceramics, an impurity in thermoelectric device material and etc. [1].

As a result of these boron carbide properties there is an interest not only in powder, but also in coating, which can be used as protective layer from aggressive environments and high temperatures in different units. One of the most perspective investigations in that sphere is obtaining thick boron carbide coatings (300-500 μm) for tokamak wall in nuclear fusion reactor. The most popular method for obtaining such coatings is vacuum plasma deposition method.

B4C can be prepared by several methods, such as laser assisted CVD, microwave CVD, filament activated CVD, sputtering and radio-frequency plasma CVD. Most of these methods have a number of disadvantages, such as process duration, coatings thinness, low adhesion, necessity of high temperatures and energy intensity as a result [2].

2. Experimental

Plasmodynamic method is a possible way of sintering boron carbide phase avoiding above mentioned problems. The method is realized on the basis of the pulsed high-current coaxial magnetoplasma accelerator [3]. The dominant part of the precursors is acquired by accelerator channel erosion due to high velocity of the plasma jet.

Powder-like electro conductive precursor compound is consisted of a black amorphous boron and amorphous carbon. Their ratio is 4:1 and total mass is 1.0 g. That precursor material is put into plasma forming structure area. It acts as electro conductive wire providing arc discharge initiation between graphitic central electrode and graphitic accelerator channel (AC) surface at the beginning. The

IOP Conf. Series: Materials Science and Engineering **66** (2014) 012051 doi:10.1088/1757-899X/66/1/012051

cylindrical AC diameter is 9.5 mm and length is 150 mm. Power source of CMPA is obtained from a capacitor, which has volume of C=6 mF. Charge voltage is about 3 kV. The plasma discharge is produced in chamber-reactor at room temperature. The chamber-reactor is filled with argon at 1 atmosphere. The volume of the chamber-reactor is 0.002 m3. These dimensions are not fixed and can be changed if there is necessity.

The energy parameters of the CMPA operational cycle are plotted in the form of voltage and current oscillograms, power and energy consumed curves in Fig.1.

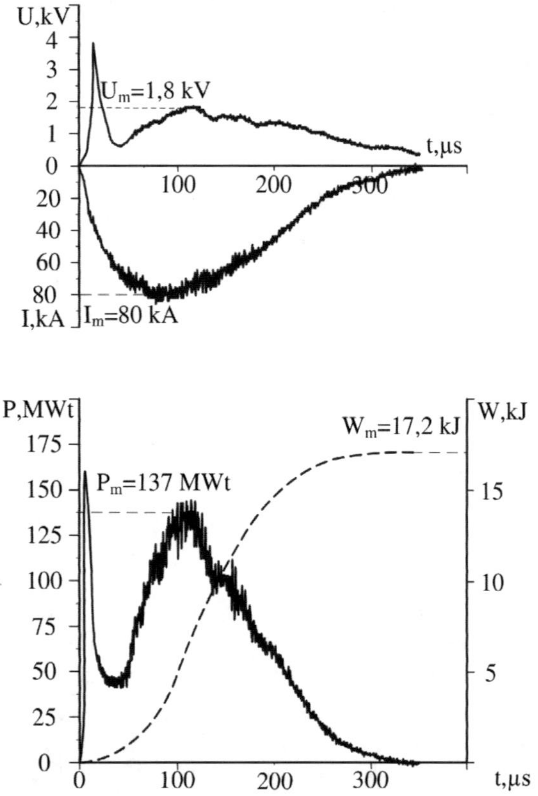

Figure 1. The typical oscillograms of the impulse current I(t) and voltage U(t) of the coaxial magneto plasma accelerator electrodes, the power discharge P(t) and energy curves W(t).

By applying a high discharge current of roughly104-105 A the temperature of the order of 103 K can be achieved between graphitic electrodes, in the limited volume of the AC, causing evaporation or initiating chemical reactions. Conductive induction electrodynamics force in a coaxial system and gas-kinetic pressure in the aggregate allow achieving on the section of the AC the velocity of the bow shock wave of supersonic plasma jet that equals about 2 km/s. The chamber-reactor opening is performed after approximately 1 hour of the plasma shot. As a result, occurs full deposition on the walls of the ultrafine synthesis product fraction.

3. Results and discussion

Analytical data of the sintered samples are received by the following methods without any preparation: X-ray diffraction (XRD) (diffractometer Shimadzu XRD 7000, 0.15406 nm); scanning electron microscopy (SEM) (Quanta 200 3D microscope and Hitachi TM3000); transmission electron microscopy (TEM) (microscope Philips CM12).

Significant reduction of amorphous carbon fractions and pure crystalline carbon phases is obtained by halving the total weight of the sum compound of the precursors to 0.5 grams. But the ratio of the precursors is the same thus precursor powder contains 80 wt. % boron and 20 wt. % carbon.

In the Fig. 2a an obtained powder is presented. As seen in the Figure 2b, the most part of the plasmodynamic synthesis has deposited on the copper substrate in the form of bulk gray material. The character of deposition and dispersion on the surface indicates that the product has turned into a liquid state and crystallized due to quick cooling. Thus gas phase precursors that have been dissociated to form molecular fragments which condense to form thin film.

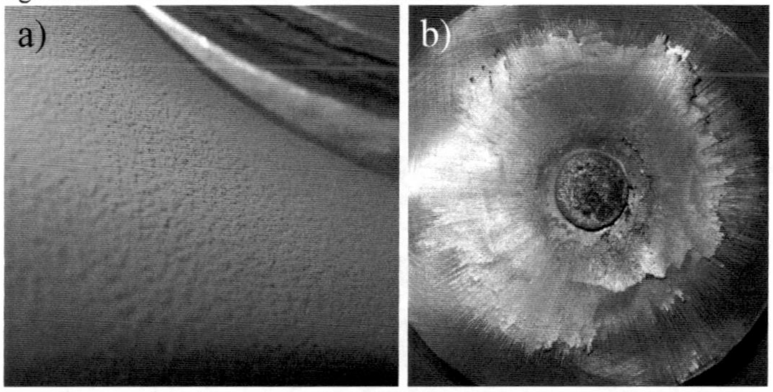

Figure 2. The chamber-reactor walls powder photograph (a) and the boron carbine coating on the copper substrate (b).

The outer surface of the deposited material and the surface of adhesion with the surface of the copper substrate have been analyzed by the XRD method. Practically identical XRD – patterns have been obtained. Fig. 3 shows the diffraction pattern taken from a front side of the deposited bulk material.

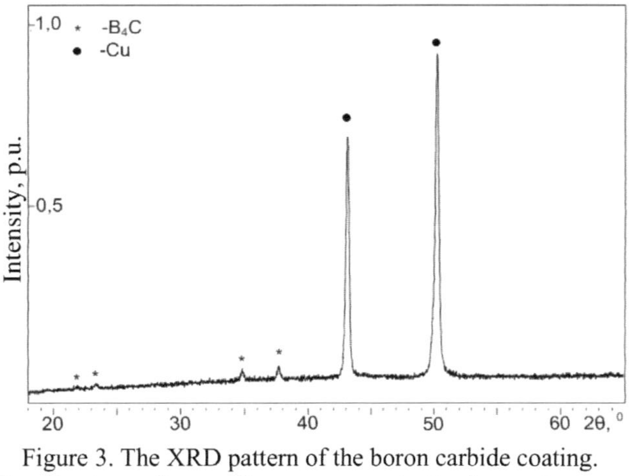

Figure 3. The XRD pattern of the boron carbide coating.

The strongest XRD-reflections of crystalline copper are in the figures because of its higher density as compared with boron carbide. In addition the crystalline boron carbide reflexes are in the XRD patterns. Fig. 4 shows the SEM images in detection mode, back scattered electrons, giving contrast material density. In the overview photograph (a) at low magnification x1000 surface irregularity of the bulk material precipitated is visible.

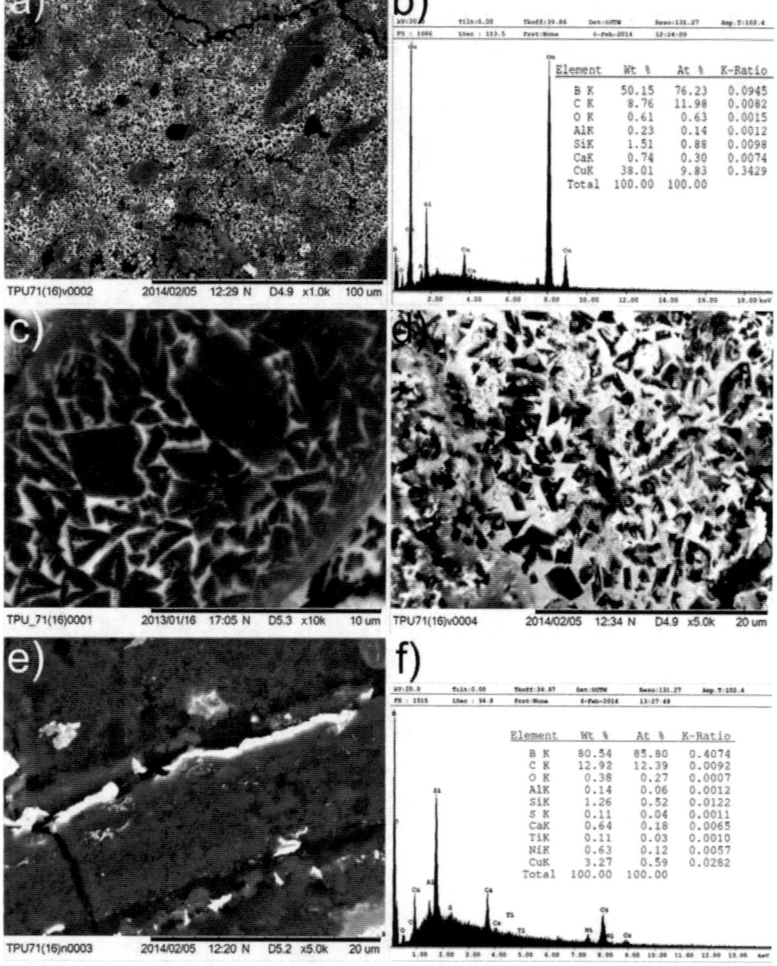

Figure 4. SEM images of the boron carbide coating. The right side of the coating (a); the element analysis data (b) of right side of the coating; the right sides of the coating (c), (d); the backside of the boron carbide coating (e); the elemental analysis of the backside (f).

Large gray contrast areas, mainly consisting of low density material, as well as areas with granular surface structure, mainly consisting of high-density material, are clearly distinguished. SEM images of fragments of these regions with magnification x10000 (c) and x5000 (d) allow concluding that in both areas the solid coating is a composite. Its basis is gray or very dark grains with geometrically irregular shapes of low-density material being in the matrix of the high-density material. In gray areas contrast (Fig. 4c) reveals grains in light matrix more closely, and their material can prevail over the light matrix material in volume and weight. In the second case (Fig. 4d) the structure with matrix material dominance takes place. Fig. 4b shows energy dispersive analysis of the surface and it gives the integral content of elements shown in Fig. 4a. The energy dispersive analysis results conform well to the visual assessment and X-ray analysis of the deposited bulk material results.

The layer thickness of the material is not uniform and does not exceed 100 microns on average. Besides XRD pattern of the back-side contacting copper substrate is identical to the coating diffraction pattern (Fig. 3). However, as shown in Fig. 4e, the material of the back-side substrate is low density and porous. The data of the energy-dispersive analysis of the front and back surfaces of the sample reflecting the integral ratio of the elements are presented in Fig. 10b, f.

It is seen that on the front surface there are about ~ 40 % of copper and about 60 % of the B-C material in the ratio which is close to B13C2 stoichiometric boron carbide. On the back side the copper is practically absent in the material of the gray contrast. And the copper content is about 3.0% presumably due to the presence of very large fragments peeled off from the copper surface. Dominant B-C material has also the ratio of elements close to the stoichiometry of the boron carbide B13C2.
Fig. 5 shows a precipitated material cross cleavage separated from the copper substrate.

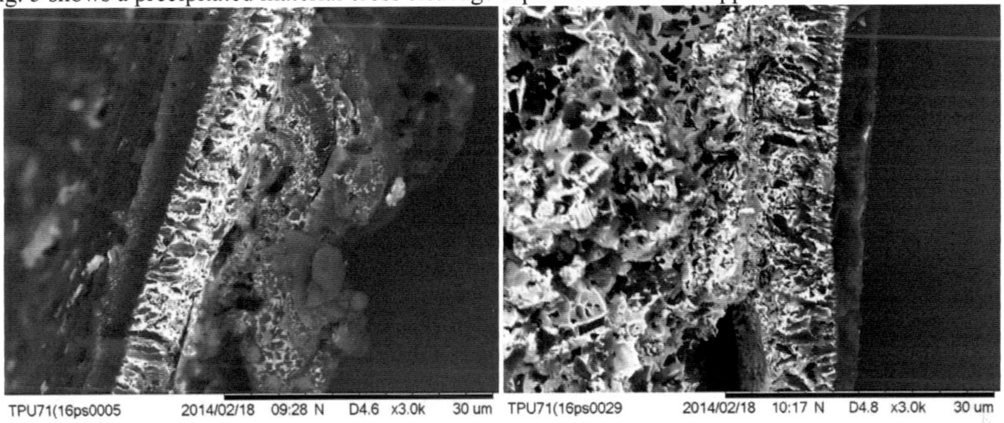

Figure 5. SEM micrographs of the cross-section of the boron carbide film deposited on copper.

On the border with the copper substrate uniform layer of gray material is visible, which is boron carbide. Above this layer a composite layer of boron carbide and copper is formed. Such distribution of the main phases of the bulk material thickness can be interpreted as follows. Boron carbide is synthesized in the discharge plasma in the accelerating channel at a high temperature. After that liquid state boron carbide is carried out of the AC and covers still "cold" surface of the copper substrate. The temperature of the liquid state boron carbide is not less than 2350 °C, which is considerably higher than the melting point of copper. The surface of the copper layer melts by a large temperature gradient meanwhile the carbide melt temperature decreases and the crystallization starts with the formation of solid carbide layer ejecting still hot copper melt to the surface layers. While it is cooling due to heat shrinkage the layer is cracking and dividing into a geometrically irregular grain shapes.
In microgaps formed between the grains by capillary action comes copper, which is still in a liquid state, and is in near-surface micro- layer of material layers. The composite material is formed.

4. Introduction

Despite the superlative properties and the wide variety of potential applications of boron carbide, relatively little effort has been put. Boron carbide is one of the hardest materials. It has a high melting point, high resistance to chemical agents, low density and lightweight. It can be applied in many areas as abrasive powders and coatings, in nuclear applications as neutron radiation absorbent, ceramics, an impurity in thermoelectric device material and etc. [1].

As a result of these boron carbide properties there is an interest not only in powder, but also in coating, which can be used as protective layer from aggressive environments and high temperatures in different units. One of the most perspective investigations in that sphere is obtaining thick boron carbide coatings (300-500 μm) for tokamak wall in nuclear fusion reactor. The most popular method for obtaining such coatings is vacuum plasma deposition method.

B4C can be prepared by several methods, such as laser assisted CVD, microwave CVD, filament activated CVD, sputtering and radio-frequency plasma CVD. Most of these methods have a number of disadvantages, such as process duration, coatings thinness, low adhesion, necessity of high temperatures and energy intensity as a result [2].

IOP Conf. Series: Materials Science and Engineering **66** (2014) 012051 doi:10.1088/1757-899X/66/1/012051

5. Summary

Thus, this scheme of supersonic jet electro plasma sputtering on the substrate allows forming a coating based on boron carbide. Low adhesion is due to the high degree of crystallinity and a low temperature substrate [4].

References

[1] Thvenot F 1990 Boron carbide - a comprehensive review *J. Eur. Ceram. Soc.* **6** 205-25

[2] Sezer A, Brand J I 2001 Chemical vapor deposition of boron carbide *J. Mater. Sci. Eng.* B79 191-202

[3] Сивков А and Пак А Патент № 2431947 РФ Коаксиальный магнитоплазменный ускоритель

[4] Postel O and Heberlein J 1998 Deposition of boron carbide thin film by supersonic plasma jet CVD with secondary discharge *J. Surf. Coat. Technol.* **108** 247-52

IOP Conf. Series: Materials Science and Engineering **66** (2014) 012052 doi:10.1088/1757-899X/66/1/012052

Biomass recycling heat technology and energy products

R B Tabakaev[1] P S Gergelizhiu[1] A V Kazakov[1] and A S Zavorin[1]
[1] Tomsk Polytechnic University, Tomsk, Russia

Email: TabakaevRB@tpu.ru

Abstract. Relevance is determined by necessity of utilizing of local low-grade fuels by energy equpment. Most widespread Tomsk oblast (Russian Federation region) low-grade fuels are described and listed. Capability of utilizing is analysed. Mass balances of heat-technology conversion materials and derived products are described. As a result, recycling capability of low-grade fuels in briquette fuel is appraised.

Introduction

The modern view on the problems in the power engineering industry development is directed to gradual increase in the percentage of renewable energy sources (RES) in the total power balance: governmental programs on involving RES in the fuel and energy balance have been adopted in more than 73 countries [1]. In turn, the Government of the Russian Federation has elaborated the Federal Law "On the Electrical Power Engineering" and "Strategy in Russia's Power Engineering Development by 2030", according to which it is required to put power generation systems based on RES with capacity more than 20 GW into operation by the end of 2030.

One of the RES types, in particular, organic biomass, has been studied within the framework of this research. The organic biomass proves difficult for the use as an energy source due to its high humidity and low combustion heat, which makes its direct combustion in the existing boiler equipment inefficient and inexpedient.

Biomass processed thermally into refined products, such as peat briquettes [2] and high-energy hydrogen-containing gas [3], is considered quite promising. However, prior to evaluating the possibility and expedience of such processing, data on material balance and properties of raw and final products should be obtained.

The main aim of this paper is to obtain information on the material balance at low-temperature pyrolysis of the Tomsk Region biomass, as well as on the properties of final refined products.

Biomass of Tomsk region (Russia)

The following major types of Tomsk Region biomass have been sampled (Fig. 1): the peat bogs of the Bakcharsky, Kolpashevsky, and Tomsky districts, wood wastes of the forest-industry complex, and sapropel of the Kolpashevsky district. Burning characteristics of the fuels under study are given in the Table 1.

The Sukhovsky and Arkadevsky peat bogs are filled with one and the same type of deposits - the lowland deposit referring to the group of decomposed peat, which was drained for industrial use. The Sukhovsky peat is compacted and can be characterised by high yield of volatile materials, high humidity and moderate ash content, so combustion proceeds at low temperature values. The Arkadevsky peat is loose and can be characterised by high yield of volatile materials, moderate humidity and ash content, which also results in low heat of combustion.

Figure 1. Researched types of Tomsk Region biomass
a) – Sukhovsky peat; b) – Arkadevsky peat; c) – Kandinsky peat; d) – wood chips; e) – Karasevsky sapropel

Table 1. Burning characteristics of fuels under study

Burning characteristics	Peat			Karasevsky sapropel	Wood chips
	Sukhovsky	Arkadevsky	Kandinsky		
Ash content on dry weight basis A^d, %	39.5	31.5	9.1	38.4	0.6
Yield of volatile materials on dry ash-free basis V^{daf}, %	69.3	71.0	71.6	84.8	91.8
Heat of combustion on dry ash-free basis Q^{daf}, MJ/kg	12.8	15.2	19.8	12.5	19.1

The Kandinsky peat is brown and in its native state, can be characterised by high yield of volatile materials and high humidity. Contrasted to other types of peat, the Kandinsky peat has low ash content, and quite high (for a peat) combustion heat.

The sapropel samples were delivered from the Karasevoe bog situated near the Chazhemto village of the Kolpashevsky district. Sapropel is a viscous mass of high humidity; whose dry ash content is quite high, so combustion proceeds at low temperature values. Besides, its yield of volatile materials is about 95 %.

Wood chips of different timber species from the CJSC "Lesopererabatyvauschy kombinat "Partner-Tomsk" were used in the research as wood wastes of the forest-industry complex.

Heat-technology recycling of biomass

The samples of biomass listed above were subjected to low-temperature pyrolysis at a special laboratory facility for heat processing (Fig. 2). The processing temperature reached 450 °C with a maximum heater capacity (750 W) of the given facility, which enabled to obtain the highest yield of pyrolysis condensate. Material balance was comprised on the experiment completion (Table 2).

Figure 2. Photo of the laboratory facility for heat processing [3]

Table 2. Material balance in thermal technological processing of Tomsk Region biomass

Low-grade fuels	Yield of pyrolysis products on dry weight basis		
	Carbon residue	Pyrolysis condensate	Hydrogen-containing gas
Sukhovsky peat	60.7 ± 7.9	20.3 ± 1.6	19.0 ± 6.9
Arkadevsky peat	74.8 ± 1.9	12.0 ± 0.7	13.2 ± 1.5
Kandinsky peat	43.4 ± 0.5	26.2 ± 2.4	30.4 ± 2.9
Wood chips	31.1 ± 3.3	23.6 ± 4.4	45.3 ± 3.0
Karasevsky sapropel	67.4 ± 1.8	14.1 ± 3.1	18.5 ± 4.3

The composition of the gases emitted was registered over the range of temperature values from 200 to 450 °C with an interval of 50 C. The Khromatek-Kristall 5000.2 chromatograph was used for measurements. According to recommendations [4], combustion heat was calculated based on the defined composition. Table 3 comprises the averaged values of fuel gas composition and its combustion heat over the temperature value range 200-450 °C.

Table 3. Characteristics of hydrogen-containing gas

Low-grade fuels	Fuel gas composition, %					Averaged value of heat of combustion of fuel gas, MJ/m^3
	H_2	CO	CH_4	C_mH_n	CO_2+N_2	
Sukhovsky peat	25.9	9.4	10.8	11.8	42.2	16.3
Arkadevsky peat	27.6	9.1	9.7	10.6	43.0	15.2
Kandinsky peat	23.8	9.8	12.9	12.8	40.7	17.6
Wood chips	6.5	26.0	19.3	11.5	36.8	19.1
Karasevsky sapropel	26.8	7.6	10.2	12.8	42.5	16.6

Table 4 contains burning characteristics of carbon residue gained at low-temperature pyrolysis.

Table 4. Burning characteristics of carbon residue

Burning characteristics	Sukhovsky peat	Arkadevsky peat	Kandinsky peat	Karasevsky sapropel	Wood chips
Moisture of fuel as received W^r_t, %	0.0	0.0	0.0	0.0	0.0
Ash content on dry weight basis A^d, %	40.6	61.9	22.9	56.5	3.5
Yield of volatile materials on dry ash-free basis V^{daf}, %	25.1	18.6	23.3	19.8	15.7
Heat of combustion on dry ash-free basis Q^{daf}, MJ/kg	17.0	16.3	27.9	23.0	33.0

Analysis of Tables 2-4 enables to draw a conclusion that thermal processing of wood chips and the Kandinsky peat by means of low-temperature pyrolysis can be carried out both for the purpose of gaining high-energy fuel gas (high yield and high combustion heat) and producing peat briquettes from carbon residue (an extremely high combustion heat) [2].

The characteristics of the Sukhovsky peat pyrolysis products allow to consider them relating to power generation only after thoroughly studying the technical and economic feasibility of using it under specific circumstances.

High ash content of carbon residue leads to lowering the combustion heat and increasing the operating costs. For this reason, thermal technological processing of the Arkadevsky peat and Karasevsky sapropel into peat briquettes is not expedient. Low yield of gas makes them inefficient for fuel gas generation as well.

Summary

1. Wood wastes and the Kandinsky peat can be considered promising raw material for thermal technological processing both into fuel gas and peat briquettes.
2. The Sukhovsky peat may be used for power generation only after conducting a thorough investigation and feasibility study of its use in certain conditions.
3. Using the Karasevsky sapropel and Arkadevsky peat is considered unpromising due to low ash content of raw material and final products of thermal technological processing.

The reported study was partially supported by the Ministry of education and science of The Russian Federation (Government Order No. 2.1322.2014).

References

[1] Popel O S Reutov B F and Antropov A P 2010 *Thermal engineering* **11** pp. 909-918.
[2] Zavorin A S Kazakov A V Makeev A A and Podorov S V 2012 *Thermal engineering* **1** pp. 77-82.
[3] Zavorin A S Kazakov A V and Tabakaev R B 2012 *Bulletin of Tomsk Polytechnic University* **4** pp.18-22.
[4] *Teplovoj raschet kotlov (Normativny metod)*. Saint-Petersburg: Energiya, 1998. 256 p. [in Russian]

AUTHOR INDEX

AUTHOR INDEX

Institute of Physics
Dirac House, Temple Back
Bristol BS1 6BE UK

CURRAN ASSOCIATES INC
proceedings.com